Cytogenetics of Domestic Animals: Clinical, Molecular and Evolutionary Aspects

Cytogenetics of Domestic Animals: Clinical, Molecular and Evolutionary Aspects

Editors

**Leopoldo Iannuzzi
Pietro Parma**

MDPI • Basel • Beijing • Wuhan • Barcelona • Belgrade • Manchester • Tokyo • Cluj • Tianjin

Editors
Leopoldo Iannuzzi
National Research Council
(CNR) of Italy
Portici
Italy

Pietro Parma
Milan University
Milan
Italy

Editorial Office
MDPI
St. Alban-Anlage 66
4052 Basel, Switzerland

This is a reprint of articles from the Special Issue published online in the open access journal *Animals* (ISSN 2076-2615) (available at: https://www.mdpi.com/journal/animals/special_issues/cytogenetic_animals_clinical_molecular_evolutionary).

For citation purposes, cite each article independently as indicated on the article page online and as indicated below:

LastName, A.A.; LastName, B.B.; LastName, C.C. Article Title. *Journal Name* **Year**, *Volume Number*, Page Range.

ISBN 978-3-0365-7588-9 (Hbk)
ISBN 978-3-0365-7589-6 (PDF)

Cover image courtesy of Leopoldo Iannuzzi and Pietro Parma

© 2023 by the authors. Articles in this book are Open Access and distributed under the Creative Commons Attribution (CC BY) license, which allows users to download, copy and build upon published articles, as long as the author and publisher are properly credited, which ensures maximum dissemination and a wider impact of our publications.

The book as a whole is distributed by MDPI under the terms and conditions of the Creative Commons license CC BY-NC-ND.

Contents

About the Editors vii

Preface to "Cytogenetics of Domestic Animals: Clinical, Molecular and Evolutionary Aspects" ix

Alessandra Iannuzzi, Pietro Parma and Leopoldo Iannuzzi
Chromosome Abnormalities and Fertility in Domestic Bovids: A Review
Reprinted from: *Animals* **2021**, *11*, 802, doi:10.3390/ani11030802 1

Izabela Szczerbal and Marek Switonski
Clinical Cytogenetics of the Dog: A Review
Reprinted from: *Animals* **2021**, *11*, 947, doi:10.3390/ani11040947 29

Monika Bugno-Poniewierska and Terje Raudsepp
Horse Clinical Cytogenetics: Recurrent Themes and Novel Findings
Reprinted from: *Animals* **2021**, *11*, 831, doi:10.3390/ani11030831 45

Brendan Donaldson, Daniel A. F. Villagomez and W. Allan King
Classical, Molecular, and Genomic Cytogenetics of the Pig, a Clinical Perspective
Reprinted from: *Animals* **2021**, *11*, 1257, doi:10.3390/ani11051257 71

Miluse Vozdova, Svatava Kubickova, Halina Cernohorska, Jan Fröhlich and Jiri Rubes
Anchoring the CerEla1.0 Genome Assembly to Red Deer (*Cervus elaphus*) and Cattle (*Bos taurus*) Chromosomes and Specification of Evolutionary Chromosome Rearrangements in Cervidae
Reprinted from: *Animals* **2021**, *11*, 2614, doi:10.3390/ani11092614 93

Rafael Kretschmer, Benilson Silva Rodrigues, Suziane Alves Barcellos, Alice Lemos Costa, Marcelo de Bello Cioffi, Analía del Valle Garnero, et al.
Karyotype Evolution and Genomic Organization of Repetitive DNAs in the Saffron Finch, *Sicalis flaveola* (Passeriformes, Aves)
Reprinted from: *Animals* **2021**, *11*, 1456, doi:10.3390/ani11051456 109

Sharmila Ghosh, Josefina Kjöllerström, Laurie Metcalfe, Stephen Reed, Rytis Juras and Terje Raudsepp
The Second Case of Non-Mosaic Trisomy of Chromosome 26 with Homologous Fusion 26q;26q in the Horse
Reprinted from: *Animals* **2022**, *12*, 803, doi:10.3390/ani12070803 123

Emanuele D'Anza, Francesco Buono, Sara Albarella, Elisa Castaldo, Mariagiulia Pugliano, Alessandra Iannuzzi, et al.
Chromosome Instability in Pony of Esperia Breed Naturally Infected by Intestinal Strongylidae
Reprinted from: *Animals* **2022**, *12*, 2817, doi:10.3390/ani12202817 135

Angela Perucatti, Alessandra Iannuzzi, Alessia Armezzani, Massimo Palmarini and Leopoldo Iannuzzi
Comparative Fluorescence In Situ Hybridization (FISH) Mapping of Twenty-Three Endogenous Jaagsiekte Sheep Retrovirus (enJSRVs) in Sheep (*Ovis aries*) and River Buffalo (*Bubalus bubalis*) Chromosomes
Reprinted from: *Animals* **2022**, *12*, 2834, doi:10.3390/ani12202834 145

Izabela Szczerbal, Joanna Nowacka-Woszuk, Monika Stachowiak, Anna Lukomska, Kacper Konieczny, Natalia Tarnogrodzka, et al.
XX/XY Chimerism in Internal Genitalia of a Virilized Heifer
Reprinted from: *Animals* **2022**, *12*, 2932, doi:10.3390/ani12212932 153

Halina Cernohorska, Svatava Kubickova, Petra Musilova, Miluse Vozdova, Roman Vodicka and Jiri Rubes
Supernumerary Marker Chromosome Identified in Asian Elephant (*Elephas maximus*)
Reprinted from: *Animals* **2023**, *13*, 701, doi:10.3390/ani13040701 . . . . **165**

Sebastián Demyda-Peyrás, Nora Laseca, Gabriel Anaya, Barbara Kij-Mitka, Antonio Molina, Ayelén Karlau and Mercedes Valera
Prevalence of Sex-Related Chromosomal Abnormalities in a Large Cohort of Spanish Purebred Horses
Reprinted from: *Animals* **2023**, *13*, 539, doi:10.3390/ani13030539 . . . . **177**

Alessandra Iannuzzi, Leopoldo Iannuzzi and Pietro Parma
Molecular Cytogenetics in Domestic Bovids: A Review
Reprinted from: *Animals* **2023**, *13*, 944, doi:10.3390/ani13050944 . . . . **187**

About the Editors

Leopoldo Iannuzzi

Leopoldo Iannuzzi is an accomplished researcher and academic in Agricultural Science, with expertise in animal production. He has held various prestigious positions, including Research Director and Associate of the National Research Council's Institute of Animal Production Systems in Mediterranean Environments (ISPAAM) in Italy. He has also coordinated international scientific committees focused on standardizing banded karyotypes of domestic bovids and shared his expertise as an invited speaker in 23 international and two national meetings on animal cytogenetics and gene mapping. His research interests include clinical cytogenetics, evolutionary cytogenetics, molecular cytogenetics, environmental cytogenetics, and chromosome characterization. He has explored chromosome abnormalities' relationships and reproduction in clinical cytogenetics, studied chromosome relationships among species using chromosome banding and FISH-mapping techniques in evolutionary cytogenetics, and constructed detailed cytogenetic maps in domestic animals using molecular cytogenetics. He has also examined the effects of environmental mutagens on animal chromosomes in environmental cytogenetics and worked on constructing standard banded karyotypes in bovids using high-resolution chromosome banding in chromosome characterization. Iannuzzi has published 371 papers, 210 of which were in international journals. He has 5,774 citations on Google Scholar and an H-index of 39. Since 2018, he has 1,591 citations and has a H-index of 19 for that period. His i10-index stands at 149 overall, and 39 since 2018. Overall, Iannuzzi's contributions to animal cytogenetics have been significant.

Pietro Parma

Pietro Parma is an Associate Professor at the Department of Agricultural and Environmental Sciences—Production, Landscape, Agroenergy at the University of Milano, Milano. He holds a degree in Agricultural Sciences, with a major in the Genetics of Animal Production. Parma's research primarily focuses on the Genetics of sex determination in mammals, where he identifies and characterizes genetic factors involved in the process of sex determination. Additionally, he has researched Clinical cytogenetics, identifying and characterizing chromosomal anomalies present in subjects with anomalies attributable to the reproductive aspect. Furthermore, he has also worked in the area of Molecular cytogenetics, utilizing technologies such as FISH, Array-CGH, and NGS sequencing for the molecular characterization of chromosomal abnormalities. Parma's work has received significant attention, with 1689 citations and 89 documents indexed in Scopus. His H-index stands at 21.

Preface to "Cytogenetics of Domestic Animals: Clinical, Molecular and Evolutionary Aspects"

With the discovery of 1;29 Robertsonian translocation in Swedish red cattle (Gustavson and Rockborn, 1964), which was later demonstrated to reduce fertility in the carriers and was found in more than 50 different breeds (mainly meat breeds) in various frequencies (reaching 80% of the carriers in some breeds), cytogenetics became a discipline applied in various laboratories dedicated to research on domestic animal genetics. The attention of scientists now focuses on several sectors, such as clinical cytogenetics, evolutionary cytogenetics, molecular cytogenetics, and environmental cytogenetics.

In this Special Issue, several important groups from different parts of the world provide interesting contributions to enlighten the field of domestic animal cytogenetics. Both exhaustive reviews (five papers) and original contributions (height papers) are included. The reviews focus particularly on the clinical cytogenetics of the most important domestic species, such as cattle, river buffalo, sheep, goats (Iannuzzi et al., 2021), horses (Bugno-Poniewierska and Raudsepp, 2021), dogs (Szczerbal et al., 2021), and pigs (Donaldson et al., 2021). A final review focuses on the molecular cytogenetics of domestic bovids (Iannuzzi et al., 2023).

The original contributions involve different animal species and investigate various aspects. These include (a) the clinical cytogenetics of horses (Ghosh et al., 2022; Demyda-Peyrás et al., 2022), cattle (Szczerbal et al., 2022), and Asian elephants (Cernohorska et al., 2023); (b) chromosomal instability in Esperia ponies naturally infected with intestinal Strongylidae (D'Anza et al., 2022); (c) evolutionary chromosomal rearrangements in Cervidae (Vozdova et al., 2021); (d) the karyotype evolution and genomic organization of repetitive DNA in the saffron finch, Passeriformes (Kretschmern et al., 2021); and (e) the comparative FISH mapping of twenty-three endogenous Jaagsiekte sheep retroviruses (enJSRVs) in sheep and river buffalo (Perucatti et al., 2022).

Leopoldo Iannuzzi and Pietro Parma
Editors

Review

Chromosome Abnormalities and Fertility in Domestic Bovids: A Review

Alessandra Iannuzzi [1], Pietro Parma [2] and Leopoldo Iannuzzi [1,*]

[1] Institute for Animal Production System in Mediterranean Environment, National Research Council, 80055 Portici, Italy; alessandra.iannuzzi@cnr.it

[2] Department of Agricultural and Environmental Sciences, University of Milan, 20133 Milan, Italy; pietro.parma@unimi.it

* Correspondence: leopiannuzzi949@gmail.com; Tel.: +39-(06)-499327734

Simple Summary: In domestic bovids, numerical autosome abnormalities have been rarely reported, as they present abnormal animal phenotypes quickly eliminated by breeders. However, numerical abnormalities involving sex chromosomes and structural (balanced) chromosome anomalies have been more frequently detected because they are most often not phenotypically visible to breeders. For this reason, these chromosome abnormalities, without a cytogenetic control, escape animal selection, with subsequent deleterious effects on fertility, especially in female carriers.

Abstract: After discovering the Robertsonian translocation rob(1;29) in Swedish red cattle and demonstrating its harmful effect on fertility, the cytogenetics applied to domestic animals have been widely expanded in many laboratories in order to find relationships between chromosome abnormalities and their phenotypic effects on animal production. Numerical abnormalities involving autosomes have been rarely reported, as they present abnormal animal phenotypes quickly eliminated by breeders. In contrast, numerical sex chromosome abnormalities and structural chromosome anomalies have been more frequently detected in domestic bovids because they are often not phenotypically visible to breeders. For this reason, these chromosome abnormalities, without a cytogenetic control, escape selection, with subsequent harmful effects on fertility, especially in female carriers. Chromosome abnormalities can also be easily spread through the offspring, especially when using artificial insemination. The advent of chromosome banding and FISH-mapping techniques with specific molecular markers (or chromosome-painting probes) has led to the development of powerful tools for cytogeneticists in their daily work. With these tools, they can identify the chromosomes involved in abnormalities, even when the banding pattern resolution is low (as has been the case in many published papers, especially in the past). Indeed, clinical cytogenetics remains an essential step in the genetic improvement of livestock.

Keywords: chromosome abnormality; cattle; river buffalo; sheep; goat; fertility

1. Introduction

After discovering the Robertsonian translocation rob(1;29) in the Swedish red cattle breed [1], and the demonstration of its harmful effect on fertility [2–4], the cytogenetics applied to domestic animals have been widely expanded in many laboratories in order to find relationships between chromosome abnormalities and their phenotypic effects, primarily in terms of fertility.

However, in the years immediately following this discovery, various cytogeneticists published reports on chromosome abnormalities, mostly involving sex chromosomes, underlining the importance of these types of abnormalities, often responsible for sterility, especially in females [5–11].

Numerical autosome abnormalities have been rarely reported, as they present abnormal animal phenotypes quickly eliminated in early embryo development or by breeders [12]. In contrast, numerical sex chromosome abnormalities and structural (balanced) chromosome anomalies have been more frequently detected in domestic bovids because they are most often not phenotypically visible to breeders (Table 1). For this reason, these chromosome abnormalities, without cytogenetic control, escape selection, with subsequent harmful effects on fertility (and production), especially in female carriers. Chromosome abnormalities can also be easily spread by offspring, especially when using artificial insemination, with adverse economic effects on animal breeding.

Table 1. Schematic representation of the chromosome abnormalities in domestic bovids.

Chromosome Abnormalities		
Numerical		Structural
Autosomes	Sex Chromosomes	
Very rare (the animal body conformation being abnormal; these abnormalities are eliminated directly by the breeders)	More tolerated by the species but almost all related to sterility or low fertility, especially in the females Generally not visible in the carriers (normal body conformation and external genitalia)	Deviation from the normal chromosome shape or gene order Very important for the (a) high percentage of carriers (i.e., cattle rob1;29); (b) normal body conformation; (c) because they escape the normal breeding selection They can be balanced (translocations and inversions) or unbalanced (deletions, insertions, and duplications)

The advent of chromosome-banding and FISH-mapping techniques with specific molecular markers (generally BAC clones), reviewed by [13], as well as chromosome painting probes (Zoo-FISH) [14,15], the use of CGH arrays [16], and the availability of standard chromosome nomenclatures [17], have led to the development of powerful tools for cytogeneticists in their daily work. With these tools, they can identify the chromosomes involved in abnormalities and the possible loss or gain of genetic material (especially using CGH arrays). Indeed, clinical cytogenetics remains an essential step in the genetic improvement of livestock.

In this review, we discuss the most crucial chromosome abnormalities (CA) found in domestic bovids (mainly cattle, sheep, goats, and river buffalo) by grouping most of them in tables to synthetize the data. We also suggest possible strategies for a better investigation of CA in animal populations, using efficient and simple banding and molecular techniques to speed up the analyses for the improved selection of reproductive animals.

2. Numerical Chromosome Abnormalities

2.1. Autosomes

Numerical autosome abnormalities have been rarely found in domestic bovids because they are directly eliminated in early embryo development or by breeders when severe anatomical defects occur [12]. Most trisomies reported in cattle involve multiple and heterogeneous defects, especially including those of the muscular-skeletal, cardiovascular, and urogenital systems. Table 2 summarizes the numerical autosomal abnormalities found so far in cattle. Due to the poor banding techniques available in the past, as well as the lack of the use of specific chromosome markers in the FISH-technique in most studies, the accuracy of the chromosome identification can be doubtful. An example is trisomies 22 [18,19] and 28 [20], found in the same animal, when the case was revisited some years after the previous studies, using the same animal slides, chromosome banding, and FISH-mapping technique (Table 2, Figure 1).

Table 2. Autosomal trisomies in cattle.

Chromosome Involved	Phenotype	References
Large Autosome	Male calf with extreme brachygnathia inferior	[21]
12	Anatomical defect, lethal	[22,23]
16 (TAN,1;16)	Anatomical defects	[24]
18 (?)	Anatomical defects	[25]
19	Anatomical defects (BI)	[26]
20	Sterile cow	[27]
	Malformed calf, absence of external genitalia	[28]
	Malformed fetus, cranial defects	[29]
	Fetus with pulmonary hypoplasia and anasarca syndrome (genomic analysis)	[30]
21 (?)	Anatomical defects	[31]
21	Newborn Hereford with a cleft palate, hydrocephalus, a cardiac interventricular septal defect, and arthrogryposis	[32]
22	Anatomical defects (no lethality) [33] Multiple malformations, including hypoplasia of palpebral fissures, cleft palate, kyphoscoliosis, and arthrogryposis	[32–34]
21 and 27	Fetuses	[35,36]
22 [1]	Anatomical defects	[18]
	Anatomical defects	[19]
24	Malformed heifer (slight prognathia, heart defects, slow growth rate)	[37]
26	Sterility, growth retardation	[38]
25 +;11−	Anatomical defects	[39]
28 [1]	Anatomical defects	[20]
29	Malformed female calf showing dwarfism with severe facial anomalies (genomic analysis)	[40]

[1] Same animal. ? means uncertain chromosome involved.

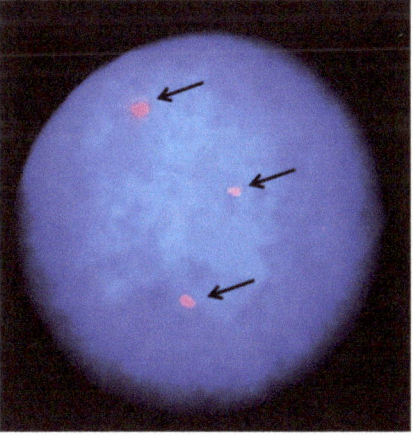

Figure 1. Interphase nucleus of a female cattle calf affected by trisomy 28. Arrows indicate the three FITC signals of the BAC clone containing the conglutinin (CGN1) gene, the official marker of BTA28 (ISCNDB2000, 2001).

Large chromosomes were no longer found to be involved in the autosomal trisomies (see Table 2), probably due to the fetus's lethal condition, which caused it to die in early embryonic life.

A particular case has been reported in a calf of the Agerolese breed (southern Italy). This animal, unable to stand up and which died a few weeks after birth, was found to be affected by partial trisomy 25 and partial monosomy 11 [39] (Table 2) due to an unbalanced meiosis of the mother cow, which had been affected by a balanced rcp(11;25) and reduced fertility [41]. Two cases of trisomy involving BTA20 and BTA29 have been found using only genomic analyses [30,40]. It should be interesting to compare this approach with cytogenetic analyses, such as chromosome banding and FISH mapping using specific chromosome markers, as recently performed in a case of tandem fusion translocation [42]. A useful approach to detecting numerical chromosome abnormalities using a FISH-mapping technique with two marker chromosomes has been applied to cattle embryos derived from in vitro production (IVP) [43]. These authors observed an increased number of mixoploid cells (diploid and polyploid) compared to in vivo embryos obtained by superovulation (72% of IVP blastocysts versus 25% in vivo). However, the authors maintain that the survival of most calves derived from IVP indicates that a considerable number of these embryos can compensate for the adverse effects of the in vitro procedures [43]. The in vitro aspect is very interesting regarding chromosomal abnormalities, especially in a breeding context. Future breeding might involve in vitro embryo production, subsequent genotyping of the embryo, and selection. In this respect, looking for structural abnormalities will be very important because they will often escape "regular" genomic selection protocols.

2.2. Sex Chromosomes

Sex chromosome abnormalities are generally better tolerated by animal species, including the bovids, because one of the X chromosomes genetically is inactivated as gene dosage compensation [44]. However, some genes escape inactivation and cause reproductive disorders involving the abnormal development of internal sex organs [45]. The sex chromosomes of domestic bovids are easily identifiable by both standard chromosome-staining and C-banding techniques. In fact, the X chromosomes of domestic bovids have a different size, shape, and C-banding pattern compared with the autosomes, in particular, (a) BTA-X is submetacentric when all autosomes are acrocentric; (b) BBU-X is the largest acrocentric chromosome, with typically one extensive centromeric C band (and an additional, proximally located C band), compared to all acrocentric autosomes; (c) OAR-X and CHI-X are acrocentric with visible p arms and negative C banding; (d) and BIN-X is submetacentric (as in BTA-X).

The Y chromosome can also be easily detected by both standard chromosome staining (cattle, sheep and goat) or C-banding techniques (river buffalo and zebu). Indeed, the Y chromosome is small and submetacentric in cattle and small and metacentric in both sheep and goat (where the other acrocentric autosomes are all acrocentric). The Y chromosome is acrocentric in both river buffalo and zebu, presenting a positive, distally located C band (C-banding patterns are centromeric in all remaining autosomes). More detailed information about sex chromosome banding is available in [46].

2.2.1. X Trisomy

X trisomy has been rarely found in domestic bovids. The few cases found have only occurred in cattle and river buffalo (Table 3).

Table 3. X-trisomy in domestic bovids.

Species	Phenotype	Reference
Cattle	Meiotic disturbances, familiar disposition, infertility	[47]
	Infertility	[48]
	Infertility	[22]
	Infertility	[49]
	Continuous estrus	[50]
	Infertility	[51]
	Infertility, 2 cases	[52]
R. Buffalo	Sterile (damages to internal sex structures)	[53]
	Sterile (damages to internal sex structures)	[54]
	Sterile (damages to internal sex structures), male traits	[55]

Generally, X-trisomic females have a normal body conformation and external genitalia, although a female river buffalo with male traits (prominent withers, tight pelvis, and large horns) has been observed (Figure 2). Carriers are generally affected by infertility (cattle) or sterility (river buffalo) due to damage to the internal sex structures, including ovarian hypoplasia, smaller uterus body, and lack of estrus. As has been established, one of two X chromosomes is randomly inactivated in these females during meiosis as gene-dosage compensation. The same inactivation occurs in X-trisomy cases where one X chromosome is active and the other two are inactivated. Still, abnormalities may result from the presence of three active X chromosomes in early embryonic development, either before X inactivation or due to X-linked genes that escape the inactivation process [56]. In humans, this syndrome is the most common sex chromosome abnormality (1/1000 births, [56]).

Figure 2. Female river buffalo, five years old, affected by X trisomy ($2n = 51$, XXX). Note the prominent withers (male trait).

2.2.2. X Monosomy

This type of chromosome abnormality is also rare in domestic bovids. Indeed, only a few cases have been recorded so far (Table 4).

Table 4. X-monosomy in domestic bovids.

Species	Phenotype	Reference
Cattle	Gonadal disgenesis (sterility)	[57]
	Gonadal disgenesis (sterility)	[58]
	Body smaller in size, the uterus and uterine tubes appeared immature and inactive.	[59]
	Infertile heifer (XY/X0/Y-isochromosome)	[60]
R. Buffalo	Gonadal disgenesis (sterility)	[61]
	Gonadal disgenesis (sterility)	[62]
	Gonadal disgenesis (sterility)	[63]
Sheep	Normal phenotype and external genitalia, no nursing of offspring	[64]
	Gonadal dygenesis in the X0/XX karyotype	[65]
	Dizygotic sheep twins with internal sex damages and mammary gland development very limited	[66]
Goat	Gonadal dysgenesis (XO/XX/XXX mixoploidy)	[67]

Generally, females carrying X monosomy (active X, Figure 3) showed gonadal dysgenesis and sterility [57–59,63,68], although in sheep, the effects on the internal sex organs can be less damaging (Table 4), [64,66]. In humans, 1 in 5000 live births is $2n = 45,X$. In addition, 45,X represents one of the most common chromosome abnormalities identified in spontaneous abortions [56]. Very probably, the same occurs in domestic bovids, complicating the cytogenetic analyses of aborted fetuses. Thus, it is difficult to know the real frequency of this chromosome abnormality in domestic bovids and its fertility effects.

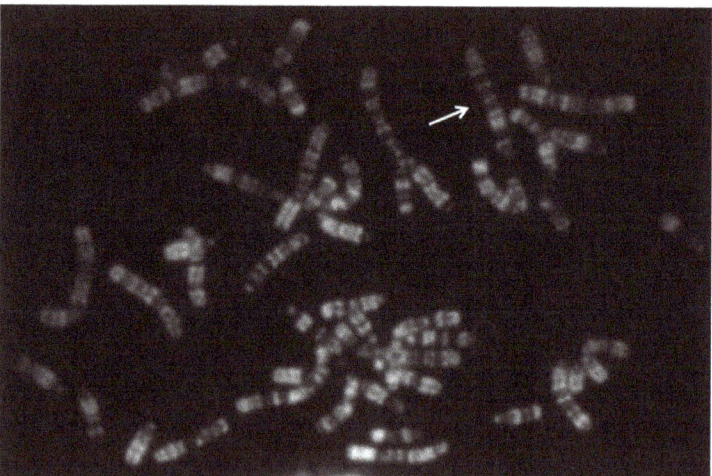

Figure 3. RBA-banding river buffalo metaphase from a female affected by X monosomy ($2n = 49,X$). The only active X chromosome (arrow) was observed in all metaphases. This female was sterile due to damage to her internal sex organs.

2.2.3. XXY Syndrome

Known in humans as Klinefelter's syndrome, this abnormality has rarely been found in males of domestic bovids (Table 5).

Table 5. XXY-syndrome in domestic bovids.

Species	Phenotype	References
Cattle	Testicular hypoplasia in a mosaicism case XY/XX/XXY	[69]
	Testicular hypoplasia	[70]
	Testicular hypoplasia	[22]
	Intersexuality in a mosaicism case XX/XXY	[71]
	Bilateral testicular hypoplasia	[72]
	Testicular hypoplasia	[11]
	Testicular hypoplasia in a mosaicism case XX/XYY	[73]
	Masculinization effects in a mosaicism case XX/XXY	[70]
	Testicular hypoplasia	[74]
	Testicular hypoplasia (XXY + rob(1;29))	[75]
	2 cases (testicular hypoplasia with degradation of seminiferous tubules in one examined case)	[76]
	Azospermic bull	[77]
	Testicular hypoplasia in a bull with mosaicism (XY/XYY)	[78]
	Testicular hypoplasia	[79]
	Testicular hypoplasia	[80]
	Testicular hypoplasia in 3 cases	[52]
	Young male excluded for reproduction being mosaic for XY/XYY	Present Study
R. Buffalo	Testicular hypoplasia in a case of $2n = 50,Y, rob(X;X)$	[81]
Sheep	2 cases in rams showing hypoplastic testis	[82]
	Ram with no particular phenotypic effects (XX/XYY mosaicism)	[83]
Goat	Testicular hypoplasia in a case of XXY/XY mosaicism	[84]
	XX/XXY fertile buck	[85]

Even when two or more X chromosomes are present, the presence of only one Y chromosome is sufficient to induce testes development. This is due to the presence of the SRY gene on the Y chromosome. Carriers are generally affected by testicular hypoplasia, as found also in several cases of mosaicism, XY/XX/XXY, XX/XXY, or XXY/XY (Table 5). An interesting XXY case has been reported in a river buffalo [81]. This male, showing gonadal dysgenesis, presented an unusual karyotype: $2n = 50,Y, rob(X;X)$. A case of mosaicism XY/XYY was found in a young male of the Chianina cattle breed intended for reproduction (Figure 4, Table 5). The animal was promptly eliminated after a karyotype analysis, and it was not possible to further investigate the case.

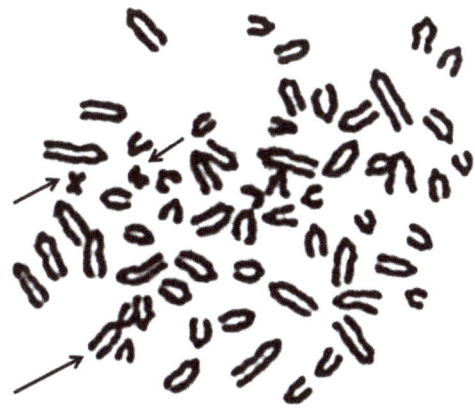

Figure 4. Normal Giemsa-staining metaphase plate of young male cattle for reproduction but promptly eliminated because it was found to be affected by XY/XYY mosaicism. The X chromosome (large arrow) and Y chromosomes (small arrows) are indicated.

2.3. Sex Reversal Syndrome

This syndrome occurs when male and female phenotypes (or gonadic sex) differ from the expected sex chromosome constitution, as in XX males and XY females. All cases found with this syndrome in domestic bovids are reported in Table 6.

Table 6. Cases with sex reversal syndrome in domestic bovids.

Species	Sex Chrom.	Phenotype/Effects on Fertility	Reference
Cattle	XY	Female (2) with reproductive defects	[86]
	XY	Female with internal sex anatomical defects and no estrus	[87]
	XY	Female with no estrus and streak gonads	[88]
	XY	Female with hypoplastic ovaries	[89]
	XY	Single birth female with normal internal sex adducts but feeble estrus	[27]
	XY	Female normal gonads and genital development with AMGY and ZFY genes present (no SRY determination)	[90]
	XY	Female with hypoplastic gonads (the right one resembled an ovary and the left one an undeveloped testis)	[91]
	XY	Females (3) with no estrus and abnormal Y (Yp-iso)	[92,93]
	XX	Male with both testis and ovotestis development	[94]
	XX	Male XX + rob(1;29) apparently with the normal reproductive parameters but eliminated for rob(1;29)	[95]
R. buffalo	XY	Females (2) sterile with abnormal internal sex adducts (one case with SRY-positive)	[55,96]
Sheep	XY	Sterile ewe with streak gonads, SRY+	[97]
	XY	Ewe with a longer ano-vulvar distance, enlarged clitoris, two testes-like structures at the inguinal level	[98]
Goat	XX	Testicular biosynthesis of testosterone	[99]
	XX	Males intersex, SRY-, Polled Intersex Syndrome (PIS)	[100–102]

2.3.1. XY Sex Reversal

Bovine XY sex reversal has been observed much more frequently than its counterpart (i.e., XX sex reversal syndrome). Several cases have been reported in this species (Table 6). When the SRY gene sequences were published [103], a test for this syndrome in animals revealed a lack of SRY gene sequences by both PCR and FISH-mapping analysis in such individuals [92,93]. Only two cases of XY sex reversal syndrome have been reported in river buffalo (Table 6). Both females were sterile with severe disruption to their internal sex organs. However, upon investigation by both FISH-mapping and gene-sequence analysis, one individual displayed the SRY gene at its expected location on the Y chromosome with its normal DNA sequence [55]. Similar cases have been reported in sheep [97]. Other authors [104] reported a case of a woman with a 46,XY karyotype and a female phenotype, including histologically normal ovaries. This phenotype, which originated from loss of function due to mutations on the CBX2 gene (human homolog of mouse gene M33), is the only known report of an XY sex reversal with ovary development.

2.3.2. XX Sex Reversal

This syndrome is very rare in domestic animals [105]. Although very rare, XX human males show a variety of clinical manifestations from a normal male phenotype to ambiguous genitalia in newborns. The syndrome is correlated to a translocation of the SRY gene from the Y chromosome to the X chromosome in about 80% of XX sex reversal cases [106,107]. An essential role in this syndrome is played by the chromosome position of the SRY gene in the Y chromosome. When it is located close to the PAR region (as in humans), there are more probabilities for translocations from the Y to X chromosomes during meiotic recombination. In domestic animals, the SRY gene is generally located far from the PAR region [108–110], thus explaining its rare occurrence in domestic animals. No documented XX sex reversal related to the SRY gene have been found so far in domestic animals [111,112]. Detailed information on sex reversal syndrome in placental animal species has been reviewed by Parma et al. [113].

2.4. XX/XY Mosaicism (Free-Martinism)

This syndrome is the most common sex chromosome abnormality found in domestic bovids in twins of different sexes. In cattle, about 90% of twins of different sexes are free-martin [80,114]. In dairy cattle, the percentage of free-martin twins is higher than that in meat breeds. It varies between 0.5% and 2.0%, with the rate of twinning in dairy breeds between 1% and 4% [115] when the male–female sex ratio is 1:1. Twin pregnancy percentages are also influenced by seasonal effects, reaching the highest levels during springtime and in older dairy cows (6%) [116]. Alterations of internal sex traits seem to be more severe in females than in males, although studies following several free-martin males also reported damage to interior male features [114]. In Italian Friesian cattle, most females with chromosome abnormalities (13%) were free-martin [80]. The presence of XX/XY mosaicism has been found also in bone marrow cattle cells [5].

Free-martin females generally show the typical body conformation and external genitalia. Still, they have pronounced gonadal dysgenesis, varying from a complete lack of internal sex organs (closed vagina) to Mullerian-duct atrophy (Figure 5). Furthermore, several studies reported that damage to the internal sex structures is not correlated with the percentage of male cells in either cattle [116] or river buffalo [117]. Indeed, in both cattle and river buffalo, aberrant internal sex organs were found even in the presence of small percentages of male cells [117]. This is essentially due to three events: (1) placental anastomosis occurring at 20–25 days of embryonic life; (2) sex differentiation occurring later (at 40–45 days) in cattle; and (3) male sex differentiation occurring one week before females [118]. For this reason, the presence of male cells, even in low percentages (and male hormones, in particular AMH), affects the development of internal female sex characteristics [118,119]. For this reason, male free-martins seem to be less prone to abnormal sex anomalies. However, some cases of reduced fertility have been reported in free-martin males [120–123]. The presence of material belonging to the Y chromosome has also been identified in female subjects with reduced reproductive efficiency [124].

Many free-martin cases are from single births (the other twin dying during early embryonic development). In river buffalo, about 90% of free-martin females were born in single births [55]. This phenomenon is essential because these females generally show normal body conformation and external genitalia, thus escaping breeding selection, unlike in twin births. In the latter case, the breeder knows that the female is probably free-martin and requires a veterinary examination by rectal palpation and cytogenetic or molecular (PCR with specific sex markers) analyses to confirm it.

In sheep and goats, although twins are frequent (but also triplets or quadruplets in some breeds), XX/XY mosaicism correlated to free-martinism occurs at very low frequencies (5–6%) in twins of different sexes, probably because sex differentiation occurs much earlier in sheep (20–25 days after fertilization) than in cattle [125]. Several cases of free-martins have been reported in both sheep [114] and goats [126–128], although the frequency of free-martinism is much lower in sheep and goats than in cattle and river buffalo. Sheep and goats carrying XX/XY mosaicism show a pronounced presence of both male and female traits, easily recognizable by breeders [98,129,130].

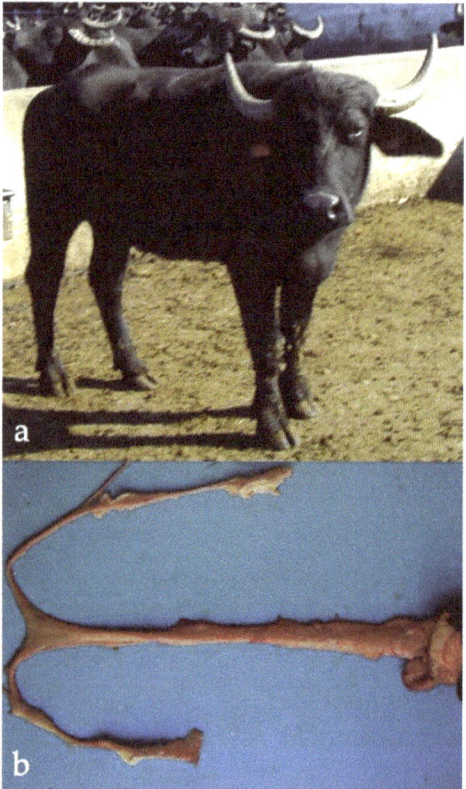

Figure 5. (**a**) River buffalo female showing normal body conformation and external genitalia but found with XX/XY mosaicism (free-martin). Note the atrophic uterine body (**b**).

2.5. Diploid-Triploid XX/XXY Mosaicism (Mixoploidy)

This syndrome is very rare in both humans and animals. In domestic bovids, only four cases have been reported of cattle with 2*n* = 60,XX and 3*n* = 90,XXY mosaicism [131]. Generally, the mixoploidy depends on the type of cell in cattle and humans, triploid cells being absent or present in lower percentages in blood lymphocytes and present in higher percentages in fibroblasts or cells of the uterine body or limbs [131–133]. In humans, the few 46,XX/69,XXY cases fall into three phenotypic groups: males with testicular development, ovo-testicular disorder of sex development (DSD), or under-virilized male DSD [134]. In cattle, the four cases reported so far showed various phenotypes, including aplasia of the vulva, a rudimentary penis, the presence of ovaries, an empty scrotum, and ovaries with corpus luteum [131].

3. Structural Chromosome Abnormalities

3.1. Reciprocal Translocations

Reciprocal translocations (rcp) have been found only in cattle and sheep (Table 7).

Table 7. Reciprocal translocations (rcp) found in cattle and sheep, with the chromosomes involved, phenotypic effects (when available), and author reference.

Species	Rcp/Chrom. Involved	Phenotype	Reference
Cattle	double rcp(2q−;20q +, 8q−;27q +)	reduced fertility	[135]
	rcp(8;15) (q21;q24)	reduced fertility	[136]
	rcp(1;8) (q44:q16)	2 males and 3 females, reduced fertility	[137]
	rcp(1;8;9) (q43;q13;q26)	subfertile bull subfertile bulls (n = 3)	[138,139]
	rcp(8;13) (q11;q24)	azoospemic bull	[140]
	rcp(20;24) (q17;q25)	subfertile bull	[141]
	rcp(X;1) (42;13)	normal female calf with mosaicism XX/XY	[142]
	rcp(12;17) (q22;q14)	subfertile bull	[143]
	rcp(1;5) (q21;q35)	azoospermic bull and its dam (reduced fertility)	[144]
	rcp(Y;9) (q12.3;q2.1)	azoospermic bull	[145]
	rcp(11;21) (q28;q12)	bull, no libido, rare spermatozoa	[146]
	rcp(9;11) (q27;q11)	male addressed to reproduction	[147]
	rcp(2;4) (q45;q34)	bull (post mortem SC-analysis)	[148]
	rcp(4;7) (q14;q28)	bull, balanced, cyto-genomic analysis (CGH-arrays)	[149]
	rcp(Y;21) (p11;q11)	bull testosterone negative	[150]
	rcp(11;25) (q24;q11)	cow with reduced fertility	[41]
	rcp(13;26)	cow with reduced fertility	[151]
	rcp(5;6) (q13;q34)	bull, balanced, cyto-genomic analysis (CGH-arrays)	[16]
	rcp(13;26) (q24;q11)	dam and calf, balanced	[152]
	rcp(12;23)	two subfertile bulls	[153]
Sheep	rcp(1p;19q)	low fertility	[154]
	rcp(13;20) (q12;q22)	low fertility	[155]
	rcp(2q;3q)	low fertility	[156,157]
	rcp(2p−;3q +)	low fertility	[80,158]
	rcp(4q;12q) (q13;q25)	low fertility	[159]
	rcp(18;23) (q14;q26)	low fertility	[160]
	rcp(13;20) (q12;q22)	poor fertility	[155]

Rcp are generally balanced, and for this reason, animal carriers show a normal body conformation. Still, they have reduced fertility due to disturbances that occurred during meiosis caused by abnormal (quadrivalent) configurations and erroneous chromosome disjunctions, which can give rise to abnormal embryos that generally die during early embryonic life [138,141,161–163]. Without a cytogenetic analysis, these abnormalities escape genetic selection and spread in the offspring, especially when using AI. However, rcp often escape cytogenetic analyses. Most animal cytogenetic labs apply routine cytogenetic analyses with only standard chromosome staining to detect robs, in particular rob(1;29). All cattle autosomes being acrocentric, only when abnormal autosomes are larger and/or shorter than BTA1 and BTA29, respectively, does the lab try to better investigate the case to identify a possible presence of rcp using chromosome-banding techniques and, more recently, chromosome-specific molecular markers (or chromosome-painting probes) by FISH-mapping techniques. For this reason, rcp have been reported with lower frequencies in cattle compared to dicentric robs. A study investigating all rcp found in cattle and correlating them to relative chromosome length concluded that the expected frequency of rcp in cattle is about four times higher than dicentric robs [164]. This estimate is based on two different approaches: (i) a mathematical approach; and (ii) a bioinformatics simulation approach. Both approaches provided similar value and therefore this estimate is believed to be reliable. However, when fertility values, such as (a) the interval between two births, (b) the return to estrus after natural or artificial insemination, and (c) a low number of calves during the reproductive life, appear abnormal, cytogenetic investigations must be done using both chromosome-banding and FISH-mapping techniques [13] to determine the presence, or lack thereof, of chromosome abnormalities like rcp. Generally, only single rcp has been found in bovids, involving only two chromosomes (Table 7). Only rarely has single rcp involved three chromosomes (Table 7) [80,138]. The only case of double rcp involving four chromosomes has been reported by De Schepper et al. [135] (Table 7). Only two rcp involved an autosome and the Y chromosome in an azoospermic bull [145] and a bull negative for testosterone (Table 7) [148].

Significant advantages for detecting rcp in domestic bovids (i.e., cattle and sheep) have been derived from improved chromosome-banding and FISH-mapping techniques with specific molecular markers (generally bovine or ovine BAC clones; Figure 6) or chromosome paint probes. Recently, a method using a panel of subtelomeric FISH-probes on a multi-hybridization device, as a means of highlighting the ends of each chromosome, has also been applied to cattle chromosomes to detect structural chromosome abnormalities [153]. However, only two studies extended the analyses using the CGH array to establish possible genetic material losses during chromosome rearrangements (Table 7) [16,149]. At least in these two latter cases, no genetic losses occurred during the rearrangements. Considering that the carriers of rcp are morphologically normal, it is possible to support the hypothesis that the rcp found so far in cattle and sheep are generally balanced.

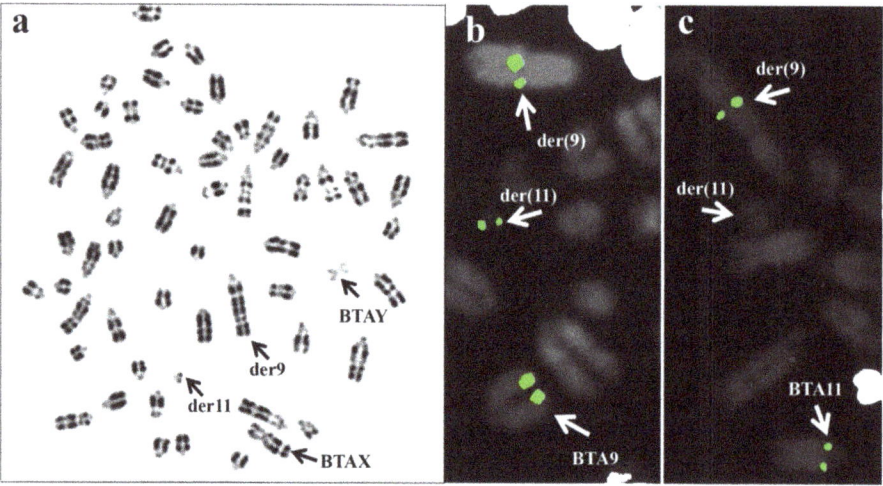

Figure 6. (a) Cattle metaphase treated for RBG banding and showing a case of rcp(9;11) (q27;q11) in a young male for reproduction. Arrows indicate the sex chromosomes der(9) and der(11). FISH mapping with two chromosome-specific BAC clones mapping on BTA9 and BTA11 confirmed the chromosomes involved in the rcp (**b,c**). Note the presence of FITC signals of a BTA9 marker in BTA9, der(9), and der(11) (**b**), as well as of FITC signals of a BTA11 marker only in BTA11 and der(9), being absent in der(11) (**c**) because the chromosome region was positioned after the break point.

In humans, the routine uses of genomic investigations allow the study of rcp. Indeed, mapping discordant mate pairs from long-insert, low-pass genome sequencing now permits efficient, cost-effective discovery and nucleotide-level resolution of rearrangement breakpoints, necessary for interpreting the etiology of clinical phenotypes in patients with rearrangements [165]. However, in domestic bovids, because breeders directly eliminate calves showing abnormal phenotypes potentially born from carriers of rcp, it is difficult to study these kinds of mating products.

A rare example has been found in a female calf with partial trisomy 11 and partial monosomy 25, which was unable to stand up and died after a few weeks (Table 2) [39]. The mother of this calf was a carrier of rcp(11;25) (Table 7) [41]. These two latter cases demonstrate that rcp cause reduced fertility by generating unbalanced embryos that die in early embryonic life or a few days after birth.

3.2. Robertsonian Translocations (rob)

Centric-fusion translocations are the most common chromosome abnormalities found in cattle. With the exception of rob(1;29), which is monocentric, all remaining robs found in cattle are dicentric (two centromeres; Table 8).

Table 8. Dicentric Robertsonian translocations reported in cattle, river buffalo, sheep, and goat.

Species	Rob/Chrom.		Breed/Country	Reference
Cattle	1	4	Czech Republic	[166]
	-	7	Not reported	[167]
	-	7	Blond D'Aquitaine, France	[80]
	-	21	Friesian	[168]
	-	22	Czech Republic	[166]
	-	23	Czech Republic	[166]
	-	25	Blonde d'Aquitaine, N.Z. Piebald cattle Germany	[169,170]
	-	26	Friesian, Japan	[171]
	-	27	British Friesian	[172]
	-	28	Czech Republic	[166]
	2	4	Friesian, England	[173]
	-	8	Friesian, England	[167]
	-	27	Not reported	[167]
	-	28	Vietnamese cattle	[174]
	3	4	Limousine, France	[175]
	-	12	Blond D'Aquitaine, France	[80]
	-	16	Montbéliarde, France	[176]
	-	27	Black spotted, Romania	[95]
	4	4	Czech Republic	[167]
	-	8	Chianina, Italy	[177]
	-	10	Blonde d'Aquitaine, France	[178]
	5	18	Simmenthal, Hungary	[179]
	-	21	Japanese Black, Japan	[167]
	-	22	Polish Red White, Poland	[180]
	-	23	Brown, Romania	[95]
	6	8	Chianina, Italy	[177,181]
	-	28	Czech Republic	[166]
	7	21	Japanese Black Cattle, Japan	[182,183]
	8	9	Brown Swiss, Switzerland	[167]
	-	23	Ukrainian Grey	[167]
	9	23	Blonde d'Aquitaine, France	[184]
	10	15	Pitangueiras, Spain	[185]
	11	16	Simmenthal, Hungary	[186]
	-	21	Brown, Romania	[95]
	-	22	Czech Republic	[167]
	12	12	Simmenthal, Germany	[167]
	-	15	Friesian, Argentina	[167]
	13	14	Friesian, Slovakia	[187]
	-	19	Marchigiana, Italy	[188]
	-	21	Friesian, Hungary	[189]
	-	24	Red &White, Poland. Not reported	[80,187,190]
	14	17	Marchigiana, Italy	[191,192]
	-	19	Braunvieh, Switzerland	[167]
	-	20	Simmenthal, Switzerland, USA. Spotted, Romania	[95,193–195]
	-	21	Simmental, Hungary	[167]
	-	24	Podolian, Italy	[196]
	-	28	Friesian, USA	[197]
	15	25	Barrosã, Portugal	[198]
	16	18	Barrosã, Portugal	[199]
	-	19	Marchigiana, Italy	[167]
	-	20	Simmenthal, Czeck Rep.	[200,201]
	-	21	RedPied, Czeck Rep.	[167]
	19	21	Friesian, France	[202]
	20	20	Simmenthal, Germany	[167]
	21	27	Blonde d'Aquitaine, France	[203]
	21	23	Maremmana, Italy	[204]
	-	29	Blonde d'Aquitaine, France	[80]
	24	27	Friesian, Egypt	[167]
	25	27	Alpine Grey, Italy	[139]
	26	29	Alpine Grey, Italy	[139,181,205]
	27	29	Guernsey, Canada	[206]

Table 8. Cont.

Species	Rob/Chrom.		Breed/Country	Reference
R. buffalo	1p	23	Ital. Mediterranean, Italy	[207]
	1p	18	Ital. Mediterranean, Italy	[208]
	X	X	Murrah, India	[81]
Sheep	6	24	(t1) New Zeland Romney, NZ	[209,210]
	9	10	(t2) New Zeland Romney, NZ	[210,211]
	7	25	(t3) New Zeland Romney, New Zeland	[210,211]
	5	8	(t4) New Zeland Romney, New Zeland	[212]
	8	22	(t5) New Zeland Romney, New Zeland	[212]
	1	20	Undefined Race, Germany	[213]
	8	11	Churra da Terra Quente, Portugal	[214]
Goat	2	13	Undefined Race, France	[215]
	3	7	-	[161]
	5	15	Saanen, Scotland. Saanen, Brazil	[216,217]
	6	17	Saanen, Switzerland. Saanen, Germany	[218,219]
	6	15	Saanen, Italy. Saanen, France. Saanen, Brazil	[220–222]
	10	12	Malaguena, Spain	[223]

The dicentric translocations reported so far in cattle have generally been found in single cases. Two exceptions are rob(14;20), reported in Simmenthal cattle in both Switzerland and the USA [193–195], and rob(26;29), reported in Alpine Grey cattle [139,181,205], where several carriers were found, probably due to the use of AI from bull carriers.

Generally, dicentric robs disappear after some generations, being unstable due to the presence of two active centromeres and restabilizing to the normal diploid number. In contrast, rob(1;29) is monocentric, showing one (and large) C-banding block particularly present in the q arm (Figure 7). Although this abnormality was discovered a long time ago [1,2], and various studies tried to show the origin of this translocation, only recently and with the use of cytogenetic (high-resolution chromosome banding and FISH-mapping techniques) and genomic (CGH array) analyses, was it possible to establish the origin and evolution of this frequent chromosome abnormality. Indeed, a chromosome segment of about 5 Mb translocated from the proximal region of BTA29 to the proximal region of BTA1, with inversion during the evolution of rob(1;29) [224]. A loss of constitutive heterochromatin (C bands) and of some SAT DNA also was observed on rob(1;29) [225,226].

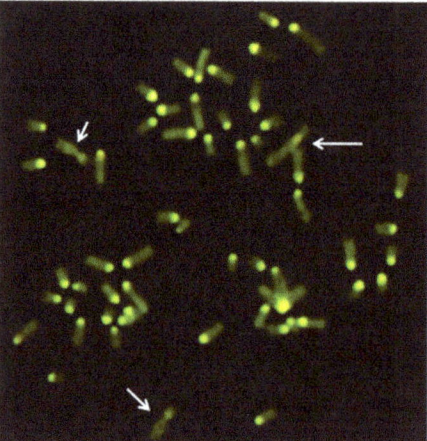

Figure 7. Female cattle metaphase treated for CBA banding in a heterozygous carrier of rob(1;29) (2n = 59,XX). Note the single C-band block in the rob(1;29), especially present on the q arms (large arrow). Small arrows indicate X chromosomes.

Rob(1;29), first found in Swedish red cattle [1,2], has been found widely in several breeds (more than 50) [227], mainly in meat breeds. Thus, cytogenetic investigations are particularly focused on meat breeds rather than on dairy cattle breeds, where rob(1;29) has rarely been found, probably because the genetic selection is more strictly applied to dairy breeds. Another hypothesis is that the lower frequency is due to the attempt to reduce the meat breeds' diploid number from $2n = 60$ to $2n = 58$ to gain genetic advantages derived from this new genetic linkage between the two chromosomes. The frequency of this translocation varies among cattle breeds, reaching high values in several breeds, in particular in the Barrosa (Portugal), where the frequency of rob(1;29) carriers has been observed at 70%, of which 53.2% were heterozygous carriers ($2n = 59$) and 16.6% were homozygous ($2n = 58$) carriers [228]. This abnormality reduces fertility in the carriers due to the presence of abnormal trivalent meiotic configurations [2,229] originating in unbalanced gametes that give rise to abnormal embryos that die in early embryonic life. The cow returns to estrus but with some delay compared to the normal interval due to the service's failure after AI [205]. The reduction in reproductive value in cow rob(1;29) carriers is around 8–9% [80], while in the male carriers it appears to be lower. Indeed, meiotic studies by sperm-FISH in two bulls carrying rob(1;29) revealed a lower percentage (around 2%) of abnormal and unbalanced sperm [230] than those achieved in oocytes of four female carriers of rob(1;29), which showed 21.83% diploid oocytes and 4.06% chromosomally unbalanced sets, with significant variation among carriers. However, these studies should be applied to a larger number of carriers (at least to males) to better establish the real reproductive value of bulls carrying the translocation in terms of unbalanced gametes. Sperm-FISH analyses also should be performed not only on the total sperm fraction but primarily on the motile sperm fraction (i.e., the effective sperm which fertilize the oocytes), as demonstrated in a river buffalo bull sperm carrying a rob(1p;18) translocation [231]. A possible effect of bulls carrying robs(16;20) and (14;20) on the development of bovine oocytes fertilized and matured in vitro was assessed on the basis of embryo yield and blastocyst formation [232]. The study demonstrated that, in bulls carrying the 16;20 and 14;20 translocations, in vitro preimplantation embryo development was reduced (compared to fertilization by a bull with a normal karyotype), probably due to genetically unbalanced spermatozoa [232].

A chromosome-specific marker for rob(1;29) has been found, making it possible to directly detect the presence of this translocation on sperm [233]. This marker, and sperm-FISH with specific chromosome markers, could be particularly useful in males bred for reproduction when no karyotype analyses are applied.

In river buffalo, in addition to the five biarmed pairs originating from centric-fusion translocations during the karyotype evolution [234], three more robs have been found so far as chromosome abnormalities in this species (Table 8). Two of them originated from a complex chromosome mechanism: fission of BBU1 and subsequent centric-fusion translocation between BBU1p and BBU23 in a cow with reduced fertility [207], and later with BBU18 in a very famous Italian bull (named Magnifico) of the Mediterranean Italian breed [208]. Since rob(1p;18) was also found in the bull's offspring [208], the bull was excluded from reproduction by the Italian buffalo breeder association. Analyses in both total and motile sperm fractions of carrier bulls, by triple-color FISH analysis with a pool of specific BAC probes, revealed that normal sperm were 27% and 69% in the total sperm fraction and motile sperm fraction, respectively [231].

The third case of centric-fusion translocation, rob(X;X), found in river buffalo (Table 8) was reported in a case of an XXY bull with testicular hypoplasia (Table 5) [81].

These studies suggested the necessity of applying cytogenetic investigations in this important species, particularly for all males bred for reproduction and all females with reproductive disturbances, in order to increase the fertility and economic value of river buffalo.

The normal karyotype of sheep (*Ovis aries*, $2n = 54$) has three biarmed pairs (OAR1, OAR2, and OAR3), which originated from centric-fusion translocations on chromosomes homologous to cattle (and goat, ancestral bovid) 1–3, 2–8, and 5–11, respectively [17]. In addition to these normal biarmed pairs, six centric-fusion translocations, as chromosome abnormalities, were found in sheep, of which five were named t1, t2, t3, t4, and t5, and involving goat-cattle homologous chromosomes 6–24, 9–10, 7–25, 5–8, and 8–22, respectively (Table 8) [209–212]. More recently, rob(8;11) was found in the Churra da Terra Quente sheep breed (Portugal) [214]. Except for the t4 translocation, which disappeared, and the most recent rob(8;11), found in a single case, the remaining four robs (t1, t2, t3, and t5) remained in New Zealand sheep flocks. Homozygous carriers ($2n = 48$ and $2n = 46$) were later found in these same sheep flocks [235]. At least for t1, t2, and t3, no particular effects on reproduction seemed to be present in the carriers [236].

Several Robertsonian translocations have also been reported in goats (Table 8). Very probably, some robs, like rob(5;15), rob(6;17), and rob(6;15), reported in Saanen goats, are identical [220,221]. As has generally occurred in other bovids, the translocations were reported in single cases, except for those found in the offspring of males carrying the translocation [217]. The authors performed cytogenetic and genealogical analyses on 205 goats, which were descendants of a sire imported from Switzerland. They reported 29.7% and 4.9% heterozygous and homozygous carriers of rob (5;15), respectively.

3.3. Simple Translocation

This chromosome abnormality consists of a chromosome segment region translocated from one chromosome to another. It has been rarely reported. A case of a Y;17 translocation was found in a cattle bull, phenotypically normal (normal reproductive organs and testicular function), but with slight pathospermia (oligozoospermia and asthenozoospermia), However, the portions of the Y chromosome with TDF and AZF were not lost [237]. A case of X-autosome translocation was reported involving almost all of chromosome 23 translocating to the p- arms of the X chromosome of a cow [238]. The same translocation was later found in a bull, which showed malformed spermatozoa [162]. Five cases of 1;8 simple translocation (two males and three females), including a carrier of rob(1;29), were reported by [137].

A case of 2q−;5p+ translocation mosaicism has been reported in a bull, identified by chromosome painting using probes generated by conventional microdissection [239]. Its fertility could not be estimated since the owner culled it before reproduction.

3.4. Pericentric Inversion

Few cases of pericentric inversions have been reported in cattle. Popescu [240] found a pericentric inversion involving BTA14 in a female bovine showing reduced fertility. Switonsky [241] found a pericentric inversion involving one of the two X chromosomes in a female with reduced fertility. Iannuzzi et al. [242] found a pericentric inversion in the Y chromosome of 12 male offspring (Podolian breed), of which one had a female-shaped head with reduced horn size, signs of udders, a significantly reduced scrotum, and an atrophic penis. Once slaughtered, an atrophic penis, absence of testis, sign of prostate, and absence of internal female organs were observed. All the remaining carriers of the chromosome abnormality showed normal phenotypes.

De Lorenzi et al. [243] found a possible case of pericentric inversion in the autosomes of a young male cattle. Still, after a detailed FISH-mapping analysis, the authors demonstrated that a centromere repositioning had occurred in BTA17. Subsequent CGH and SNP arrays indicated no loss or gain had occurred in the centromeric region of BTA17 or other BTA17 regions [243].

3.5. Tandem Fusion (TAN)

The TANs found so far are centromere–telomere (with two active centromeres as revealed by C-banding techniques) and were rarely found in domestic bovids. Hansen [244] found a case of TAN in the red Holstein breed, while two cases of TAN were found in a male and female of Romanian cattle [95], demonstrating the maternal origin of this abnormality by genealogical investigations. The female carrier of TAN showed a lower non-returned rate and had only two offspring, of which one had a normal karyotype and the other carried the same TAN. The evolution of male carriers was fascinating because the first two analyses revealed a large percentage of mitosis with TAN. Subsequent investigations in four examinations revealed a decreasing number of mitosis with TAN until a total lack of TAN occurred. Indeed, six descendants of this bull showed normal karyotypes [95]. A particular case of TAN (1;16) has been found in a Brown Swiss bull affected by anatomical defects with the simultaneous presence of both TAN(1;16) and trisomy 16 [24]. A case of TAN (4;21) was found in a new-born Holstein-Friesian heifer, which was also XX/XY mosaic (free-martin) [245].

A recent TAN case has been found in a female calf affected by hypospadias, growth retardation, and ventricular septal defects [42]. The TAN involved BTA18 and BTA27 with an accompanying loss of genomic sequences, as demonstrated by chromosome banding, FISH mapping, and genome sequencing [42].

3.6. Cytogenetically Detectable Deletions and Duplications

Genetic deletions and duplications have been reported in several studies using genomic approaches and have rarely been reported as chromosome abnormalities. This is probably due to the harmful effects of large genomic losses (deletions) or gains (duplications). These conditions can cause the death of embryos in early embryonic life, especially chromosome deletions. Among the few reported cases of chromosome deletions, only two involved an autosome: the first one in an infertile cow [246] and another one, more recently, in a female calf with several anatomic defects (head asymmetry, relocation of the frontal sinus and eye orbits, hypoplastic thymus without neck part, ductus Botalli, unfinished obliteration in umbilical arteries, and a bilateral series of tooth germs in the temporal region) [247]. In this case, mosaic cells were observed, of which 92% were normal ($2n = 60$, XX) and 8% abnormal ($2n = 60$, XX+ mar) due to the presence of a small marker chromosome showing only the centromere and a proximal part due to the deletion of the remaining material [247].

The remaining cases of deletions involve the X chromosome (generally the inactive and late-replicating X). Indeed, chromosome abnormalities are often found on sex chromosomes because they are more tolerated by the species (for gene inactivation in one of the two Xs) and easily discovered for both shape and C-banding, which are different from the autosomes. A Swiss Holstein bovine, affected by hypotrichosis and oligodontia, was found affected by Xq deletion [248]. A large Xq-arm deletion has been found in a cow carrying rob(1;29) [249]. An interesting case of Xp deletion ($2n = 60$, XX) has been found in a young cow of the Marchigiana breed (central Italy) with normal body conformation and external genitalia [250]. Detailed cytogenetic investigation by both C- and R-banding and FISH-mapping techniques showed that almost all the p arms of the late-replicating (inactive) X chromosome were absent. A CGH-array analysis showed that the deletion involved the Xp arm from the telomere to around 39.5 Mb, referring to the BosTau6 cattle genome assembly. This abnormality deletes about 40 Mb of the X-chromosome sequences, but none of them are programmed to escape from inactivation despite the large number of genes deleted, explaining the normal phenotype of the female. However, this carrier gave rise to a female carrying the same deletion, which later would not remain pregnant after several services and was then eliminated from the farm. The second female carrier gave birth to two calves, both females, of which one was normal and another one carried the same deletion. Later, after several failed services the mother carrier was eliminated from the farm [251]. Both female carries had essentially similar reproductive problems.

Only two cases of chromosome duplications correlating to abnormal phenotypes have been reported in cattle. A possible duplication of a survival motor neuron gene (SMN) has been demonstrated in a calf affected by arthrogryposis (a disease characterized by congenital contractures in the limbs having different origins) using extended-chromosome fiber-FISH [252]. Another chromosome duplication of about 99 Kb has been found in BTA18 using a CGH array on an XY female cattle (SRY positive) affected by a disorder of sex development (DSD), although the authors could not demonstrate its relationship with the phenotype [253].

4. Conclusions

As shown in this review, there is a strict relationship between chromosome abnormalities and fertility problems in domestic bovids. In particular, numerical abnormalities have been found very rarely because of their phenotypical visibility, resulting in elimination by breeders. On the other hand, numerical sex chromosome abnormalities often escape selection, as the body conformation and external genitalia are generally normal, but are responsible for sterility in most of cases, including free-martinism, or lower fertility. Structural chromosome abnormalities are usually related to lower fertility compared to normal-karyotyped animals. However, centric-fusion translocations are often present in high percentages in meat breeds, particularly rob(1;29). For this reason, many breeder associations required karyotype analyses for males bred for reproduction, especially for AI, only in meat breeds. This choice is only partially correct because animals belonging to dairy breeds are generally not examined. This could cause reproductive problems in animals, as has occurred in the Italian Friesian breed, where 16.2% of the investigated animals (males and females showing reproductive problems) were found to be carriers of sex chromosome abnormalities, especially of XX/XY mosaicism (see [80]). Finally, only with a good collaboration between breeders, veterinary doctors, and cytogeneticists, as well as between different labs that use genomic and/or cytogenetic approaches, is it possible to correctly investigate the presence of chromosome abnormalities and their effects on fertility in domestic animals in order to better select reproductive animals to improve both their genetic and economic value.

Author Contributions: Conceptualization, L.I. and P.P.; writing—original draft preparation, L.I., A.I. and P.P.; writing—review and editing, L.I. and A.I. All authors have read and agreed to the published version of the manuscript.

Funding: The study has been supported by the project PON1_486 GENOBU.

Institutional Review Board Statement: The Institutional Review Board statement was not required.

Data Availability Statement: Data sharing is not applicable to this article as no new data were created or analyzed in this study.

Conflicts of Interest: The authors declare no conflict of interest.

Abbreviations

BTA	*Bos taurus* chromosome, $2n = 60$
BBU	*Bubalus bubalis* chromosome, $2n = 50$
OAR	*Ovis aries* chromosome, $2n = 54$
CHI	*Capra hircus* chromosome, $2n = 60$
BIN	*Bos indicus* chromosome, $2n = 60$
FISH	fluorescence in situ hybridization
Fiber-FISH	extended chromatin fiber-FISH
CGH-array	comparative genomic hybridization array
DSD	disorder sex development

References

1. Gustavsson, I.; Rockborn, G. Chromosome abnormality in three cases of lymphatic leukemia in cattle. *Nature* **1964**, *203*, 990. [CrossRef]
2. Gustavsson, I. Cytogenetics, distribution and phenotypic effects of a translocation in swedish cattle. *Hereditas* **2009**, *63*, 68–169. [CrossRef]
3. Gustavsson, I. Distribution of the 1/29 translocation in the A.I. Bull population of Swedish Red and White cattle. *Hereditas* **1971**, *69*, 101–106. [CrossRef] [PubMed]
4. Dyrendahl, I.; Gustavsson, I. Sexual functions, semen characteristics and fertility of bulls carrying the 1/29 chromosome translocation. *Hereditas* **1979**, *90*, 281–289. [CrossRef] [PubMed]

5. Kanagawa, H.; Kawata, K.; Ishikawa, T.; Odajima, T.; Inoue, T. Chromosome studies on heterosexual twins in cattle. 3. Sex-chromosome chimerism (XX/XY) in bone marrow specimens. *Jpn. J. Vet. Res.* **1966**, *14*, 123–126.
6. Mcfeely, R.; Hare, W.; Biggers, J.; Diggers, J. Chromosome Studies in 14 Cases of Intersex in Domestic Mammals. *Cytogenet. Genome Res.* **1967**, *6*, 242–253. [CrossRef] [PubMed]
7. Mcfeely, R.A. Chromosome abnormalities in early embryos of the pig. *J. Reprod. Fertil.* **1967**, *13*, 579–581. [CrossRef] [PubMed]
8. Basrur, P.; Kosaka, S.; Kanagawa, H. Blood Cell Chimerism and Freemartinism in Heterosexual Bovine Quadruplets. *J. Hered.* **1970**, *61*, 15–18. [CrossRef]
9. Fechheimer, N.S. A cytogenetic survey of young bulls in the U.S.A. *Vet. Rec.* **1973**, *93*, 535–536. [CrossRef] [PubMed]
10. Sysa, P.; Bernacki, Z.; Kunska, A. Intersexuality in cattle—A case of male pseudohermaphroditismus with a 60.XY karotype. *Vet. Rec.* **1974**, *94*, 30–31. [CrossRef]
11. Dunn, H.O.; Lein, D.H.; McEntee, K. Testicular hypoplasia in a Hereford bull with 61,XXY karyotype: The bovine counterpart of human Klinefelter's syndrome. *Cornell Vet.* **1980**, *70*, 137–146. [PubMed]
12. King, W. Chromosome variation in the embryos of domestic animals. *Cytogenet. Genome Res.* **2008**, *120*, 81–90. [CrossRef]
13. Iannuzzi, L.; Di Berardino, D. Tools of the trade: Diagnostics and research in domestic animal cytogenetics. *J. Appl. Genet.* **2008**, *49*, 357–366. [CrossRef] [PubMed]
14. Hayes, H. Chromosome painting with human chromosome-specific DNA libraries reveals the extent and distribution of conserved segments in bovine chromosomes. *Cytogenet. Cell Genet.* **1995**, *71*, 168–174. [CrossRef]
15. Iannuzzi, A.; Pereira, J.; Iannuzzi, C.; Fu, B.; Ferguson-Smith, M. Pooling strategy and chromosome painting characterize a living zebroid for the first time. *PLoS ONE* **2017**, *12*, e0180158. [CrossRef] [PubMed]
16. De Lorenzi, L.; Rossi, E.; Gimelli, S.; Parma, P. De novo Reciprocal Translocation t(5;6)(q13;q34) in Cattle: Cytogenetic and Molecular Characterization. *Cytogenet. Genome Res.* **2013**, *142*, 95–100. [CrossRef] [PubMed]
17. Cribiu, E.P.; Di Berardino, D.; Di Meo, G.P.; Gallagher, D.S.; Hayes, H.; Iannuzzi, L.; Popescu, C.P.; Rubes, J.; Schmutz, S.; Stranzinger, G.; et al. International System for Chromosome Nomenclature of Domestic Bovids (ISCNDB 2000). *Cytogenet. Genome Res.* **2001**, *95*, 283–299. [CrossRef]
18. Agerholm, J.S.; Christensen, K. Trisomy 22 in a calf. *J. Vet. Med. Ser. A* **1993**, *40*, 576–581. [CrossRef] [PubMed]
19. Christensen, K.; Juul, L. A Case of Trisomy 22 in a Live Hereford Calf. *Acta Vet. Scand.* **1999**, *40*, 85–88. [CrossRef]
20. Iannuzzi, L.; Meo, G.P.; Leifsson, P.S.; Eggen, A.; Christensen, K. A Case of Trisomy 28 in Cattle Revealed by both Banding and FISH-Mapping Techniques. *Hereditas* **2001**, *134*, 147–151. [CrossRef]
21. Dunn, H.O.; Johnson, R.H., Jr. A 61,XY Cell Line in a Calf with Extreme Brachygnathia. *J. Dairy Sci.* **1972**, *55*, 524–526. [CrossRef]
22. Herzog, A.; Höhn, H.; Rieck, G. Survey of recent situation of chromosome pathology in different breeds of german cattle. *Ann. Genet. Sel. Anim.* **1977**, *9*, 471–491. [CrossRef] [PubMed]
23. Herzog, A.; Hoehn, H. Uber zwei weitere Fälle von autosomaler Trisomie, 61,XY, + 12 und 61,XX, + 12, beim Rind [Two ad-ditional cases of autosomal trisomy, 61,XY, + 12 and 61,XX, + 12, in cattle]. *Cytogenet. Cell Genet.* **1991**, *57*, 211–213. [CrossRef] [PubMed]
24. Kovács, A.; Foote, R.H.; Lein, D.H. 1;16 tandem translocation with trisomy 16 in a Brown Swiss bull. *Vet. Rec.* **1990**, *127*, 205. [PubMed]
25. Tschudi, P.; Ueltschi, G.; Martig, J.; Küpfer, U. Autosomale Trisomie als Ursache eines hohen Ventrikelseptumdefekts bei einem Kalb der Simmentalerrasse [Autosomal trisomy as the cause of a high ventricular septal defect in a calf of the Simmental breed]. *Schweiz. Arch. Tierheilkd.* **1975**, *117*, 335–340. [PubMed]
26. Kulikova, S.G.; Petukhov, V.L.; Grafodatskiĭ, A.S. Novyĭ sluchaĭ trisomii u krupnogo rogatogo skota [A new case of trisomy in cattle]. *Tsitol. Genet.* **1991**, *25*, 28–31. [PubMed]
27. Murakami, R.; Miyake, Y.; Kaneda, Y. Cases of XY female, single-birth freemartin and trisomy (61, XX, +20) observed in cy-togenetical studies on 18 sterile heifers. *Nihon Juigaku Zasshi* **1989**, *51*, 941–945. [CrossRef]
28. Lioi, M.B.; Scarfì, M.R.; Di Berardino, D. An autosomal trisomy in cattle. *Genet. Sel. Evol.* **1995**, *27*, 473–476. [CrossRef]
29. Gallagher, D.S., Jr.; Lewis, B.C.; De Donato, M.; Davis, S.K.; Taylor, J.F.; Edwards, J.F. Autosomal trisomy 20 (61,XX, + 20) in a malformed bovine fetus. *Vet. Pathol.* **1999**, *36*, 448–451. [CrossRef] [PubMed]
30. Häfliger, I.M.; Agerholm, J.S.; Drögemüller, C. Constitutional trisomy 20 in an aborted Holstein fetus with pulmonary hy-poplasia and anasarca syndrome. *Anim. Genet.* **2020**, *51*, 988–989. [CrossRef] [PubMed]
31. Long, S.E. Autosomal trisomy in a calf. *Vet. Rec.* **1984**, *115*, 16–17. [CrossRef] [PubMed]
32. Schmutz, S.M.; Moker, J.S.; Clark, E.G.; Orr, J.P. Chromosomal Aneuploidy Associated with Spontaneous Abortions and Neonatal Losses in Cattle. *J. Vet. Diagn. Investig.* **1996**, *8*, 91–95. [CrossRef] [PubMed]
33. Mayr, B.; Krutzler, H.; Auer, H.; Schleger, W.; Sasshofer, K.; Glawischnig, E. A viable calf with trisomy 22. *Cytogenet. Genome Res.* **1985**, *39*, 77–79. [CrossRef] [PubMed]
34. Mayr, B.; Schellander, K.; Auer, H.; Tesarik, E.; Schleger, W.; Sasshofer, K.; Glawischnig, E. Offspring of a trisomic cow. *Cytogenet. Genome Res.* **1987**, *44*, 229–230. [CrossRef] [PubMed]
35. Coates, J.; Rousseaux, C.; Schmutz, S. Multiple defects in an aborted bovine foetus associated with chromosomal trisomy. *N. Z. Vet. J.* **1987**, *35*, 173–174. [CrossRef] [PubMed]
36. Schmutz, S.M.; Coates, J.W.; Rousseaux, C.G. Chromosomal Trisomy in an Anomalous Bovine Fetus. *Can. Vet. J.* **1987**, *28*, 61–62. [PubMed]
37. Mäkinen, A.; Alitalo, I.; Alanko, M. Autosomal Trisomy in a Heifer. *Acta Vet. Scand.* **1987**, *28*, 1–8. [CrossRef]

38. Ducos, A.; Seguela, A.; Pinton, A.; Berland, H.; Brun-Baronnat, C.; Darre, R.; Manesse, M.; Darre, A. Trisomy 26 mosaicism in a sterile Holstein-Friesian heifer. *Vet. Rec.* **2000**, *146*, 163–164. [CrossRef]
39. Iannuzzi, A.; Genualdo, V.; Perucatti, A.; Pauciullo, A.; Varricchio, G.; Incarnato, D.; Matassino, D.; Iannuzzi, L. Fatal Outcome in a Newborn Calf Associated with Partial Trisomy 25q and Partial Monosomy 11q, 60,XX,der(11)t(11;25)(q11;q14~21). *Cytogenet. Genome Res.* **2015**, *146*, 222–229. [CrossRef] [PubMed]
40. Häfliger, I.M.; Seefried, F.; Drögemüller, C. Trisomy 29 in a stillborn Swiss Original Braunvieh calf. *Anim. Genet.* **2020**, *51*, 483–484. [CrossRef] [PubMed]
41. Perucatti, A.; Genualdo, V.; Iannuzzi, A.; De Lorenzi, L.; Matassino, D.; Parma, P.; Di Berardino, D.; Di Meo, G. A New and Unusual Reciprocal Translocation in Cattle: Rcp(11;25)(q11;q14–21). *Cytogenet. Genome Res.* **2011**, *134*, 96–100. [CrossRef] [PubMed]
42. Iannuzzi, A.; Braun, M.; Genualdo, V.; Perucatti, A.; Reinartz, S.; Proios, I.; Heppelmann, M.; Rehage, J.; Hülskötter, K.; Beineke, A.; et al. Clinical, cytogenetic and molecular genetic characterization of a tandem fusion translocation in a male Holstein cattle with congenital hypospadias and a ventricular septal defect. *PLoS ONE* **2020**, *15*, e0227117. [CrossRef] [PubMed]
43. Hyttel, P.; Viuff, D.; Laurincik, J.; Schmidt, M.; Thomsen, P.; Avery, B.; Callesen, H.; Rath, D.; Niemann, H.; Rosenkranz, H.; et al. Risks of in-vitro production of cattle and swine embryos: Aberrations in chromosome numbers, ribosomal RNA gene activation and perinatal physiology. *Hum. Reprod.* **2000**, *15*, 87–97. [CrossRef] [PubMed]
44. Lyon, M.F. Gene Action in the X-chromosome of the Mouse (*Mus musculus* L.). *Nat. Cell Biol.* **1961**, *190*, 372–373. [CrossRef]
45. Burgoyne, P.S.; Ojarikre, O.A.; Turner, J.M.A. Evidence that postnatal growth retardation in XO mice is due to haploinsufficiency for a non-PAR X gene. *Cytogenet. Genome Res.* **2002**, *99*, 252–256. [CrossRef]
46. Iannuzzi, L.; King, W.; Di Berardino, D. Chromosome Evolution in Domestic Bovids as Revealed by Chromosome Banding and FISH-Mapping Techniques. *Cytogenet. Genome Res.* **2009**, *126*, 49–62. [CrossRef] [PubMed]
47. Rieck, G.W.; Höhn, H.; Herzog, A. X-Trisomie beim Rind mit Anzeichen familiärer Disposition für Meiosestörungen [X-trisomy in cattle with signs of familial disposition for meiotic disturbances]. *Cytogenetics* **1970**, *9*, 401–409. (In German) [CrossRef] [PubMed]
48. Norberg, H.S.; Refsdal, A.O.; Garm, O.N.; Nes, N. A case report on X-trisomy in cattle. *Hereditas* **2009**, *82*, 69–72. [CrossRef]
49. Buoen, L.C.; Seguin, B.E.; Weber, A.F.; Shoffner, R.N. X-trisomy karyotype and associated infertility in a Holstein heifer. *J. Am. Vet. Med. Assoc.* **1981**, *179*, 808–811.
50. Moreno-Millan, M.; Bermejo, J.V.D.; Garcia, F.A. X-trisomy in Friesian cow with continuous oestrus. *Vet. Rec.* **1987**, *121*, 167–168. [CrossRef]
51. Pinheiro, L.; Almeida, I.; Garcia, J.; Basrur, P. Trisomy X and translocation in infertile heifers. *Theriogenology* **1987**, *28*, 891–898. [CrossRef]
52. Citek, J.; Rubeš, J.; Hájková, J. Short communication: Robertsonian translocations, chimerism, and aneuploidy in cattle. *J. Dairy Sci.* **2009**, *92*, 3481–3483. [CrossRef] [PubMed]
53. Yadav, B.R.; Balakrishnan, C.R. Trisomy of the X chromosome in a Murrah buffalo. *Vet. Rec.* **1982**, *111*, 184–185. [CrossRef] [PubMed]
54. Prakash, B.; Balain, D.S.; Lathwal, S.S.; Malik, R.K. Trisomy-X in a sterile river buffalo. *Vet. Rec.* **1994**, *134*, 241–242. [CrossRef] [PubMed]
55. Iannuzzi, L.; Di Meo, G.P.; Perucatti, A.; Incarnato, D.; Palo, R.D.; Zicarelli, L. Reproductive disturbances and sex chromosome abnormalities in two female river buffaloes. *Vet. Rec.* **2004**, *154*, 823–824. [CrossRef]
56. Powell, C. Sex chromosomes and sex chromosome abnormalities. In *The Principles of Clinical Cyto-Genetics*; Gersen, S., Keagle, M., Eds.; Humana Press: Totowa, NJ, USA, 1999; pp. 229–258.
57. Prakash, B.; Balain, D.S.; Lathwal, S.S.; Malik, R.K. Infertility associated with monosomy-X in a crossbred cattle heifer. *Vet. Rec.* **1995**, *137*, 436–437. [CrossRef] [PubMed]
58. Romano, J.E.; Raussdepp, T.; Mulon, P.Y.; Villadóniga, G.B. Non-mosaic monosomy 59,X in cattle: A case report. *Anim. Reprod. Sci.* **2015**, *156*, 83–90. [CrossRef] [PubMed]
59. Berry, D.P.; Wolfe, A.; O'Donovan, J.; Byrne, N.; Sayers, R.G.; Dodds, K.G.; McEwan, J.C.; O'Connor, R.E.; McClure, M.; Purfield, D.C. Characterization of an X-chromosomal non-mosaic monosomy (59, X0) dairy heifer detected using routinely available single nucleotide polymorphism genotype data. *J. Anim. Sci.* **2017**, *95*, 1042–1049.
60. Pinheiro, L.E.L.; Mikich, A.B.; Bechara, G.H.; Almeida, I.L.; Basrur, P.K. Isochromosome Y in an infertile heifer. *Genome* **1990**, *33*, 690–695. [CrossRef] [PubMed]
61. Yadav, B.R.; Kumar, P.; Tomer, O.S.; Kumar, S.; Balain, D.S. Monosomy X and gonadal dysgenesis in a buffalo heifer (*Bubalus bubalis*). *Theriogenology* **1990**, *34*, 99–105. [CrossRef]
62. Prakash, B.; Balain, D.S.; Lathwal, S.S. A 49, XO sterile murrah buffalo (*Bubalus bubalis*). *Vet. Rec.* **1992**, *130*, 559–560. [CrossRef]
63. Iannuzzi, L.; Di Meo, G.P.; Perucatti, A.; Zicarelli, L. Sex chromosome monosomy (2n = 49,X) in a river buffalo (Bubalus bu-balis). *Vet. Rec.* **2000**, *147*, 690–691.
64. Zartman, D.L.; Hinesley, L.L.; Gnatkowski, M.W. A 53, X female sheep (Ovis aries). *Cytogenet. Genome Res.* **1981**, *30*, 54–58. [CrossRef] [PubMed]
65. Baylis, M.; Wayte, D.; Owen, J. An XO/XX mosaic sheep with associated gonadal dysgenesis. *Res. Vet. Sci.* **1984**, *36*, 125–126. [CrossRef]
66. Berry, D.P.; O'Brien, A.; O'Donovan, J.; McHugh, N.; Wall, E.; Randles, S.; McDermott, K.; O'Connor, R.E.; Patil, M.A.; Ho, J.; et al. Aneuploidy in dizygotic twin sheep detected using genome-wide single nucleotide polymorphism data from two commonly used commercial vendors. *Animal* **2018**, *12*, 2462–2469. [CrossRef] [PubMed]
67. Bhatia, S.; Shanker, V. A case report on XO/XX/XXX mixoploidy in a goat. *Vet. Rec.* **1990**, *126*, 312–313. [PubMed]
68. Switonski, M.; Szczerbal, I. Chromosome Abnormalities in Domestic Animals as Causes of Disorders of Sex Development or Impaired Fertility. In *Insights from Animal Reproduction*; Intech Open Science: London, UK, 2016.

69. Rieck, G.W.; Höhn, H.; Herzog, A. Hypogonadismus, intermittierender Kryptorchismus und segmentäre Aplasie der Ductus Wolffii bei einem männlichen Rind mit XXY–Gonosomen-Konstellation bzw. XXY–XX–XY-Gonosomen-Mosaik [Hy-pogonadism, intermittent cryptorchism and segmentary aplasia of the Wolffian duct in a bull with XXY gonosome constel-lation or XXY–XX–XY gonosome mosaic]. *Dtsch. Tierarztl. Wochenschr.* **1969**, *76*, 133–138.
70. Rieck, G.W.; Höhn, H.; Schmidt, I. Vulvaaplasie und Urethra masculina: Maskulinisierungseffekte des Sinus urogenitalis bei genetisch weiblichen Rindern durch Chimärismen mit dem XXY-Gonosomen-Komplement [Aplastic vulva and masculine urethra: Masculinization effects of the urogenital sinus in genetically female cattle due to chimerism with the XXY gonosomal complement]. *Berl. Munch. Tierarztl. Wochenschr.* **1982**, *95*, 181–185.
71. Dain, A.R.; Bridge, P.S. A chimaeric calf with XY/XXY mosaicism and intersexuality. *J. Reprod. Fertil.* **1978**, *54*, 197–201. [CrossRef]
72. Logue, D.N.; Harvey, M.J.; Munro, C.D.; Lennox, B. Hormonal and histological studies in a 61XXY bull. *Vet. Rec.* **1979**, *104*, 500–503. [CrossRef]
73. Miyake, Y.; Ishikawa, T.; Kanagawa, H.; Sato, K. A first case of XY/XYY mosaic bull. *Jpn. J. Vet. Res.* **1981**, *29*, 94–96. [PubMed]
74. Alam, M.G.S.; Hurtado, R. Testicular hypoplasia syndrome due to chromosomal aberration (Trisomy 61/XXY). *Indian Vet. J.* **1982**, *2*, 55–60.
75. Schmutz, S.M.; Barth, A.D.; Moker, J.S. A Klinefelter bull with a 1;29 translocation born to a fertile 61,XXX cow. *Can. Vet. J.* **1994**, *35*, 182–184.
76. Molteni, L.; Macchi, A.D.G.; Meggiolaro, D.; Sironi, G.; Enice, F.; Popescu, P. New cases of XXY constitution in cattle. *Anim. Reprod. Sci.* **1999**, *55*, 107–113. [CrossRef]
77. Joerg, H.; Janett, F.; Schlatt, S.; Mueller, S.; Graphodatskaya, D.; Suwattana, D.; Asai, M.; Stranzinger, G. Germ cell transplan-tation in an azoospermic Klinefelter bull. *Biol. Reprod.* **2003**, *69*, 1940–4194. [CrossRef] [PubMed]
78. Jaszczak, K.; Parada, R.; Wardecka, B.; Niemczewski, C. A note on analysis of chromosome constitution in tissues, quality of semen and DNA microsatellite loci in bull with 60,XY/61,XXY karyotype. *J. Anim. Feed. Sci.* **2003**, *12*, 521–527. [CrossRef]
79. Słota, E.; Kozubska-Sobocińska, A.; Kościelny, M.; Danielak-Czech, B.; Rejduch, B. Detection of the XXY trisomy in a bull by using sex chromosome painting probes. *J. Appl. Genet.* **2003**, *44*, 379–382.
80. Ducos, A.; Revay, T.; Kovacs, A.; Hidas, A.; Pinton, A.; Bonnet-Garnier, A.; Molteni, L.; Slota, E.; Switonski, M.; Arruga, M.V.; et al. Cytogenetic screening of livestock populations in Europe: An overview. *Cytogenet. Genome Res.* **2008**, *120*, 26–41. [CrossRef]
81. Patel, R.; Singh, K.; Soni, K.; Chauhan, J. Novel cytogenetic finding: An unusual X;X-translocation in Mehsana buffalo (*Bubalus bubalis*). *Cytogenet. Genome Res.* **2006**, *115*, 186–188. [CrossRef]
82. Bruere, A.N.; Marshall, R.B.; Ward, D.P.J. Testicular hypoplasia and xxy sex chromosome complement in two rams: The ovine counterpart of klinefelter's syndrome in man. *J. Reprod. Fertil.* **1969**, *19*, 103–108. [CrossRef]
83. Moraes, J.C.; Mattevi, M.S.; Ferreira, J.M. Chromosome studies in Brazilian rams. *Vet. Rec.* **1980**, *107*, 489–490. [CrossRef]
84. Takebayashi, S.G.; Jorg, W. Testicular hypoplasia in a horned goat with 61, XXY/60,XY karyotype. *Jpn. J. Genet.* **1986**, *61*, 177–181. [CrossRef]
85. Bhatia, S.; Shanker, V. First report of a XX/XXY fertile goat buck. *Vet. Rec.* **1992**, *130*, 271–272. [CrossRef] [PubMed]
86. Henricson, B.; Åkesson, A. Two Heifers with Gonadal Dysgenesis and the Sex Chromosomal Constitution XY. *Acta Vet. Scand.* **1967**, *8*, 262–272. [CrossRef]
87. Chapman, H.; Bruère, A.; Jaine, P. XY gonadal dysgenesis in a Charolais heifer. *Anim. Reprod. Sci.* **1978**, *1*, 9–18. [CrossRef]
88. Sharma, A.K.; Vijaykumar, N.K.; Khar, S.K.; Verma, S.K.; Nigam, J.M. XY gonadal dysgenesis in a heifer. *Vet. Rec.* **1980**, *107*, 328–330. [CrossRef]
89. Macmillan, K.L.; Fielden, E.D.; McNatty, K.P.; Henderson, H.V. LH concentrations in two cattle with XY gonadal dysgenesis. *J. Reprod. Fertil.* **1984**, *71*, 525–531. [CrossRef] [PubMed]
90. Kondoh, S.; Miyake, Y.; Nakahori, Y.; Nakagome, Y.; Kaneda, Y. Cytogenetical and molecular biological studies on a bovine XY female. *J. Vet. Med. Sci.* **1992**, *54*, 1077–1080. [CrossRef] [PubMed]
91. Hare, J.E.; Baird, J.D.; Duignan, P.; Saunders, J.; Floetenmeyer, R.; Basrur, P.K. XY gonadal dysgenesis and tetralogy of Fallot in an Angus calf. *Can. Vet. J.* **1994**, *35*, 510–512.
92. Kawakura, K.; Miyake, Y.I.; Murakami, R.K.; Kondoh, S.; Hirata, T.I.; Kaneda, Y. Deletion of the SRY region on the Y chro-mosome detected in bovine gonadal hypoplasia (XY female) by PCR. *Cytogenet. Cell Genet.* **1996**, *72*, 183–184. [CrossRef] [PubMed]
93. Kawakura, K.; Miyake, Y.-I.; Murakami, R.; Kondoh, S.; Hirata, T.-I.; Kaneda, Y. Abnormal structure of the Y chromosome detected in bovine gonadal hypoplasia (XY female) by FISH. *Cytogenet. Genome Res.* **1997**, *76*, 36–38. [CrossRef] [PubMed]
94. Kieffer, M.; Sorensen, A.M., Jr. Some cytogenetic aspects of intersexuality in the bovine. *J. Anim. Sci.* **1971**, *32*, 1219–1228. [CrossRef]
95. Nicolae, I.; Popescu, C.P. Cytogenetic studies on Romanian cattle breeds. *Arch. Zootec.* **2001**, *50*, 355–361.
96. Iannuzzi, L.; Di Meo, G.P.; Perucatti, A.; Di Palo, R.; Zicarelli, L. 50,XY gonadal dysgenesis (Swyer's syndrome) in a female river buffalo (*Bubalus bubalis*). *Vet. Rec.* **2001**, *148*, 634–635. [CrossRef] [PubMed]
97. Ferrer, L.; Monteagudo, L.; De Jalon, J.G.; Tejedor, M.; Ramos, J.; Lacasta, D. A Case of Ovine Female XY Sex Reversal Syndrome Not Related to Anomalies in the Sex-Determining Region Y (SRY). *Cytogenet. Genome Res.* **2009**, *126*, 329–332. [CrossRef]
98. Albarella, S.; D'Anza, E.; Galdiero, G.; Esposito, L.; De Biase, D.; Paciello, O.; Ciotola, F.; Peretti, V. Cytogenetic Analyses in Ewes with Congenital Abnormalities of the Genital Apparatus. *Animals* **2019**, *9*, 776. [CrossRef]

99. Sulimovici, S.; Weissenberg, R.; Lunenfeld, B.; Padeh, B.; Soller, M. Testicular testosterone biosynthesis in male Saanen goats with XX sex chromosomes. *Clin. Genet.* **2008**, *13*, 397–403. [CrossRef] [PubMed]
100. Pailhoux, E.; Cribiu, E.P.; Chaffaux, S.; Darre, R.; Fellous, M.; Cotinot, C. Molecular analysis of 60,XX pseudohermaphrodite polled goats for the presence of SRY and ZFY genes. *J. Reprod. Fertil.* **1994**, *100*, 491–496. [CrossRef] [PubMed]
101. Just, W.; Almeida, C.C.; Goldshmidt, B.; Vogel, W. The Male Pseudohermaphrodite XX Polled Goat is Zfy and Sry Negative. *Hereditas* **2004**, *120*, 71–75. [CrossRef] [PubMed]
102. Vaiman, D.; Koutita, O.; Oustry, A.; Elsen, J.-M.; Manfredi, E.; Fellous, M.; Cribiu, E.P. Genetic mapping of the autosomal region involved in XX sex-reversal and horn development in goats. *Mamm. Genome* **1996**, *7*, 133–137. [CrossRef] [PubMed]
103. Payen, E.J.; Cotinot, C.Y. Sequence evolution of SRY gene within Bovidae family. *Mamm. Genome* **1994**, *5*, 723–725. [CrossRef]
104. Biason-Lauber, A.; Konrad, D.; Meyer, M.; Debeaufort, C.; Schoenle, E.J. Ovaries and Female Phenotype in a Girl with 46,XY Karyotype and Mutations in the CBX2 Gene. *Am. J. Hum. Genet.* **2009**, *84*, 658–663. [CrossRef]
105. Pailhoux, E.; Vigier, B.; Vaiman, D.; Schibler, L.; Vaiman, A.; Cribiu, E.; Nezer, C.; Georges, M.; Sundström, J.; Pelliniemi, L.J.; et al. Contribution of domestic animals to the identification of new genes involved in sex determination. *J. Exp. Zoo.* **2001**, *290*, 700–708. [CrossRef] [PubMed]
106. Pepene, C.E.; Coman, I.; Mihu, D.; Militaru, M.; Duncea, I. Infertility in a new 46, XX male with positive SRY confirmed by fluorescence in situ hybridization: A case report. *Clin. Exp. Obstet. Gynecol.* **2008**, *35*, 299–300.
107. Nebesio, T.D.; Torres-Martinez, W.; Rink, R.C.; Eugster, E.A. Spurious Case of XX Maleness in A Patient With A History of Wiskott-Aldrich Syndrome. *Endocr. Pract.* **2011**, *17*, e1–e3. [CrossRef]
108. Di Meo, G.P.; Perucatti, A.; Floriot, S.; Incarnato, D.; Rullo, R.; Jambrenghi, A.C.; Ferretti, L.; Vonghia, G.; Cribiu, E.; Eggen, A.; et al. Chromosome evolution and improved cytogenetic maps of the Y chromosome in cattle, zebu, river buffalo, sheep and goat. *Chromosom. Res.* **2005**, *13*, 349–355. [CrossRef]
109. Das, P.; Chowdhary, B.; Raudsepp, T. Characterization of the Bovine Pseudoautosomal Region and Comparison with Sheep, Goat, and Other Mammalian Pseudoautosomal Regions. *Cytogenet. Genome Res.* **2009**, *126*, 139–147. [CrossRef] [PubMed]
110. Raudsepp, T.; Das, P.; Avila, F.; Chowdhary, B. The Pseudoautosomal Region and Sex Chromosome Aneuploidies in Domestic Species. *Sex. Dev.* **2012**, *6*, 72–83. [CrossRef] [PubMed]
111. Villagómez, D.; Parma, P.; Radi, O.; Di Meo, G.; Pinton, A.; Iannuzzi, L.; King, W. Classical and Molecular Cytogenetics of Disorders of Sex Development in Domestic Animals. *Cytogenet. Genome Res.* **2009**, *126*, 110–131. [CrossRef]
112. Favetta, L.; Villagómez, D.; Iannuzzi, L.; Di Meo, G.; Webb, A.; Crain, S.; King, W. Disorders of Sexual Development and Abnormal Early Development in Domestic Food-Producing Mammals: The Role of Chromosome Abnormalities, Environment and Stress Factors. *Sex. Dev.* **2012**, *6*, 18–32. [CrossRef]
113. Parma, P.; Veyrunes, F.; Pailhoux, E. Sex Reversal in Non-Human Placental Mammals. *Sex. Dev.* **2016**, *10*, 326–344. [CrossRef]
114. Padula, A. The freemartin syndrome: An update. *Anim. Reprod. Sci.* **2005**, *87*, 93–109. [CrossRef] [PubMed]
115. Komisarek, J.; Dorynek, Z. Genetic aspects of twinning in cattle. *J. Appl. Genet.* **2002**, *43*, 55–68.
116. Greene, W.; Dunn, H.; Foote, R. Sex-chromosome ratios in cattle and their relationship to reproductive development in freemartins. *Cytogenet. Genome Res.* **1977**, *18*, 97–105. [CrossRef] [PubMed]
117. Di Meo, G.; Perucatti, A.; Di Palo, R.; Iannuzzi, A.; Ciotola, F.; Peretti, V.; Neglia, G.; Campanile, G.; Zicarelli, L. Sex chromosome abnormalities and sterility in river buffalo. *Cytogenet. Genome Res.* **2008**, *120*, 127–131. [CrossRef] [PubMed]
118. Ruvinsky, A.; Spicer, L.J. Developmental genetics: Sex determination and differentiation. In *The Genetics of Cattle*; Fries, R., Ruvinsky, A., Eds.; CARI Puhlishing and CAB International: Wallingford, UK, 1999; pp. 456–461.
119. Cabianca, G.; Rota, A.; Cozzi, B.; Ballarin, C. Expression of AMH in Female Fetal Intersex Gonads in the Bovine. *Anat. Histol. Embryol.* **2006**, *36*, 24–26. [CrossRef] [PubMed]
120. Dunn, H.O.; McEntee, K.; Hall, C.E.; Johnson, R.H., Jr.; Stone, W.H. Cytogenetic and reproductive studies of bulls born co-twin with freemartins. *J. Reprod. Fertil.* **1979**, *57*, 21–30. [CrossRef] [PubMed]
121. Gustavsson, I. Chromosome aberrations and their influence on the reproductive performance of domestic animals Y: A review. *Z. Tierz. Züchtungsbiol.* **1980**, *97*, 176–195. [CrossRef]
122. Bongso, T.A.; Jainudeen, M.R.; Lee, J.Y. Testicular hypoplasia in a bull with XX/XY chimerism. *Cornell Vet.* **1981**, *71*, 376–382.
123. Seguin, B.E.; Zhang, T.Q.; Buoen, L.C.; Weber, A.F.; Ruth, G.R. Cytogenetic survey of Holstein bulls at a commercial artificial insemination company to determine prevalence of bulls with centric fusion and chimeric anomalies. *J. Am. Vet. Med. Assoc.* **2000**, *216*, 65–67. [CrossRef]
124. McDaneld, T.G.; Kuehn, L.A.; Thomas, M.G.; Snelling, W.M.; Sonstegard, T.S.; Matukumalli, L.K.; Smith, T.P.L.; Pollak, E.J.; Keele, J.W. Y are you not pregnant: Identification of Y chromosome segments in female cattle with decreased reproductive efficiency1,2,3,4. *J. Anim. Sci.* **2012**, *90*, 2142–2151. [CrossRef] [PubMed]
125. Payen, E.; Pailhoux, E.; Merhi, R.A.; Gianquinto, L.; Kirszenbaum, M.; Locatelli, A.; Cotinot, C. Characterization of ovine SRY transcript and developmental expression of genes involved in sexual differentiation. *Int. J. Dev. Biol.* **1996**, *40*, 567–575.
126. Bosu, W.T.; Basrur, P.K. Morphological and hormonal features of an ovine and a caprine intersex. *Can. J. Comp. Med.: Rev. Can. de Med. Comp.* **1984**, *48*, 402–409.
127. Smith, M.C.; Dunn, H.O. Freemartin condition in a goat. *J. Am. Vet. Med. Assoc.* **1981**, *178*, 735–737. [PubMed]
128. Bongso, T.; Robinson, E.; Fatimah, I.; Abeynayake, P. Foetal membrane fusion and its developmental consequences in goat twins. *Br. Vet. J.* **1986**, *142*, 59–64. [CrossRef]

129. Santucciu, C.; Iannuzzi, L.; Fogu, G.; Bonelli, P.; Bogliolo, L.; Rosati, I.; Ledda, S.; Zedda, M.T.; Pau, S. Clinical and cytogenetic studies in intersex ewes. *Caryologia* **2006**, *59*, 67–74. [CrossRef]
130. Di Meo, G.; Neglia, G.; Perucatti, A.; Genualdo, V.; Iannuzzi, A.; Crocco, D.; Incarnato, D.; Romano, G.; Parma, P. Numerical Sex Chromosome Aberrations and Abnormal Sex Development in Horse and Sheep. *Sex. Dev.* **2009**, *3*, 329–332. [CrossRef] [PubMed]
131. Szczerbal, I.; Komosa, M.; Nowacka-Woszuk, J.; Uzar, T.; Houszka, M.; Semrau, J.; Musial, M.; Barczykowski, M.; Lukomska, A.; Switonski, M. A Disorder of Sex Development in a Holstein–Friesian Heifer with a Rare Mosaicism (60,XX/90,XYY): A Genetic, Anatomical, and Histological Study. *Animals* **2021**, *11*, 285. [CrossRef] [PubMed]
132. Meinecke, B.; Kuiper, H.; Wohlsein, P.; Wehrend, A.; Meinecke-Tillmann, S.; Drögemüller, C.; Bürstel, D.; Ebeling, S. A Diploid-Triploid (60,XX/90,XYY) Intersex in a Holstein Heifer. *Sex. Dev.* **2006**, *1*, 59–65. [CrossRef] [PubMed]
133. Graham, J.M., Jr.; Hoehn, H.; Lin, M.S.; Smith, D.W. Diploid-triploid mixoploidy: Clinical and cytogenetic aspects. *Pediatrics* **1981**, *68*, 23–28. [PubMed]
134. Oktem, O.; Paduch, D.A.; Xu, K.; Mielnik, A.; Oktay, K. Normal Female Phenotype and Ovarian Development Despite the Ovarian Expression of the Sex-Determining Region of Y Chromosome (SRY) in a 46,XX/69,XXY Diploid/Triploid Mosaic Child Conceived after in Vitro Fertilization–Intracytoplasmic Sperm Injection. *J. Clin. Endocrinol. Metab.* **2007**, *92*, 1008–1014. [CrossRef] [PubMed]
135. De Schepper, G.G.; Aalbers, J.G.; Brake, J.H.T. Double reciprocal translocation heterozygosity in a bull. *Vet. Rec.* **1982**, *110*, 197–199. [CrossRef] [PubMed]
136. Mayr, B.; Krutzler, H.; Auer, H.; Schleger, W. Reciprocal translocation 60,XY, t(8;15) (21;24) in cattle. *J. Reprod. Fertil.* **1983**, *69*, 629–630. [CrossRef]
137. Christensen, K.; Agerholm, J.S.; Larsen, B. Dairy breed bull with complex chromosome translocation: Fertility and linkage studies. *J. Reprod. Fertil.* **2008**, *117*, 199–202. [CrossRef]
138. Kovacs, A.; Villagómez, D.; Gustavsson, I.; Lindblad, K.; Foote, R.; Howard, T. Synaptonemal complex analysis of a three-breakpoint translocation in a subfertile bull. *Cytogenet. Genome Res.* **1992**, *61*, 195–201. [CrossRef]
139. De Giovanni, A.; Succi, G.; Molteni, L.; Castiglioni, M. A new autosomal translocation in "Alpine grey cattle". *Ann. Genet. Sel. Anim.* **1979**, *11*, 115–120. [CrossRef]
140. Ansari, H.; Jung, H.; Hediger, R.; Fries, R.; König, H.; Stranzinger, G. A balanced autosomal reciprocal translocation in an azoospermic bull. *Cytogenet. Genome Res.* **1993**, *62*, 117–123. [CrossRef]
141. Villagómez, D.; Andersson, M.; Gustavsson, I.; Plöen, L. Synaptonemal complex analysis of a reciprocal translocation, rcp(20;24) (q17;q25), in a subfertile bull. *Cytogenet. Genome Res.* **1993**, *62*, 124–130. [CrossRef]
142. Mayr, B.; Korb, H.; Kiendler, S.; Brem, G. Reciprocal X;1 translocation in a calf. *Genet. Sel. Evol.* **1998**, *30*, 305–308. [CrossRef]
143. Ducos, A.; Dumont, P.; Seguela, A.; Pinton, A.; Berland, H.; Brun-Baronnat, C.; Darre, R.; Guienne, B.M.-L.; Humblot, P.; Boichard, D.; et al. A new reciprocal translocation in a subfertile bull. *Genet. Sel. Evol.* **2000**, *32*, 589–598. [CrossRef] [PubMed]
144. Iannuzzi, L.; Molteni, L.; Di Meo, G.; Perucatti, A.; Lorenzi, L.; Incarnato, D.; De Giovanni, A.; Succi, G.; Gustavsson, I. A new balanced autosomal reciprocal translocation in cattle revealed by banding techniques and human-painting probes. *Cytogenet. Cell Genet.* **2001**, *94*, 225–228. [CrossRef] [PubMed]
145. Iannuzzi, L.; Molteni, L.; Di Meo, G.; De Giovanni, A.; Perucatti, A.; Succi, G.; Incarnato, D.; Eggen, A.; Cribiu, E. A case of azoospermia in a bull carrying a Y-autosome reciprocal translocation. *Cytogenet. Cell Genet.* **2001**, *95*, 225–227. [CrossRef] [PubMed]
146. Molteni, L.; Perucatti, A.; Iannuzzi, A.; Di Meo, G.; De Lorenzi, L.; De Giovanni, A.; Incarnato, D.; Succi, G.; Cribiu, E.; Eggen, A.; et al. A new case of reciprocal translocation in a young bull: Rcp(11;21)(q28;q12). *Cytogenet. Genome Res.* **2007**, *116*, 80–84. [CrossRef] [PubMed]
147. De Lorenzi, L.; De Giovanni, A.; Molteni, L.; Denis, C.; Eggen, A.; Parma, P. Characterization of a balanced reciprocal translocation, rcp(9;11)(q27;q11) in cattle. *Cytogenet. Genome Res.* **2007**, *119*, 231–234. [CrossRef] [PubMed]
148. Switonski, M.; Andersson, M.; Nowacka-Woszuk, J.; Szczerbal, I.; Sosnowski, J.; Kopp, C.; Cernohorska, H.; Rubes, J. Identification of a new reciprocal translocation in an AI bull by synaptonemal complex analysis, followed by chromosome painting. *Cytogenet. Genome Res.* **2008**, *121*, 245–248. [CrossRef] [PubMed]
149. De Lorenzi, L.; Kopecna, O.; Gimelli, S.; Cernohorska, H.; Zannotti, M.; Béna, F.; Molteni, L.; Rubes, J.; Parma, P. Reciprocal Translocation t(4;7)(q14;q28) in Cattle: Molecular Characterization. *Cytogenet. Genome Res.* **2010**, *129*, 298–304. [CrossRef] [PubMed]
150. Switonski, M.; Szczerbal, I.; Krumrych, W.; Nowacka-Woszuk, J. A case of Y-autosome reciprocal translocation in a Holstein-Friesian bull. *Cytogenet. Genome Res.* **2011**, *132*, 22–25. [CrossRef] [PubMed]
151. Kochneva, M.L.; Zhidenova, A.N.; Biltueva, L.S.; Kiseleva, Y.T. A new cse of reciprocal translocation rcp(13;26) in cattle. *Biologiya* **2011**, *6*, 84–89.
152. Biltueva, L.; Kulemzina, A.; Vorobieva, N.; Perelman, P.; Kochneva, M.; Zhidenova, A.; Graphodatsky, A. A New Case of an Inherited Reciprocal Translocation in Cattle: Rcp(13;26)(q24;q11). *Cytogenet. Genome Res.* **2014**, *144*, 208–211. [CrossRef] [PubMed]
153. Jennings, R.L.; Griffin, D.K.; O'Connor, R.E. A New Approach for Accurate Detection of Chromosome Rearrangements That Affect Fertility in Cattle. *Animals* **2020**, *10*, 114. [CrossRef] [PubMed]
154. Glahn-Luft, B.; Schneider, H.; Schneider, J.; Wassmuth, R. Agnathie beim Schaf mit Chromosomenaberrationen und Hb-Mangel [Agnathia in the sheep associated with chromosome aberration and Hb deficiency]. *Dtsch. Tierarztl. Wochenschr.* **1978**, *85*, 472–474.
155. Anamthawat-Jonsson, K.; Long, S.; Basrur, P.; Adalsteinsson, S. Reciprocal translocation (13;20)(q12;q22) in an Icelandic sheep. *Res. Vet. Sci.* **1992**, *52*, 367–370. [CrossRef]

156. Popescu, C.P.; Tixier, M. L'incidence des anomalies chromosomiques chez les animaux de ferme et leurs conséquences économiques [The frequency of chromosome abnormalities in farm animals and their economic consequences]. *Ann. Genet.* **1984**, *27*, 69–72.
157. Popescu, C.P. Conséquences des anomalies de la structure chromosomique chez les animaux domestiques [Consequences of abnormalities of chromosome structure in domestic animals]. *Reprod. Nutr. Dev.* **1990**, *30* (Suppl. 1), 105s–116s. [CrossRef]
158. Slota, E.; Danielak, B.; Kozubska, A. Structural rearrangement of metacentric chromosomes in the ram of Polish wrzosówka breed. In Proceedings of the 7th European Colloquium on Cytogenetics Dom. Anim, Warsaw, Poland, 23–26 July 1986; p. 37.
159. Iannuzzi, A.; Perucatti, A.; Genualdo, V.; De Lorenzi, L.; Di Berardino, D.; Parma, P. Cytogenetic Elaboration of a Novel Reciprocal Translocation in Sheep. *Cytogenet. Genome Res.* **2013**, *139*, 97–101. [CrossRef] [PubMed]
160. Iannuzzi, A.; Perucatti, A.; Genualdo, V.; Pauciullo, A.; Incarnato, D.; Musilova, P.; Rubes, J.; Iannuzzi, C. The Utility of Chromosome Microdissection in Clinical Cytogenetics: A New Reciprocal Translocation in Sheep. *Cytogenet. Genome Res.* **2014**, *142*, 174–178. [CrossRef]
161. Switonski, M.; Stranzinger, G. Studies of synaptonemal complexes in farm mammals—A review. *J. Hered.* **1998**, *89*, 473–480. [CrossRef]
162. Basrur, P.K.; Koykul, W.; Baguma-Nibasheka, M.; King, W.A.; Ambady, S.; de León, F.A.P. Synaptic pattern of sex com-plements and sperm head malformation in X-autosome translocation carrier bulls. *Mol. Reprod. Dev.* **2001**, *59*, 67–77. [CrossRef]
163. Rho, G.-J.; Coppola, G.; Sosnowski, J.; Kasimanickam, R.; Johnson, W.H.; Semple, E.; Mastromonaco, G.F.; Betts, D.H.; Koch, T.G.; Weese, S.; et al. Use of Somatic Cell Nuclear Transfer to Study Meiosis in Female Cattle Carrying A Sex-Dependent Fertility-Impairing X-Chromosome Abnormality. *Cloning Stem Cells* **2007**, *9*, 118–129. [CrossRef] [PubMed]
164. De Lorenzi, L.; Morando, P.; Planas, J.; Zannotti, M.; Molteni, L.; Parma, P. Reciprocal translocations in cattle: Frequency es-timation. *J. Anim. Breed. Genet.* **2012**, *129*, 409–416. [CrossRef]
165. Wilch, E.S.; Morton, C.C. Historical and Clinical Perspectives on Chromosomal Translocations. *Adv. Exp. Med. Biol.* **2018**, *1044*, 1–14. [CrossRef] [PubMed]
166. Lojda, L.; Rubes, J.; Staisksova, M.; Havrandsova, J. Chromosomal findings in some reproductive disorders in bulls. In Proceedings of the 8th International Congress Animal Reproduction Artificial Insemination, Krakow, Poland, 12–16 July 1976; Volume 158, p. 141.
167. Garick, D.J.; Ruvinsky, A. *The Gentic of Cattle*; CAB International: Wallingford, UK, 2015.
168. Miyake, Y.; Murakami, R.K.; Kaneda, Y. Inheritance of the Robertsonian translocation (1/21) in the Holstein-Friesian cattle. I. Chromosome analysis. *J. Vet. Med. Sci.* **1991**, *53*, 113–116. [CrossRef] [PubMed]
169. Pearce, P.; Ansari, H.; Maher, D.; Amarante, M.; Monro, T.; Hendrikse, W. 1/25 translocations in Blonde d'Aquitaine cattle in New Zealand. *N. Z. Vet. J.* **1997**, *45*, 69–71. [CrossRef]
170. Stranzinger, G.F.; Förster, M. Autosomale Chromosomentranslokationen beim Fleck- und Braunvieh [Autosomal chromosometranslocation of piebald cattle and brown cattle (author's transl)]. *Experientia* **1976**, *32*, 24–27. (In German) [CrossRef]
171. Miyake, Y.; Kaneda, Y. A new type of Robertsonian translocation (1/26) in a bull with unilateral cryptorchidism, probably occurring de novo. *Nihon Juigaku Zasshi* **1987**, *49*, 1015–1019. [CrossRef] [PubMed]
172. Eldridge, F.E. High frequency of a Robertsonian translocation in a herd of British White cattle. *Vet. Rec.* **1975**, *97*, 71–73. [CrossRef]
173. Pollock, D.L.; Bowman, J.C. A Robertsonian Translocation in British Friesian Cattle. *J. Reprod. Fertil.* **1974**, *40*, 423–432. [CrossRef]
174. Tanaka, K.; Yamamoto, Y.; Amano, T.; Yamagata, T.; Dang, V.-C.; Matsuda, Y.; Namikawa, T. A Robertsonian Translocation, Rob(2;28), Found in Vietnamese Cattle. *Hereditas* **2000**, *133*, 19–23. [CrossRef]
175. Popescu, C.P. Observations sur le caryotype normal et anormal des bovins [Normal and abnormal karyotypes of cattle]. *Can. Vet. J.* **1977**, *18*, 143–149. [PubMed]
176. Barasc, H.; Mouney-Bonnet, N.; Peigney, C.; Calgaro, A.; Revel, C.; Mary, N.; Ducos, A.; Pinton, A. Analysis of Meiotic Seg-regation Pattern and Interchromosomal Effects in a Bull Heterozygous for a 3/16 Robertsonian Translocation. *Cytogenet. Genome Res.* **2018**, *156*, 197–203. [CrossRef] [PubMed]
177. Bouvet, A.; Popescu, C.P.; De Giovanni-Macchi, A.M.; Colombo, G.; Molteni, L. Synaptonemal complexes analysis in a bull carrying a 4;8 Robertsonian translocation. *Ann. Génét.* **1989**, *32*, 193–199. [PubMed]
178. Bahri-Darwich, I.; Cribiu, E.; Berland, H.; Darré, R. A new Robertsonian translocation in Blonde d'Aquitaine cattle, rob(4;10). *Genet. Sel. Evol.* **1993**, *25*, 413–419. [CrossRef]
179. Papp, M.; Kovacs, A. 5/18 dicentric Robertsonian translocation in a Simmental bull. In Proceedings of the 4th Europe Colloquium Cytogen Domestic Animals, Milano Gargnano, Italy, 7–11 June 1980; p. 51.
180. Slota, E.; Switonski, M. A new Robertsonian translocation 5;22 in cattle. Studies of banded chromosomes and synaptonemal complexes. *Genet. Pol.* **1992**, *33*, 227–231.
181. Di Meo, G.P.; Molteni, L.; Perucatti, A.; De Giovanni, A.; Incarnato, D.; Succi, G.; Schibler, L.; Cribiu, E.P.; Iannuzzi, L. Chromosomal characterization of three centric fusion translocations in cattle using G-, R- and C-banding and FISH technique. *Caryologia* **2000**, *53*, 213–218. [CrossRef]
182. Tateno, H.; Miyake, Y.I.; Mori, H.; Kamiguchi, Y.; Mikamo, K. Sperm chromosome study of two bulls heterozygous for dif-ferent Robertsonian translocations. *Hereditas* **1994**, *120*, 7–11. [CrossRef]
183. Hanada, H.; Muramatsu, S.; Abe, T.; Fukushima, T. Robertsonian chromosome polymorphism found in a local herd of the Japanese Black cattle. *Ann. Genet. Sel. Anim.* **1981**, *13*, 205–211. [CrossRef]

184. Cribiu, E.P.; Matejka, M.; Darre, R.; Durand, V.; Berland, H.M.; Bouvet, A. Identification of chromosomes involved in a Robertsonian translocation in cattle. *Genet. Sel. Evol.* **1989**, *21*, 555–560. [CrossRef]
185. Pinheiro, L.E.L.; Ferrari, L. A new type of Robertsonian translocation in cattle. In Proceedings of the 5th Encontro de Pesquisas Veterinarias, Jaboticabal, Brazil, 6–7 November 1980; p. 161.
186. Kovács, A.; Papp, M. Report on chromosomal examination of A.I. bulls in Hungary. *Ann. Génét. Sél. Anim.* **1977**, *9*, 528. [CrossRef]
187. Holecková, B.; Sutiaková, I.; Pijáková, N. Robertsonian translocation in a cattle population. *Vet. Med.* **1995**, *40*, 33–34.
188. Molteni, L.; Giovanni-Macchi, A.; Succi, G.; Cremonesi, F.; Stacchezzini, S.; Meo, G.P.; Iannuzzi, L. A New Centric Fusion Translocation in Cattle: Rob (13;19). *Hereditas* **2004**, *129*, 177–180. [CrossRef] [PubMed]
189. Kovács, A.; Mészáros, I.; Sellyei, M.; Vass, L. Mosaic centromeric fusion in a Holstein-Friesian bull. *Acta Boil. Acad. Sci. Hung.* **1973**, *24*, 215–220.
190. Slota, E.; Danielak, B.; Kozubska, A. The Robertsonian translocation in cattle quintuplets. In Proceedings of the 8th European Colloquium on Cytogenetics of Domestic Animals, Bristol, UK, 9–22 July 1988; pp. 122–124.
191. De Giovanni Macchi, A.; Molteni, L.; Parma, P.; Laurelli, A. Identification of a new Robertsonian translocation in the Marchigiana breed. In Proceedings of the 8th North American Colloquium on Domestic Animal Cytogenetics and Gene Mapping, Guelph, ON, Canada, 13–16 July 1993; p. 175.
192. De Lorenzi, L.; Molteni, L.; De Giovanni, A.; Parma, P. A new case of rob(14;17) in cattle. *Cytogenet. Genome Res.* **2008**, *120*, 144–146. [CrossRef] [PubMed]
193. Logue, D.N.; Harvey, M.J. A 14/20 Robertsonian translocation in Swiss Simmental cattle. *Res. Vet. Sci.* **1978**, *25*, 7–12. [CrossRef]
194. Schmutz, S.; Moker, J.; Pawlyshyn, V.; Haugen, B.; Clark, E. Fertility effects of the 14;20 Robertsonian translocation in cattle. *Theriogenology* **1997**, *47*, 815–823. [CrossRef]
195. Weber, A.F.; Buoen, L.C.; Zhang, T.; Ruth, G.R. Prevalence of the 14/20 centric fusion chromosomal aberration in US Simmental cattle. *J. Am. Vet. Med. Assoc.* **1992**, *200*, 1216–1219. [PubMed]
196. Di Berardino, D.; Iannuzzi, L.; Ferrara, L.; Matassino, D. A new case of Robertsonian translocation in cattle. *J. Hered.* **1979**, *70*, 436–438. [CrossRef]
197. Ellsworth, S.M.; Paul, S.R.; Bunch, T.D. A 14/28 dicentric Robertsonian translocation in a Holstein cow. *Theriogenology* **1979**, *11*, 165–171. [CrossRef]
198. Iannuzzi, L.; Rangel-Figueiredo, T.; Di Meo, G.; Ferrara, L. A new Robertsonian translocation in cattle, rob(15;25). *Cytogenet. Genome Res.* **1992**, *59*, 280–283. [CrossRef]
199. Iannuzzi, L.; Rangel-Figueiredo, T.; Meo, G.P.; Ferrara, L. A New Centric Fusion Translocation in Cattle, Rob(16;18). *Hereditas* **2004**, *119*, 239–243. [CrossRef] [PubMed]
200. Rubes, J.; Musilová, P.; Borkovec, L.; Borkovcová, Z.; Svecová, D.; Urbanová, J. A new Robertsonian translocation in cattle, rob(16;20). *Hereditas* **1996**, *124*, 275–279. [CrossRef]
201. Rybar, R.; Horakova, J.; Machatkova, M.; Hanzalova, K.; Rubes, J. Embryos produced in vitro from bulls carrying 16;20 and 1;29 Robertsonian translocations: Detection of translocations in embryos by fluorescence in situ hybridization. *Zygote* **2005**, *13*, 31–34. [CrossRef] [PubMed]
202. Pinton, A.; Ducos, A.; Berland, H.; Séguéla, A.; Blanc, M.F.; Darré, A.; Mimar, S.; Darré, R. A new Robertsonian translocation in Holstein-Friesian cattle. *Genet. Sel. Evol.* **1997**, *29*, 523–526. [CrossRef]
203. Berland, H.M.; Cribiu, E.P.; Darre, R.; Boscher, J.; Popescu, C.P.; Sharma, A. A New Case of Robertsonian Translocation in Cattle. *J. Hered.* **1988**, *79*, 33–36. [CrossRef]
204. De Lorenzi, L.; Molteni, L.; Denis, C.; Eggen, A.; Parma, P. A new case of centric fusion in cattle: Rob(21;23). *Anim. Genet.* **2008**, *39*, 454–455. [CrossRef] [PubMed]
205. Molteni, L.; Meggiolaro, D.; Macchi, A.D.G.; De Lorenzi, L.; Crepaldi, P.; Stacchezzini, S.; Cremonesi, F.; Ferrara, F. Fertility of cryopreserved sperm in three bulls with different Robertsonian translocations. *Anim. Reprod. Sci.* **2005**, *86*, 27–36. [CrossRef]
206. Bongso, A.; Basrur, P.K. Chromosome anomalies in Canadian Guernsey bulls. *Cornell Vet.* **1976**, *66*, 476–489. [PubMed]
207. Di Meo, G.; Perucatti, A.; Genualdo, V.; Iannuzzi, A.; Sarubbi, F.; Jambrenghi, A.C.; Incarnato, D.; Peretti, V.; Vonghia, G.; Iannuzzi, L. A Rare Case of Centric Fission and Fusion in a River Buffalo (*Bubalus bubalis*, 2n = 50) Cow with Reduced Fertility. *Cytogenet. Genome Res.* **2011**, *132*, 26–30. [CrossRef]
208. Albarella, S.; Ciotola, F.; Coletta, A.; Genualdo, V.; Iannuzzi, L.; Peretti, V. A new translocation t(1p;18) in an Italian Medi-terranean river buffalo (*Bubalus bubalis*, 2n = 50) bull: Cytogenetic, fertility and inheritance studies. *Cytogenet. Genome Res.* **2013**, *139*, 17–21. [CrossRef] [PubMed]
209. Bruère, A.; Mills, R.A. Observations on the incidence of Robertsonian translocations and associated testicular changes in a flock of New Zealand Romney sheep. *Cytogenet. Genome Res.* **1971**, *10*, 260–272. [CrossRef] [PubMed]
210. Ansari, H.A.; Pearce, P.D.; Maher, D.W.; Malcolm, A.A.; Broad, T.E. Resolving ambiguities in the karyotype of domestic sheep (Ovis aries). *Chromosoma* **1993**, *102*, 340–347. [CrossRef] [PubMed]
211. Bruère, A.; Chapman, H.M.; Wyllie, D.R. Chromosome polymorphism and its possible implications in the select Drysdale breed of sheep. *Cytogenet. Genome Res.* **1972**, *11*, 233–246. [CrossRef]
212. Pearce, P.D.; Ansari, H.A.; Maher, D.W.; Malcolm, A.A.; Stewart-Scott, I.A.; Broad, T.E. New Robertsonian translocation chromosomes in domestic sheep (Ovis aries). *Cytogenet. Genome Res.* **1994**, *67*, 137–140. [CrossRef]

213. Glahn-Luft, B.; Wassmuth, R. The influence of 1/20 translocation in sheep on the efficiency of reproduction. In Proceedings of the 31st Annual Meeting of the European Association for Animal Production, München, Germany, 1–4 September 1980.
214. Chaves, R.; Adega, F.; Wienberg, J.; Guedes-Pinto, H.; Heslop-Harrison, J.S. Molecular cytogenetic analysis and centromeric satellite organization of a novel 8;11 translocation in sheep: A possible intermediate in biarmed chromosome evolution. *Mamm. Genome* **2003**, *14*, 706–710. [CrossRef] [PubMed]
215. Popescu, C.P. The mode of transmission of a centric fusion to the offspring of a buck (Capra hircus L.). *Ann. Genet. Sel. Anim.* **1972**, *4*, 355–361. [PubMed]
216. Evans, H.J.; Buckland, R.A.; Sumner, A.T. Chromosome homology and heterochromatin in goat, sheep and ox studied by banding techniques. *Chromosoma* **1973**, *42*, 383–402. [CrossRef] [PubMed]
217. Gonçalves, H.; Jorge, W.; Cury, P. Distribution of a Robertsonian translocation in goats. *Small Rumin. Res.* **1992**, *8*, 345–352. [CrossRef]
218. Elminger, B.; Stranzinger, C. Identification of a centromeric fusion in the G-banding karyotype of a Saanen goat. In Proceedings of the 5th European Colloquium on Cytogenetic of Domestic Animals, Milano Gargano, Italy, 7–11 June 1982; pp. 407–409.
219. Dolf, J.; Hediger, R. Comparison of centric fusion in a Toggenburg and Saanen goat. In Proceedings of the 6th European Colloquium on Cy-togenetic of Domestic Animals, Zürich, Switzerland, 16–20 July 1984; pp. 311–312.
220. Burguete, I.; Di Berardino, D.; Lioi, M.B.; Taibi, L.; Matassino, D. Cytogenetic observations on a Robertsonian translocation in Saanen goats. *Genet. Sel. Evol.* **1987**, *19*, 391–398. [CrossRef] [PubMed]
221. Guillemot, E.; Gary, F.; Berland, H.M.; Berthelot, X.; Durand, V.; Darre, R.; Cribiu, E.P. Effects of 6/15 Robertsonian translo-cation in Saanen goats. *Reprod. Domest. Anim.* **1993**, *28*, 28–32. [CrossRef]
222. Da Mota, L.S.L.S.; da Silva, R.A.B. Centric fusion in goats (Capra hircus): Identification of a 6/15 translocation by high resolution chromosome banding. *Genet. Mol. Biol.* **1998**, *21*. [CrossRef]
223. Moreno-Millan, M.; Rodero-Franganillo, A. A new Robertsonian translocation in an intersex goat: Morphometric determi-nation. *Arch. Zootec.* **1990**, *39*, 263–270.
224. De Lorenzi, L.; Genualdo, V.; Gimelli, S.; Rossi, E.; Perucatti, A.; Iannuzzi, A.; Zannotti, M.; Malagutti, L.; Molteni, L.; Iannuzzi, L.; et al. Genomic analysis of cattle rob(1;29). *Chromosom. Res.* **2012**, *20*, 815–823. [CrossRef]
225. Di Meo, G.P.; Perucatti, A.; Chaves, R.; Adega, F.; De Lorenzi, L.; Molteni, L.; De Giovanni, A.; Incarnato, D.; Guedes-Pinto, H.; Eggen, A.; et al. Cattle rob(1;29) originating from complex chromosome rearrangements as revealed by both banding and FISH-mapping techniques. *Chromosom. Res.* **2006**, *14*, 649–655. [CrossRef]
226. Chaves, R.; Heslop-Harrsion, J.S.; Guedes-Pinto, H. Centromeric heterochromatin in the cattle rob(1;29) translocation: Al-phasatellite I sequences, in-situ MspI digestion patterns, chromomycin staining and C-bands. *Chromosom. Res.* **2000**, *8*, 621–626. [CrossRef] [PubMed]
227. Popescu, C.P.; Pech, A. Une bibliographie sur la translocation 1/29 de bovins dans le monde (1964–1990). *Ann. Zootech.* **1991**, *40*, 271–305. [CrossRef]
228. Iannuzzi, A.; Di Meo, G.; Jambrenghi, A.C.; Vonghia, G.; Iannuzzi, L.; Rangel-Figueiredo, T. Frequency and distribution of rob(1;29) in eight Portuguese cattle breeds. *Cytogenet. Genome Res.* **2008**, *120*, 147–149. [CrossRef]
229. Świtoński, M.; Gustavsson, I.; Plöen, L. The nature of the 1;29 translocation in cattle as revealed by synaptonemal complex analysis using electron microscopy. *Cytogenet. Genome Res.* **1987**, *44*, 103–111. [CrossRef]
230. Bonnet-Garnier, A.; Pinton, A.; Berland, H.M.; Khireddine, B.; Eggen, A.; Yerle, M.; Darré, R.; Ducos, A. Sperm nuclei analysis of 1/29 Robertsonian translocation carrier bulls using fluorescence in situ. *Cytogenet. Genome Res.* **2006**, *112*, 241–247. [CrossRef] [PubMed]
231. Di Dio, C.; Longobardi, V.; Zullo, G.; Parma, P.; Pauciullo, A.; Perucatti, A.; Higgins, J.; Iannuzzi, A. Analysis of meiotic segregation by triple-color fish on both total and motile sperm fractions in a t(1p;18) river buffalo bull. *PLoS ONE* **2020**, *15*, e0232592. [CrossRef]
232. Rubes, J.; Machatková, M.; Jokesová, E.; Zudová, D. A potential relationship between the 16;20 and 14;20 Robertsonian translocations and low in vitro embryo development. *Theriogenology* **1999**, *52*, 171–180. [CrossRef]
233. Vozdova, M.; Kubíčková, S.; Cernohorska, H.; Rubeš, J. Detection of translocation rob(1;29) in bull sperm using a specific DNA probe. *Cytogenet. Genome Res.* **2008**, *120*, 102–105. [CrossRef] [PubMed]
234. Iannuzzi, L. Standard karyotype of the river buffalo (Bubalus bubalis L., 2n = 50). Report of the committee for the standardization of banded karyotypes of the river buffalo. *Cytogenet. Cell Genet.* **1994**, *67*, 102–113. [CrossRef] [PubMed]
235. Broad, T.E.; McLean, N.; Grimaldi, D.; Monk, N. A novel occurrence in sheep (Ovis aries) of four homozygous Robertsonian translocations. *Hereditas* **2000**, *132*, 165–166. [CrossRef]
236. Broad, T.E.; Hayes, H.; Long, S.E. Cytogenetics: Physical maps. In *The Genetics of Sheep*; Piper, L., Ruvinsky, A., Eds.; CAB International: Wallingford, UK, 1997; pp. 241–296.
237. Vallenzasca, C.; Martignoni, M.; Galli, A. Finding of a bull with Y;17 translocation. *Hereditas* **1990**, *113*, 63–67. [CrossRef]
238. Gallagher, D.S., Jr.; Basrur, P.K.; Womack, J.E. Identification of an autosome to X chromosome translocation in the domestic cow. *J. Hered.* **1992**, *83*, 451–453. [CrossRef]
239. Pinton, A.; Ducos, A.; Yerle, M. Chromosomal rearrangements in cattle and pigs revealed by chromosome microdissection and chromosome painting. *Genet. Sel. Evol.* **2003**, *35*, 1–96. [CrossRef] [PubMed]
240. Popescu, C.; Boscher, J. New data on pericentric inversion in cattle (Bos taurus L). *Ann. Génét. Sél. Anim.* **1976**, *8*, 443. [CrossRef]

241. Switoński, M. A pericentric inversion in an X chromosome in the cow. *J. Hered.* **1987**, *78*, 58–59. [CrossRef] [PubMed]
242. Iannuzzi, L.; Di Meo, G.; Perucatti, A.; Eggen, A.; Incarnato, D.; Sarubbi, F.; Cribiu, E. A pericentric inversion in the cattle Y chromosome. *Cytogenet. Cell Genet.* **2001**, *94*, 202–205. [CrossRef] [PubMed]
243. De Lorenzi, L.; Iannuzzi, A.; Rossi, E.; Bonacina, S.; Parma, P. Centromere Repositioning in Cattle (Bos taurus) Chromosome 17. *Cytogenet. Genome Res.* **2017**, *151*, 191–197. [CrossRef]
244. Hansen, K.M. Bovine tandem fusion and fertility. *Hereditas* **1969**, *63*, 453–454.
245. Pinheiro, L.E.; Carvalho, T.B.; Oliveira, D.A.; Popescu, C.P.; Basrur, P.K. A 4/21 tandem fusion in cattle. *Hereditas* **1995**, *122*, 99–102. [CrossRef]
246. Halnan, C.R. Autosomal deletion and infertility in cattle. *Vet. Rec.* **1972**, *91*, 572. [CrossRef] [PubMed]
247. Uzar, T.; Szczerbal, I.; Serwanska-Leja, K.; Nowacka-Woszuk, J.; Gogulski, M.; Bugaj, S.; Switonski, M.; Komosa, M. Congenital Malformations in a Holstein-Fresian Calf with a Unique Mosaic Karyotype: A Case Report. *Animals* **2020**, *10*, 1615. [CrossRef] [PubMed]
248. Braun, U.; Ansari, H.A.; Hediger, R.; Süss, U.; Ehrensperger, F. Hypotrichose und Oligodontie, verbunden mit einer Xq-Deletion, bei einem Kalb der Schweizerischen Fleckviehrasse [Hypotrichosis and oligodontia, combined with an Xq-deletion, in a calf of the Swiss Holstein breed]. *Tierarztl. Prax.* **1988**, *16*, 39–44. [PubMed]
249. Sharshov, A.A.; Grafodatskiĭ, A.S. Novaia perestroĭka khromosom u krupnogo rogatogo skota (A new chromosomal rearrangement in cattle). *Tsitol. Genet.* **1990**, *24*, 30–33.
250. De Lorenzi, L.; Rossi, E.; Genualdo, V.; Gimelli, S.; Lasagna, E.; Iannuzzi, A.; Parma, P.; Perucatti, A. Molecular Characterization of Xp Chromosome Deletion in a Fertile Cow. *Sex. Dev.* **2012**, *6*, 298–302. [CrossRef] [PubMed]
251. Lasagna, E.; (University of Perugia, Perugia, Italy). Personal communication, 2020.
252. Iannuzzi, L.; Di Meo, G.; Perucatti, A.; Rullo, R.; Incarnato, D.; Longeri, M.; Bongioni, G.; Molteni, L.; Galli, A.; Zanotti, M.; et al. Comparative FISH-mapping of the survival of motor neuron gene (SMN) in domestic bovids. *Cytogenet. Genome Res.* **2003**, *102*, 39–41. [CrossRef]
253. De Lorenzi, L.; Arrighi, S.; Rossi, E.; Grignani, P.; Previderè, C.; Bonacina, S.; Cremonesi, F.; Parma, P. XY (SRY-positive) Ovarian Disorder of Sex Development in Cattle. *Sex. Dev.* **2018**, *12*, 196–203. [CrossRef]

Review

Clinical Cytogenetics of the Dog: A Review

Izabela Szczerbal and Marek Switonski *

Department of Genetics and Animal Breeding, Poznan University of Life Sciences, 60-637 Poznan, Poland; izabela.szczerbal@up.poznan.pl
* Correspondence: marek.switonski@up.poznan.pl

Simple Summary: The cytogenetic analysis of dogs is mainly focused on the diagnosis of disorders of sex development (DSD) and cancers. Unfortunately, the study of canine chromosomes is a challenging task due a high chromosome number (2n = 78) and the one-arm morphology of all autosomes. For years, the application of conventional cytogenetic techniques, Giemsa staining and G and DAPI (4′,6-diamidino-2-phenylindole) bandings, allowed the identification of sex chromosome aneuploidies and centric fusions. An advanced clinical cytogenetic diagnosis is also needed due to the fact that the dog is a valuable animal model in biomedical research. The application of hybridization methods, such as fluorescence in situ hybridization (FISH) and array comparative genome hybridization (aCGH), facilitated the detection of other chromosomal rearrangements. It can be foreseen that a wide use of modern molecular techniques (e.g., SNP microarray and next generation sequencing) will substantially extend the knowledge on canine chromosome mutations.

Abstract: The dog is an important companion animal and has been recognized as a model in biomedical research. Its karyotype is characterized by a high chromosome number (2n = 78) and by the presence of one-arm autosomes, which are mostly small in size. This makes the dog a difficult subject for cytogenetic studies. However, there are some chromosome abnormalities that can be easily identified, such as sex chromosome aneuploidies, XX/XY leukocyte chimerism, and centric fusions (Robertsonian translocations). Fluorescence in situ hybridization (FISH) with the use of whole-chromosome painting or locus-specific probes has improved our ability to identify and characterize chromosomal abnormalities, including reciprocal translocations. The evaluation of sex chromosome complement is an important diagnostic step in dogs with disorders of sex development (DSD). In such cases, FISH can detect the copy number variants (CNVs) associated with the DSD phenotype. Since cancers are frequently diagnosed in dogs, cytogenetic evaluation of tumors has also been undertaken and specific chromosome mutations for some cancers have been reported. However, the study of meiotic, gamete, and embryo chromosomes is not very advanced. Knowledge of canine genome organization and new molecular tools, such as aCGH (array comparative genome hybridization), SNP (single nucleotide polymorphism) microarray, and ddPCR (droplet digital PCR) allow the identification of chromosomal rearrangements. It is anticipated that the comprehensive use of chromosome banding, FISH, and molecular techniques will substantially improve the diagnosis of chromosome abnormalities in dogs.

Keywords: aneuploidy; cancer cytogenetics; centric fusion; chimerism; disorder of sex development; freemartinism; intersexualism; mosaicism; reciprocal translocation

Citation: Szczerbal, I.; Switonski, M. Clinical Cytogenetics of the Dog: A Review. *Animals* **2021**, *11*, 947. https://doi.org/10.3390/ani11040947

Academic Editors: Leopoldo Iannuzzi and Pietro Parma

Received: 27 February 2021
Accepted: 22 March 2021
Published: 27 March 2021

Publisher's Note: MDPI stays neutral with regard to jurisdictional claims in published maps and institutional affiliations.

Copyright: © 2021 by the authors. Licensee MDPI, Basel, Switzerland. This article is an open access article distributed under the terms and conditions of the Creative Commons Attribution (CC BY) license (https://creativecommons.org/licenses/by/4.0/).

1. Introduction

The dog is the most important companion animal species, and one for which extreme interbreed phenotypic variation has arisen over the last 200 years [1]. One side effect of this intensive breeding, caused by genetic drift, is the preservation of undesired mutations in the gene pool. About 400 DNA variants that cause hereditary diseases in dogs have been described (Online Mendelian Inheritance in Animals—OMIA, https://omia.org/home/, (accessed on 20 December 2020)).

Our knowledge of canine chromosomal mutations is less advanced, as the chromosome set of this species is a very difficult analytic subject. The diploid number of chromosomes is high (2n = 78), but the genome size of this species (2.4 Gb) is similar to other domestic mammals. The majority of chromosomes are thus small, and their banding patterns do not allow for unambiguous recognition of all homologs. Sex chromosomes are biarmed and easily recognizable, but all autosomes are acrocentric, and only the largest pair of chromosome 1 (CFA1, Canis Familiaris chromosome 1) is distinctly different from the other autosomes.

There have been several attempts to arrange a reference banded karyotype of the dog (review by [2]). In 1993, the DogMap consortium, which was focused on the development of the canine marker genome map, suggested that a commonly accepted chromosome nomenclature for this species be established by a group of cytogeneticists experienced in canine chromosomes research. The international committee agreed that, due to the similarity of G-banding patterns of the small autosomes, only the largest 21 autosome pairs and the sex chromosome pair can be recognized with certainty. As a result of this work, a partial standard karyotype was developed [3]. An important step in characterizing the canine chromosomes was the use of a set of chromosome-specific painting probes for all autosomes and sex chromosomes [4]. Unfortunately, these probes are not available for diagnostic purposes. BAC (Bacterial Artificial Chromosome) probes with known chromosomal localization are very useful tools in clinical cytogenetics. Canine BAC clones can be purchased from the BAC libraries: CHORI-82 Canine boxer (Canis familiaris) BAC library and RPCI-81 Canine BAC Library (https://bacpacresources.org/clones.htm, accessed on 2 March 2021). Information on the localization of the CHORI-82 BAC clones is available in the CanFam3.1 assembly in NCBI - National Center for Biotechnology Information (Genome Data Viewer, accessed on 2 March 2021). Moreover, chromosome specific BAC clones from the RPCI-81 library were cytogenetically assigned [5,6]. Such probes can prove very helpful in recognizing all chromosomes, including the small autosomes (pairs 22–38) that are not included in the standard partial karyotype. However, it should be pointed out that conventional Giemsa staining is sufficient to identify sex chromosome aneuploidies and XX/XY leukocyte chimerism. Detecting centric fusions is also easy on Giemsa stained metaphase spreads, but identifying the autosomes involved requires the use of chromosome banding or fluorescence in situ hybridization (FISH).

The progress on various aspects of canine cytogenetics was reviewed by Breen [2] and by Reimann-Berg et al. [7]. In this article, we focus on the importance of chromosome analysis in diagnosing clinical cases.

2. Sex Chromosome Aneuploidies

Although sex chromosome aneuploidies are an important cause of infertility and sterility, they have rarely been observed in dogs (Table 1). X monosomy has been reported in five female dogs to date, usually presenting abnormal estrus cycle and small ovaries, with no evidence of ovarian follicle development. Other abnormalities, such as small stature, juvenile appearance, or excessive skin in the ventrum of the neck (typical of Turner syndrome in women), were observed only in dogs with a pure monosomy, 77,X [8,9]. The vertical septum in the vagina, observed in a single case, could be coincidental [10]. The frequency of the monosomic cell line in individuals with mosaic karyotypes (77,X/78,XX) varied over a wide range, from 5% [10] to 95% [11]. This shows that a large number of metaphase spreads need to be evaluated in infertile female dogs suspected of X monosomy.

Table 1. Cases of X monosomy reported in dogs.

Karyotype	No. of Cells Analyzed	Breed	Characteristic Feature of Phenotype	Reference
77,X	Lack of information	Doberman Pinscher	Small stature, excessive skin in the ventrum of the neck, no signs of estrus, small ovaries consisting primarily of interstitial-type cells and solid epithelial cords	[8]
77,X	60	Miniature American Eskimo	Juvenile appearance, signs of proestrus, small and fibrous ovaries, no evidence of ovarian follicle development or corpora lutea	[9]
77,X[95%]/78,XX[5%]	40	Toy Poodle	Abnormal estrus cycle and apparently persistent follicles, gonadal dysgenesis	[11]
77,X[5%]/78,XX[95%]	220	Munsterlander	Infertility, vertical septum in vagina	[10]
77,X[6%]/78,XX[94%]	473	Bearded Collie	Infertility, irregular and poorly manifested estrus cycles	[10]

The low incidence of X monosomy in female dogs is probably associated with the large size of the pseudoautosomal region (PAR), estimated at 6.4 Mb, which is more rich in genes than human or equine PAR [12,13]. The loss of one X chromosome is associated with the lack of a long PAR, leading to haploinsufficiency in a long genomic segment responsible for embryonic mortality. No correlation between PAR size and X trisomy was observed, which indicates that the overdose of PAR-located genes has no effect on the phenotype of X trisomy carriers [12].

Only a few females with X trisomy have been described in dogs, and they usually had abnormal estrous cycles and hypoplastic ovaries (Table 2). Interestingly, among the six reported cases with X trisomy, three females showed behavioral problems, such as fearfulness, lack of barking, or coprophagy. The majority of these cases had only a single cell line, 79,XXX. A mosaic 79,XXX/78,XX karyotype was incidentally diagnosed in a female dog with normal estrus [14]. It is worth mentioning that trisomic cell lines occurred in a low frequency (5%) of cells, which suggests that the frequency of the mosaic karyotype (79,XXX/78,XX)—associated with normal fertility or subfertility—may be underestimated.

Table 2. Cases of X trisomy reported in dogs.

Karyotype	Breed	Characteristic Feature of Phenotype	Reference
79,XXX	Airedale Terrier	Primary anestrus, ovaries with solid epithelial cords and large masses of interstitial cells, lack of follicles and corpora lutea	[15]
79,XXX	Mixed breed	Infertility, normal reproductive organs, ovaries with primary follicles and corpora lutea, dental anomalies, abnormal behavior (lack of barking and fearfulness)	[16]
79,XXX	Labrador Retriever	Primary anestrus, chronic dermatitis, abnormal behavior (coprophagy)	[17]
79,XXX	Silky Terrier	Infertility, abnormal estrous cycles, hypoplastic ovaries, absence of normal follicular structures, shy and timid behavior	[18]
79,XXX	Labrador Retriever	Infertility, abnormal estrous cycles, hypoplastic ovaries, absence of normal follicular structures	[18]
79,XXX/78,XX	Boston Terrier	Estrus symptoms occurred once, ovary with corpora lutea	[14]

The XXY complement has been found in six dogs, including two cases of XX/XXY mosaicism (Table 3). This abnormality is known as a cause of hyperplastic testes and sterility. However, other abnormalities, such as congenital heart disease and bilateral cryptorchidism, have also been described [19–21]. In some XXY dogs, testicular tumors were diagnosed [20,22] and the feminization of such dogs has also been reported [23].

Table 3. Cases of the XXY complement reported in dogs.

Karyotype	Breed	Characteristic Feature of Phenotype	Reference
79,XXY	German Shorthair Pointer	Testicular hypoplasia, lack of spermatogenesis, ventricular septal defect, congenital heart abnormalities	[19]
79,XXY	Great Dane	Female external and internal genitalia, structure reminiscent of a vestigial scrotal sac	[23]
79,XXY	Norwich Terrier	Testicular dysgenesis, azoospermia	[24]
79,XXY	West Highland White Terrier	High stature, rugae of the dermis and hypodermis, low level of testosterone, Sertoli cell tumor	[22]
79,XXY/78,XY	Miniature Schnauzer	Alopecia, gynecomastia, bilateral cryptorchidism, Sertoli cell tumor	[20]
79,XXY[18%]/78,XY[82%]	Poodle	Bilateral cryptorchidism, testes with vacuolation of the seminal cells and small nests of Leydig cells, total absence of sperm cells	[21]

3. Leukocyte Chimerism XX/XY

The XX/XY leukocyte chimerism, which is caused by the formation of anastomoses between the placentas of heterosexual fetuses, is associated with freemartinism, a form of disorder of sex development (DSD). The anastomoses enable the exchange of hematopoietic cells and molecules involved in sex differentiation between the fetuses [25]. The masculinizing factors (Sex Determining Region Y - SRY, which is a transcription factor; anti-Mullerian hormone and testosterone) produced by the fetal testes alter the sexual differentiation of the female fetus. This syndrome is well known in ruminants [26], but is also observed in other species, including dogs (Table 4). Between-species differences in the frequency of the chimerism are associated with the type of placenta: the high incidence of anastomoses in ruminants is associated with a cotyledonary organization of the placenta, while in carnivores the incidence is much lower due to the zonary organization of the placenta. It also seems that diffused placentas may be associated with an elevated risk of anastomoses in litters with a large number of fetuses, as has been observed in highly prolific lines of pigs [27]. This may suggest that in dogs, too, a greater number of puppies in a litter may be associated with a higher incidence of freemartinism.

The appearance of external genitalia is a major criterion for identifying DSD in dogs, but is less useful in freemartin dogs. Some freemartins present almost normal female genitalia [28], while others have a normal male appearance [29,30]; in sporadic cases there are ambiguous external genitalia [31].

It is important to point out that virilization can be caused by freemartinism or by testicular or ovotesticular XX DSD; a correct diagnosis should thus be made using cytogenetic analysis. A comprehensive study of six French bulldogs with ambiguous external genitalia revealed that five were testicular or ovotesticular XX DSD, while one was a freemartin [32]. Using the nomenclature of DSD dogs, XX/XY leukocyte chimerism can be considered testicular, ovarian, or ovotesticular DSD [33]. These forms are also observed in dogs and, as in ruminants, there is no correlation between the percentage of XY cells and the extent of the virilization. In the reported cases, the proportion of the XY cell line ranged from 15% to 80% (Table 4).

Table 4. Leukocyte chimerism XX/XY reported in dogs.

Proportion of XX and XY Cell Lines [%]	Breed	Phenotypic Sex Considered by Owners	Characteristic Feature of Phenotype	Reference
Lack of information	Schipperke	female	Enlarged clitoris, testis and ovotestis, uterus,	[34]
43/57	Pug	female	Enlarged clitoris, hypospadias, no signs of estrus, testis and ovotestis	[23]
45/55	Dachshund	male	Abnormal urogenital tract, hematuria, ovaries	[35]
Lack of information	Spaniel × Papillon	unknown	Ovaries	[36]
Lack of information	American Eskimo	female	Normal vulva and clitoris, ovotestis	[37]
Lack of information	Spaniel	unknown	Small penis, empty rudimentary scrotum, uterus, ovaries with reduced number of follicles	[38]
85/15	Belgium Shepherd	male	Aggressive behavior, intersexuality, abdominal testes, underdeveloped penis, urethra ended under the anus, vas deferens connected to an oviduct, blind uterus	[29]
Lack of information	Fila Brasileiro	male	Prepuce-like structure located closer to the anus, testicles with an immature epididymides	[39]
43/57	Border Terrier	male	Undeveloped penis, ovarian-like structure	[30]
70/30	Wirehaired Pointing Griffon	female	Primary anestrus, juvenile vulva, enlarged clitoris, testis	[28]
78/22	Shih Tzu	ambiguous	Residual penis with a prepuce located in a position typical of a male, prostate, gonads remained undetectable	[31]
20/80	French Bulldog	female	Enlarge clitoris, ovotestes	[32]
30/70	Great Dane	female	Underdeveloped internal reproductive organs, rudimentary testicles	[40]
54/46	Great Dane	female	Undeveloped foreskin	[14]

The diagnosis of XX/XY leukocyte chimerism also requires the analysis of another tissue, such as hair follicles or buccal epithelial cells, in order to distinguish between leukocyte chimerism and whole-body chimerism. Moreover, it facilitates the establishment of concordance between phenotypic and chromosomal sex. Since external genitalia of freemartins are often ambiguous, it cannot be excluded that some cases are incorrectly considered by owners as males. In Table 4, two cases were described as males, but this was not confirmed by cytogenetic or molecular studies of other tissue.

A common diagnostic strategy involves the cytogenetic analysis of leukocytes (Figure 1), the molecular detection of Y-linked genes (e.g., *SRY* and *ZFY*), or microsatellite markers in DNA samples derived from the blood and the second tissue [31]. Droplet digital PCR (ddPCR) has recently been demonstrated to be a fast and reliable method for detecting XX/XY leukocyte chimerism in cattle and pigs [27,41]. This method can also be recommended for DSD diagnosis in dogs.

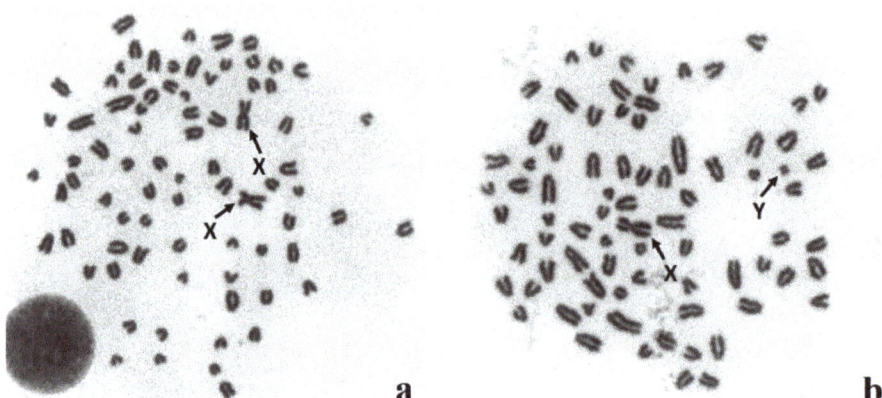

Figure 1. Identification of (**a**) 78,XX and (**b**) 78,XY Giemsa-stained metaphase spreads from in vitro cultured leukocytes obtained from a DSD (disorder of sex development) dog. Sex chromosomes are indicated with arrows.

4. Structural Chromosome Rearrangements

Robertsonian translocations (centric fusions) have been reported quite often in dogs, probably due to their ease of identification. Such mutations lead to a reduction in the diploid chromosome number to 77 and the formation of a biarmed derivative (der) chromosome (Figure 2). The first Robertsonian translocations in dogs were described in the 1960s, and over a dozen different translocations have now been reported in total (Table 5). Different chromosomes are involved in these mutations, but in several cases no attempt was undertaken to indicate the autosomes involved. Two autosomes, named CFA13 and CFA23 (Canis Familiaris chromosome 13 and 23), were identified more often, but this finding should be taken with caution due to the difficulties in identifying small autosomes when only chromosome banding was used. There is only a single report on the use of FISH with locus-specific probes to describe a centric fusion, which found rob(5;23), as described by Switonski et al. [42].

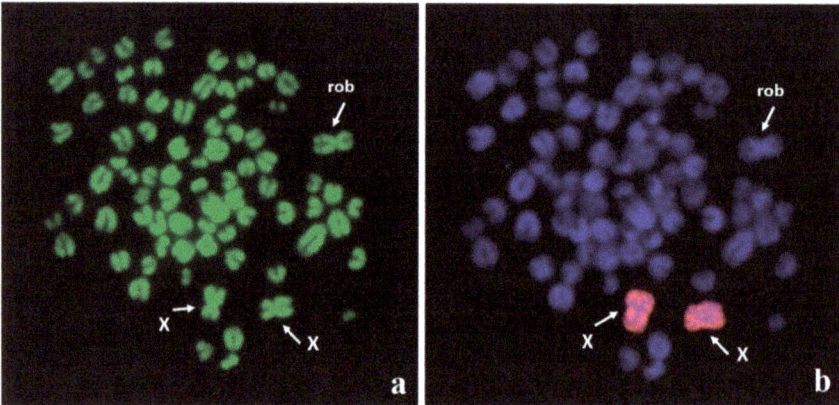

Figure 2. Robertsonian translocation, 77,XX, rob(5;23) in an infertile DSD female dog: (**a**) Q-banded metaphase spread with three biarmed chromosomes; (**b**) the same metaphase spread after fluorescence in situ hybridization (FISH) using whole X chromosome painting probe to facilitate recognition of X chromosomes from the fused chromosome (rob). The autosomes involved in the translocation were identified by FISH with locus-specific probes (for details, see [42]).

Dogs with Robertsonian translocations present normal phenotype, but there is often a small decrease in fertility. However, this abnormality has also been diagnosed in infertile bitches [43–45], which either showed a lack of estrus or were unsuccessfully mated many times. Moreover, two dogs with persistent Müllerian duct syndrome (PMDS) were found to have centric fusions, but coincidentally [46,47]. An interesting case of centric fusion in a testicular/ovotesticular XX DSD (*SRY*-negative) female dog with an enlarged clitoris and uterus was described by Switonski et al. [42]. CFA5 and CFA23 were involved in this rearrangement and the fusion led to a pericentromeric fragment of CFA23 being deleted. It was hypothesized that this could cause the deletion of regulatory sequences for genes that are important in ovarian development located in CFA23, such as *PISRT1*, *FOXL2*, and *CTNNB1*.

Table 5. Robertsonian translocations reported in dogs.

Chromosome Involved in the Fusion	Breed (Number of Cases)	Characteristic Feature of Phenotype	Reference
Not identified	Mixed terrier (1)	Cardiac defect	[48]
Not identified	Miniature Poodle (1)	Bone chondrodysplasia	[49]
Not identified	Setter–Retriever cross (1)	Phenotypically and clinically normal female	[50]
13 and 17	Golden Retriever cross (1)	Normal, fertile female	[51]
13 and 23	Golden Retriever-type (1 + 11 offspring)	Normal phenotype with the exception of congenital inguinal hernia in two female homozygotes in progeny	[52]
1 and 31	Poodle (6, including 1 homozygote male)	Normal phenotype	[53]
21 and 33	Walker Hound (1 + sister and 4 offspring)	Narrow vulva, absence of estrus	[43]
Not identified	Mixed breed (1)	Infertile female	[44]
8 and 14	West Highland White Terrier (1)	Infertile female, normal reproductive organs	[45]
5 and 23	Bernese Mountain Dog (1)	XX DSD, *SRY*-negative enlarged clitoris, testicle, ovotestis, uterus	[42]
Not identified	Miniature Schnauzer (1)	XY DSD, PMDS (Persistent Müllerian Duct Syndrome)	[46]
1 and unidentified	Miniature Schnauzer (1)	XY DSD, PMDS	[47]
Not identified	American Staffordshire Terrier (1)	XX DSD, *SRY*-negative (Sex Determining Region Y) enlarged clitoris, ovotestis,	[14]

Reciprocal translocations have rarely been diagnosed in dogs, probably due to difficulties in the recognition of one-armed autosomes. Until now, only three X/autosome mutations have been found, and this was possible because the translocation chromosome derived from the X had the abnormal morphology. The first mutation was identified in a male-to-female sex-reversed Yorkshire terrier [54]. The dog had two cell lines—a normal 78,XY and a line with X-autosome translocation. The mutation was identified using a whole X-chromosome painting probe which showed the hybridization signals on the X chromosome and unidentified autosome. Recently, two new cases of such rearrangement were observed in two female dogs with abnormalities of the genitourinary system [14]. In one of the female dogs, a pure X/autosome translocation, 78,X,t(X;2), was found, while in the second case a mosaicism of 78,X,t(X;A)/78,XX was observed. The cell line with the translocation occurred with a low incidence and it was not possible to identify the autosome involved.

It can be foreseen that the detection of canine chromosome translocations could be more efficient if a more sophisticated tool, such as multihybridization slides with a set of canine subtelomeric probes, were available, as has been recently developed for the chromosomes of pigs [55,56] and cattle [57].

5. Cytogenetic Characterization of Other Forms of DSD Cases

Cytogenetic analysis is a crucial step in classifying DSD [33,58]. Some DSD dogs may have chromosome abnormalities (sex chromosome DSD), as described above, but the majority of cases have a normal chromosome set described as XX DSD or XY DSD. Sex chromosomes are usually identified in such cases by Giemsa staining, karyotyping of R-banded chromosomes, or FISH with chromosome-specific probes (Figure 3) [32,59–61].

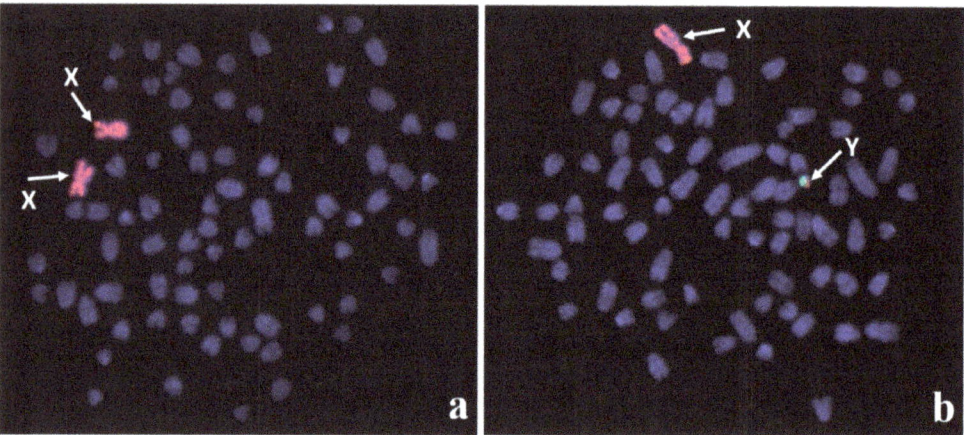

Figure 3. Identification of sex chromosomes by FISH with the use of painting probes (X: red) and (Y: green): (**a**) 78,XX, (**b**) 78,XY, with visible signals in pseudoautosomal region (PAR).

Cytogenetic analysis is also helpful for visualizing copy number variation (CNV). It has been shown that, in some dogs, CNVs in the *SOX9* gene region (CFA9) are associated with XX DSD phenotype. FISH with BAC probes specific to this region was used to identify duplication or multiplication (Figure 4) [62–64].

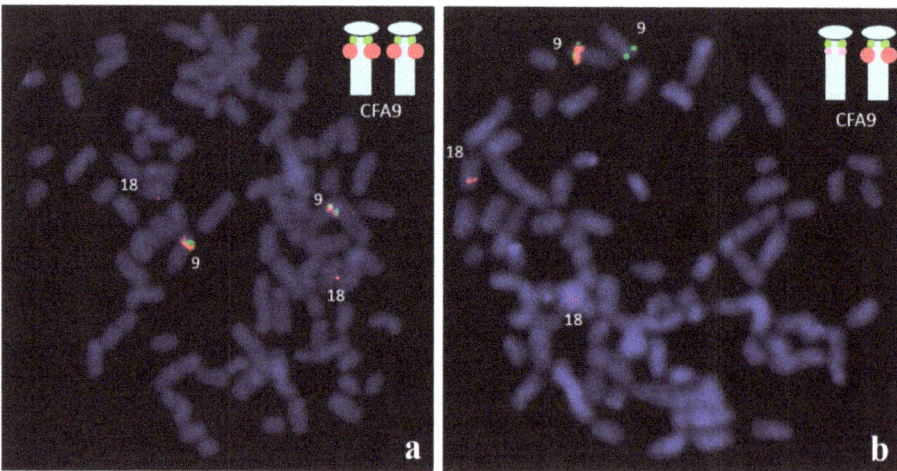

Figure 4. Identification of the copy number variation (CNV) in the region of *SOX9*, located on CFA9. Two BAC (Bacterial Artificial Chromosome) clones were used: the green signals are specific to the *SOX9* gene and the red signals are specific to the upstream CNV. The red probe also presents homology to CFA18. (**a**) The two large red signals on CFA9 indicate multiplication of the CNV region. (**b**) Another example of the variation - the single large, red signal is visible on one CFA9 chromosome, only.

Since the resolution of the hybridization signals on metaphase chromosomes is not sufficient to detect *SOX9* gene triplication, interphase nuclei were examined (Figure 5) [64]. It should be mentioned that molecular techniques such as MLPA (Multiplex Ligation-dependent Probe Amplification) and a-CGH (array Comparative Genome Hybridization) have also been employed for the identification of this CNV [60–62].

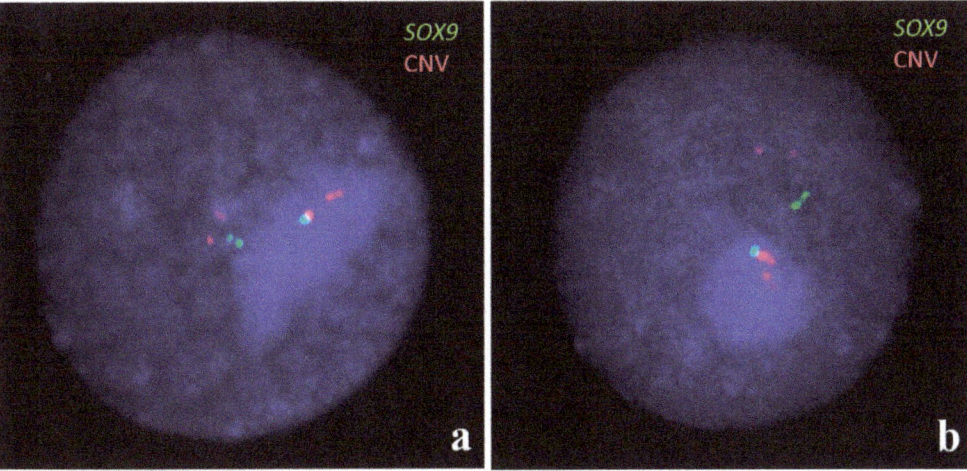

Figure 5. FISH for identification of three copies of *SOX9* gene (green signals) in two interphase nuclei (**a,b**). Moreover, multiple copies at CNV region upstream *SOX9*, as well as homologous region of CFA18 (red signals), are visible. For details, see [64].

6. Sperm Cytogenetics

The dog is a valuable large animal model in studies of human reproduction and development [65]. The segregation of sperm into two fractions, rich in, respectively, X-chromosome or Y-chromosome bearing cells, is an assisted reproductive biotechnology that has been developed for dogs. The effectiveness of this technology can be validated using the FISH technique with molecular probes specific to sex chromosomes on the segregated sperm samples. Dual color FISH has shown that the sorting of dog sperm by flow cytometry is very efficient, and the purity of the sorted samples was high at about 90% [66]. Komaki et al. [67] also used FISH to evaluate sex chromosome aneuploidy in sperm. The authors performed three color FISH with probes for chromosomes CFAX, CFAY and CFA1. Altogether, the sperms from eight dogs were analyzed and the mean frequencies of aneuploidy were: 0.016% (XX), 0.024% (YY), 0.08% (XY), and 0.176% (lack of sex chromosomes but with CFA1).

There has to date been no information on chromosome abnormalities in oocytes or embryos, despite the fact that studies of in vitro embryo production in dogs are quite advanced [68]. It can be foreseen that more such studies will be undertaken due to the increase in interest in biomedical research using induced pluripotent somatic cells [69].

7. Cancer Cytogenetics

Cancer is a genetic disease caused by gene or chromosomal mutations, classified as germline (inherited) or somatic. The somatic mutations can cause the disease or can have an effect on its development. Knowledge of the germline mutations responsible for cancer development in domestic animals, including dogs, is scarce [70]. Studies to identify mutations in cancer cells are also poorly advanced in domestic animals, though canine cancers have been considered more frequently than others [71].

Cytogenetic studies of canine cancer have a long history, with the first papers being published almost sixty years ago [72]. Early reports of chromosome abnormalities in canine tumors should be taken with caution due to the difficulties in recognizing autosomes. Since the partial international standard karyotype of the dog was agreed on in 1996, and since chromosome-specific molecular probes for FISH became available in the late 1990s, we focused this review on reports published after 1996.

Sex chromosome and autosomal aneuploidies, as well as centric fusions, are easy to identify as it was already mentioned earlier; however, the identification of small autosomes involved in such abnormalities is challenging. There are several reports showing a clonal predominance of specific aneuploidies in canine cancer cells. Analysis of G-banded chromosomes of cells derived from thyroid adenomas showed that the trisomy of chromosome 18 (CFA18) was predominant in the studied metaphase spreads [73]. Another study of in vitro cultured lymphocytes derived from the bone marrow of two dogs suffering from acute myeloid leukemia revealed two clonal aberrations: a trisomy of chromosome 1 (CFA1) and a chromosome translocation t(X;8) [74]. These aberrations were identified using G-banding. Polysomy of chromosome 13 (CFA13), caused by centric fusion between these chromosomes, was observed with an elevated frequency in cells derived from the prostate carcinomas of two dogs [75,76]. Interestingly, aberrations of this chromosome have also been observed in other dog cancers [7].

The introduction of molecular techniques into chromosome analysis was a very important step for canine cancer cytogenetics. Establishing the canine BAC library allowed the identification of BAC clones harboring 25 candidate genes for different cancers, which could be used in FISH analysis of cancer cells [77]. Researchers have searched for BAC clones in the canine genome library to use as FISH probes. Using this approach, it was shown by Breen and Modiano [78] that the well-known somatic chromosome rearrangements associated with some human cancers are also present in canine counterparts. These researchers examined the canine counterparts of three human cancers: chronic myelogenous leukemia (CML) associated with *BCR* and *ABL* fusion, sporadic Burkitt lymphoma (BL) associated with *MYC-IgH* fusion, and chronic lymphocytic leukemia/small lymphocytic lymphoma (CLL) associated with a hemizygous deletion harboring the *RB1* gene. They found that approximately 25% of the metaphase spreads or interphase nuclei of cancer cells they studied carried similar chromosome rearrangements. This study confirmed that the dog is a valuable animal model for studies of human cancerogenesis. The colocalization of *BCR-ABL* was also detected by FISH in dogs suffering from chronic monocytic leukemia (CMoL) [79] and acute myeloblastic leukemia without maturation (AML-M1) [80]. Canine BAC clones and whole chromosome painting probes were used by Vozdova et al. [81], who studied canine cutaneous mast cell tumors. Among different clonal aneuploidies and structural rearrangements, the most common was trisomy of CFA11.

Complex chromosome rearrangements causing genomic imbalances (loss or gain of genetic material) can be efficiently analyzed by comparative genomic hybridization (CGH). However, the classic CGH approach requires the reliable identification of banded chromosomes. The first attempt to use CGH to analyze canine cancer cells was by Dunn et al. [82], who studied a glial tumor cell line. Unfortunately, difficulties with chromosome recognition meant it was not possible to present a detailed characterization of the imbalances. The study showed that the only abnormality observed in all metaphase spreads was CFA1 trisomy. To overcome problems with identifying chromosomes using banding techniques, a molecular version of the CGH, called array CGH (aCGH), was developed. The first canine aCGH for 87 canine BAC clones was presented by Thomas et al. [83]. Soon after, two advanced aCGH microarrays were developed. One included 1158 canine BAC clones harboring canine genome fragments distributed along all chromosomes, with an average interval of 2 Mb [84]. In the second, the distribution of the BAC clones was approximately 10 Mb [85]. These molecular tools have replaced classical cytogenetic techniques in studies of complex chromosome rearrangements in cancer cells.

8. Conclusions and Perspectives

Although great progress has been achieved in studies of the organization of the canine genome, analysis of its chromosomes remains challenging. It is not surprising that the majority of abnormalities identified so far are sex chromosome abnormalities and centric fusions, as these can be identified by conventional Giemsa staining. The identification of sex chromosomes in DSD dogs plays a very important role in elucidating the DSD background, so classical analysis of chromosome preparations should be a common diagnostic approach. It facilitates the identification of sex chromosome abnormalities (e.g., X monosomy and XXY trisomy) or of abnormal sets of sex chromosomes in leukocytes (XX/XY leukocyte chimerism). The unequivocal identification of structural chromosome rearrangements in which small autosomes are involved usually requires the use of the FISH technique with probes derived from the canine BAC library. A promising perspective is related with the application of synthetic oligonucleotide probes (oligos) designed with the use of computational tools. The oligonucleotide libraries can be a valuable source of probes specific for a given chromosome, its region or a single gene [86–88].

It can also be expected that, in the near future, molecular techniques will play an important role in animal clinical cytogenetics. One of such techniques is digital droplet PCR (ddPCR), which allows the determination of the number of X and Y chromosome copies and the detection of sex chromosome aneuploidies and XX/XY leukocyte chimerism in a rapid, reliable manner. Other molecular techniques such as arrayCGH, SNP-microarray, MLPA, and NGS (next generation sequencing) are already very useful in human clinical cytogenetics in detecting structural rearrangements [89,90]. Moreover, the application of BioNano technologies offers the detection of chromosomal abnormalities, CNVs and structural variants [89] with a high resolution [91]. A very recent update of the canine genome reference sequence [92] should facilitate the successful use of the sequencing technologies. Taken together, it is expected that the spectrum of traditional cytogenetic techniques used in clinical diagnosis will be replaced by advanced DNA-based technologies, which are named "cytogenomics" [93].

The development of canine cytogenetics/cytogenomics also depends on the interest of veterinarians and dog breeders, who should be aware of the importance of such testing. Since the dog is an important biomedical animal, it may be expected that new diagnostic tools will be developed to overcome the difficulties of chromosome identification.

Author Contributions: The authors contributed equally in the preparation of the manuscript. Conceptualization, I.S. and M.S.; writing, I.S. and M.S.; and visualization, I.S. and M.S. Both authors have read and agreed to the published version of the manuscript.

Funding: This study was financed by the National Science Centre, Poland, project no. 2016/23/B/NZ9/03424, and by the statutory fund of the Department of Genetics and Animal Breeding (no. 506.534.04.00), Poznan University of Life Sciences, Poland.

Institutional Review Board Statement: The Institutional Review Board statement was not required.

Data Availability Statement: Data sharing is not applicable to this article as no new data were created or analyzed in this study.

Conflicts of Interest: The authors declare no conflict of interest.

References

1. Jung, C.; Pörtl, D. How Old Are (Pet) Dog Breeds? *Pet Behav. Sci.* **2019**, 29–37. [CrossRef]
2. Breen, M. Canine Cytogenetics — from Band to Basepair. *Cytogenet. Genome Res.* **2008**, *120*, 50–60. [CrossRef]
3. Świtoński, M.; Reimann, N.; Bosma, A.A.; Long, S.; Bartnitzke, S.; Pieńkowska, A.; Moreno-Milan, M.M.; Fischer, P. Report on the Progress of Standardization of the G-Banded Canine (Canis Familiaris) Karyotype. *Chromosome Res.* **1996**, *4*, 306–309. [CrossRef] [PubMed]
4. Breen, M.; Bullerdiek, J.; Langford, C.F. The DAPI banded karyotype of the domestic dog (Canis familiaris) generated using chromosome-specific paint probes. *Chromosome Res.* **1999**, *7*, 401–406. [CrossRef] [PubMed]
5. Breen, M.; Hitte, C.; Lorentzen, T.D.; Thomas, R.; Cadieu, E.; Sabacan, L.; Scott, A.; Evanno, G.; Parker, H.G.; Kirkness, E.F.; et al. An Integrated 4249 Marker FISH/RH Map of the Canine Genome. *BMC Genom.* **2004**, *5*, 65. [CrossRef] [PubMed]

6. Thomas, R.; Smith, K.C.; Ostrander, E.A.; Galibert, F.; Breen, M. Chromosome Aberrations in Canine Multicentric Lymphomas Detected with Comparative Genomic Hybridisation and a Panel of Single Locus Probes. *Br. J. Cancer* **2003**, *89*, 1530–1537. [CrossRef]
7. Reimann-Berg, N.; Bullerdiek, J.; Escobar, H.; Nolte, I. Chromosome analyses in dogs. *Tierärztl. Prax. Ausg. K Kleintiere Heimtiere* **2012**, *40*, 191–196. [CrossRef]
8. Smith, F.W.K.; Buoen, L.C.; Weber, A.F.; Johnston, S.D.; Randolph, J.F.; Waters, D.J. X-Chromosomal Monosomy (77, XO) in a Doberman Pinscher With Gonadal Dysgenesis. *J. Vet. Intern. Med.* **1989**, *3*, 90–95. [CrossRef] [PubMed]
9. Löfstedt, R.M.; Buoen, L.C.; Weber, A.F.; Johnston, S.D.; Huntington, A.; Concannon, P.W. Prolonged Proestrus in a Bitch with X Chromosomal Monosomy (77,XO). *J. Am. Vet. Med. Assoc.* **1992**, *200*, 1104–1106.
10. Switonski, M. Two Cases of Infertile Bitches with 78,XX/77,X Mosaic Karyotype: A Need for Cytogenetic Evaluation of Dogs With Reproductive Disorders. *J. Hered.* **2003**, *94*, 65–68. [CrossRef]
11. Mayenco-Aguirre, A.M.; Padilla, J.A.; Flores, J.M.; Daza, M.A. Canine Gonadal Dysgenesis Syndrome: A Case of Mosaicism (77,XO-78,XX). *Vet. Rec.* **1999**, *145*, 582–584. [CrossRef] [PubMed]
12. Raudsepp, T.; Das, P.J.; Avila, F.; Chowdhary, B.P. The Pseudoautosomal Region and Sex Chromosome Aneuploidies in Domestic Species. *Sex. Dev.* **2012**, *6*, 72–83. [CrossRef] [PubMed]
13. Raudsepp, T.; Chowdhary, B.P. The Eutherian Pseudoautosomal Region. *Cytogenet. Genome Res.* **2015**, *147*, 81–94. [CrossRef]
14. Szczerbal, I.; Nizanski, W.; Dzimira, S.; Nowacka-Woszuk, J.; Stachecka, J.; Biezynski, J.; Ligocka, Z.; Jagodka, D.; Fabian-Kurzok, H.; Switonski, M. Chromosome Abnormalities in Dogs with Disorders of Sex Development (DSD). *Anim. Reprod. Sci.* **2021**. submitted.
15. Johnston, S.D.; Buoen, L.C.; Weber, A.F.; Madl, J.E. X Trisomy in an Airedale Bitch with Ovarian Dysplasia and Primary Anestrus. *Theriogenology* **1985**, *24*, 597–607. [CrossRef]
16. Switonski, M.; Godynicki, S.; Jackowiak, H.; Piekowska, A.; Turczuk-Bierla, I.; Szymas, J.; Golinski, P.; Bereszynski, A. Brief Communication. X Trisomy in an Infertile Bitch: Cytogenetic, Anatomic, and Histologic Studies. *J. Hered.* **2000**, *91*, 149–150. [CrossRef]
17. Goldschmidt, B.; Paulino, F.O.; Sauza, L.M.; Gomes, H.F. Infertility Related to X-Trisomy in a Labrador Retriever Bitch. *J. Israeli Vet. Med. Assoc.* **2003**, *58*, 123–124.
18. O'Connor, C.L.; Schweizer, C.; Gradil, C.; Schlafer, D.; Lopate, C.; Prociuk, U.; Meyers-Wallen, V.N.; Casal, M.L. Trisomy-X with Estrous Cycle Anomalies in Two Female Dogs. *Theriogenology* **2011**, *76*, 374–380. [CrossRef]
19. Clough, E.; Pyle, R.L.; Hare, W.C.D.; Kelly, D.F.; Patterson, D.F. An XXY Sex-Chromosome Constitution in a Dog with Testicular Hypoplasia and Congenital Heart Disease. *Cytogenet. Genome Res.* **1970**, *9*, 71–77. [CrossRef]
20. Marshall, L.S.; Oehlert, M.L.; Haskins, M.E.; Selden, J.R.; Patterson, D.F. Persistent Müllerian Duct Syndrome in Miniature Schnauzers. *J. Am. Vet. Med. Assoc.* **1982**, *181*, 798–801.
21. Goldschmidt, B.; El-Jaick, K.B.; Souza, L.M.; Carvalho, E.C.Q.; Moura, V.L.S.; Benevides Filho, I.M. Cryptorchidism Associated with 78,XY/79,XXY Mosaicism in Dog. *Israel J. Vet. Med.* **2001**, *56*, 56e8.
22. Reimann-Berg, N.; Escobar, H.M.; Nolte, I.; Bullerdiek, J. Testicular Tumor in an XXY Dog. *Cancer Genet. Cytogenet.* **2008**, *183*, 114–116. [CrossRef]
23. Bosu, W.T.; Chick, B.F.; Basrur, P.K. Clinical, Pathologic and Cytogenetic Observations on Two Intersex Dogs. *Cornell Vet.* **1978**, *68*, 375–390.
24. Nie, G.J.; Johnston, S.D.; Hayden, D.W.; Buoen, L.C.; Stephens, M. Theriogenology Question of the Month. Azoospermia Associated with 79,XXY Chromosome Complement (Canine Klinefelter's Syndrome). *J. Am. Vet. Med. Assoc.* **1998**, *212*, 1545–1547.
25. Biason-Lauber, A. Control of Sex Development. *Best Pract. Res. Clin. Endocrinol. Metab.* **2010**, *24*, 163–186. [CrossRef]
26. Esteves, A.; Bage, R.; Payan-Carreira, R. Freemartinism in cattle. In *Ruminants: Anatomy, Behavior and Diseases*; Mendes, R.E., Ed.; Nova Science Publishers Inc.: Hauppauge, NY, USA, 2012; pp. 99–120.
27. Szczerbal, I.; Nowacka-Woszuk, J.; Dzimira, S.; Matuszczyk, A.; Iskrzak, P.; Switonski, M. Elevated Incidence of Freemartinism in Pigs Detected by Droplet Digital PCR and Cytogenetic Techniques. *Livest. Sci.* **2019**, *219*, 52–56. [CrossRef]
28. Beccaglia, M.; Ronchese, M.; Grieco, V.; Parma, P.; Luvoni, G.C. XX/XY Chimaerism/Mosaicism in a Phenotypically Female Wirehaired Pointing Griffon Dog. In Proceedings of the 7th International Symposium on Canine and Feline Reproduction, Whistler, BC, Canada, 26–29 July 2012.
29. Genero, E.R.; Moreno-Millán, M.; Ocaña-Quero, J.M. XX/XY Chromosome Chimaerism in an Intersex Dog. *Vet. Rec.* **1998**, *142*, 340. [CrossRef]
30. Kuiper, H.; Distl, O. Intersexuality in dogs: Causes and genetics. *DTW Dtsch. Tierarztl. Wochenschr.* **2004**, *111*, 251–258.
31. Szczerbal, I.; Nowacka-Woszuk, J.; Nizanski, W.; Salamon, S.; Ochota, M.; Dzimira, S.; Atamaniuk, W.; Switonski, M. A Case of Leucocyte Chimerism (78,XX/78,XY) in a Dog with a Disorder of Sexual Development. *Reprod. Domest. Anim.* **2014**, *49*, e31–e34. [CrossRef] [PubMed]
32. Szczerbal, I.; Nowacka-Woszuk, J.; Nizanski, W.; Dzimira, S.; Ligocka, Z.; Jastrzebska, A.; Kabala, B.; Biernacik, M.; Przadka, P.; Switonski, M. Disorders of Sex Development Are an Emerging Problem in French Bulldogs: A Description of Six New Cases and a Review of the Literature. *Sex. Dev.* **2019**, *13*, 205–211. [CrossRef]
33. Poth, T.; Breuer, W.; Walter, B.; Hecht, W.; Hermanns, W. Disorders of Sex Development in the Dog—Adoption of a New Nomenclature and Reclassification of Reported Cases. *Anim. Reprod. Sci.* **2010**, *121*, 197–207. [CrossRef]
34. Hare, W.C. Intersexuality in the Dog. *Can. Vet. J. Rev. Veterinaire Can.* **1976**, *17*, 7–15.

35. Weaver, A.D.; Harvey, M.J.; Munro, C.D.; Rogerson, P.; McDonald, M. Phenotypic Intersex (Female Pseudohermaphroditism) in a Dachshund Dog. *Vet. Rec.* **1979**, *105*, 230–232. [CrossRef] [PubMed]
36. Holt, P.E.; Long, S.E.; Gibbs, C. Disorders of Urination Associated with Canine Intersexuality. *J. Small Anim. Pract.* **1983**, *24*, 475–487. [CrossRef]
37. Johnston, S.D. Premature Gonadal Failure in Female Dogs and Cats. *J. Reprod. Fertil. Suppl.* **1989**, *39*, 65–72. [PubMed]
38. Chaffaux, S.; Cribiu, E. Clinical, Histological and Cytogenetic Observations on Nine Intersex Dogs. *Genet. Sel. Evol.* **1991**, *23*, S81. [CrossRef]
39. Meyers-Wallen, V.N. Inherited abnormalities of sexual development in dogs and cats. In *Recent Advances in Small Animal Reproduction*; Concannon, P.W., England, G., Ver-stegen, J., Eds.; International Veterinary Information Service, USA, 2001. Available online: https://www.ivis.org/library/recent-advances-small-animal-reproduction/inherited-abnormalities-of-sexual-development (accessed on 10 January 2021).
40. Sumner, S.M.; Case, J.B.; Regier, P.J.; Oliveira, L.; Abbott, J.R. Laparoscopic Gonadectomy in a Dog with 78,XX/78,XY Chimerism and Underdeveloped Reproductive Organs. *J. Am. Vet. Med. Assoc.* **2021**, *258*, 80–84. [CrossRef]
41. Szczerbal, I.; Nowacka-Woszuk, J.; Albarella, S.; Switonski, M. Technical Note: Droplet Digital PCR as a New Molecular Method for a Simple and Reliable Diagnosis of Freemartinism in Cattle. *J. Dairy Sci.* **2019**, *102*, 10100–10104. [CrossRef] [PubMed]
42. Switonski, M.; Szczerbal, I.; Nizanski, W.; Kociucka, B.; Bartz, M.; Dzimira, S.; Mikolajewska, N. Robertsonian Translocation in a Sex Reversal Dog (XX, *SRY* Negative) May Indicate That the Causative Mutation for This Intersexuality Syndrome Resides on Canine Chromosome 23 (CFA23). *Sex. Dev.* **2011**, *5*, 141–146. [CrossRef]
43. Stone, D.M.; Mickelsen, W.D.; Jacky, P.B.; Prieur, D.J. A Novel Robertsonian Translocation in a Family of Walker Hounds. *Genome* **1991**, *34*, 677–680. [CrossRef] [PubMed]
44. Switonski, M.; Slota, E.; Pietrzak, A.; Klukowska, J. Chimerism 78,XX/77,XX, Rb in a bitch revealed by chromosome and microsatellite studies. *Vet. Med. Czech.* **2000**, *45*, 296–298.
45. Switonski, M.; Szczerbal, I.; Skorczyk, A.; Yang, F.; Antosik, P. Robertsonian Translocation (8;14) in an Infertile Bitch (Canis Familiaris). *J. Appl. Genet.* **2003**, *44*, 525–527. [PubMed]
46. Dzimira, S.; Wydooghe, E.; Van Soom, A.; Van Brantegem, L.; Nowacka-Woszuk, J.; Szczerbal, I.; Switonski, M. Sertoli Cell Tumour and Uterine Leiomyoma in Miniature Schnauzer Dogs with Persistent Müllerian Duct Syndrome Caused by Mutation in the AMHR2 Gene. *J. Comp. Pathol.* **2018**, *161*, 20–24. [CrossRef] [PubMed]
47. Nogueira, D.M.; Armada, J.L.A.; Penedo, D.M.; Tannouz, V.G.S.; Meyers-Wallen, V.N. Persistent Mullerian Duct Syndrome in a Brazilian Miniature Schnauzer Dog. *An. Acad. Bras. Ciênc.* **2019**, *91*, e20180752. [CrossRef]
48. Shive, R.J.; Hare, W.C.D.; Patterson, D.F. Chromosome Studies in Dogs with Congenital Cardiac Defects. *Cytogenet. Genome Res.* **1965**, *4*, 340–348. [CrossRef]
49. Hare, W.C.; Wilkinson, J.S.; McFeely, R.A.; Riser, W.H. Bone Chondroplasia and a Chromosome Abnormality in the Same Dog. *Am. J. Vet. Res.* **1967**, *28*, 583–587.
50. Ma, N.S.F.; Gilmore, C.E. Chromosomal Abnormality in a Phenotypically and Clinically Normal Dog. *Cytogenet. Genome Res.* **1971**, *10*, 254–259. [CrossRef] [PubMed]
51. Larsen, R.E.; Dias, E.; Cervenka, J. Centric Fusion of Autosomal Chromosomes in a Bitch and Offspring. *Am. J. Vet. Res.* **1978**, *39*, 861–864.
52. Larsen, R.E.; Dias, E.; Flores, G.; Selden, J.R. Breeding Studies Reveal Segregation of a Canine Robertsonian Translocation along Mendelian Proportions. *Cytogenet. Cell Genet.* **1979**, *24*, 95–101. [CrossRef]
53. Mayr, B.; Krutzler, J.; Schleger, W.; Auer, H. A New Type of Robertsonian Translocation in the Domestic Dog. *J. Hered.* **1986**, *77*, 127. [CrossRef]
54. Schelling, C.; Pieńkowska, A.; Arnold, S.; Hauser, B.; Switoński, M. A Male to Female Sex-Reversed Dog with a Reciprocal Translocation. *J. Reprod. Fertil. Suppl.* **2001**, *57*, 435–438. [PubMed]
55. O'Connor, R.E.; Fonseka, G.; Frodsham, R.; Archibald, A.L.; Lawrie, M.; Walling, G.A.; Griffin, D.K. Isolation of Subtelomeric Sequences of Porcine Chromosomes for Translocation Screening Reveals Errors in the Pig Genome Assembly. *Anim. Genet.* **2017**, *48*, 395–403. [CrossRef] [PubMed]
56. O'Connor, R.E.; Kiazim, L.G.; Rathje, C.C.; Jennings, R.L.; Griffin, D.K. Rapid Multi-Hybridisation FISH Screening for Balanced Porcine Reciprocal Translocations Suggests a Much Higher Abnormality Rate Than Previously Appreciated. *Cells* **2021**, *10*, 250. [CrossRef] [PubMed]
57. Jennings, R.L.; Griffin, D.K.; O'Connor, R.E. A New Approach for Accurate Detection of Chromosome Rearrangements That Affect Fertility in Cattle. *Animals* **2020**, *10*, 114. [CrossRef] [PubMed]
58. Meyers-Wallen, V.N. Gonadal and Sex Differentiation Abnormalities of Dogs and Cats. *Sex. Dev.* **2012**, *6*, 46–60. [CrossRef] [PubMed]
59. Switonski, M.; Payan-Carreira, R.; Bartz, M.; Nowacka-Woszuk, J.; Szczerbal, I.; Colaço, B.; Pires, M.A.; Ochota, M.; Nizanski, W. Hypospadias in a Male (78,XY; *SRY*-Positive) Dog and Sex Reversal Female (78,XX; *SRY*-Negative) Dogs: Clinical, Histological and Genetic Studies. *Sex. Dev.* **2012**, *6*, 128–134. [CrossRef]
60. Rossi, E.; Radi, O.; De Lorenzi, L.; Vetro, A.; Groppetti, D.; Bigliardi, E.; Luvoni, G.C.; Rota, A.; Camerino, G.; Zuffardi, O.; et al. Sox9 Duplications Are a Relevant Cause of Sry-Negative XX Sex Reversal Dogs. *PLoS ONE* **2014**, *9*, e101244. [CrossRef]
61. Albarella, S.; Lorenzi, L.D.; Rossi, E.; Prisco, F.; Riccardi, M.G.; Restucci, B.; Ciotola, F.; Parma, P. Analysis of XX SRY-Negative Sex Reversal Dogs. *Animals* **2020**, *10*, 1667. [CrossRef]

62. Marcinkowska-Swojak, M.; Szczerbal, I.; Pausch, H.; Nowacka-Woszuk, J.; Flisikowski, K.; Dzimira, S.; Nizanski, W.; Payan-Carreira, R.; Fries, R.; Kozlowski, P.; et al. Copy Number Variation in the Region Harboring SOX9 Gene in Dogs with Testicular/Ovotesticular Disorder of Sex Development (78,XX; SRY-Negative). *Sci. Rep.* **2015**, *5*, 14696. [CrossRef]
63. Szczerbal, I.; Nowacka-Woszuk, J.; Dzimira, S.; Atamaniuk, W.; Nizanski, W.; Switonski, M. A Rare Case of Testicular Disorder of Sex Development in a Dog (78,XX; *SRY*-Negative) with Male External Genitalia and Detection of Copy Number Variation in the Region Upstream of the *SOX9* Gene. *Sex. Dev.* **2016**, *10*, 74–78. [CrossRef]
64. Nowacka-Woszuk, J.; Szczerbal, I.; Stachowiak, M.; Szydlowski, M.; Nizanski, W.; Dzimira, S.; Maslak, A.; Payan-Carreira, R.; Wydooghe, E.; Nowak, T.; et al. Association between Polymorphisms in the SOX9 Region and Canine Disorder of Sex Development (78,XX; SRY-Negative) Revisited in a Multibreed Case-Control Study. *PLoS ONE* **2019**, *14*, e0218565. [CrossRef]
65. Wright, S.J. Spotlight on reproduction in domestic dogs as a model for human reproduction. In *Animals Model and Human Reproduction*; Schatten, H., Constantinescu, G.M., Eds.; John Wiley & Sons, Inc.: Hoboken, NJ, USA, 2017; pp. 247–358.
66. Oi, M.; Yamada, K.; Hayakawa, H.; Suzuki, H. Sexing of Dog Sperm by Fluorescence In Situ Hybridization. *J. Reprod. Dev.* **2012**. [CrossRef]
67. Komaki, H.; Oi, M.; Suzuki, H. Detection of Sex Chromosome Aneuploidy in Dog Spermatozoa by Triple Color Fluorescence in Situ Hybridization. *Theriogenology* **2014**, *82*, 652–656. [CrossRef]
68. Nagashima, J.B.; Travis, A.J.; Songsasen, N. The Domestic Dog Embryo: In Vitro Fertilization, Culture, and Transfer. In *Comparative Embryo Culture*; Herrick, J.R., Ed.; Methods in Molecular Biology; Springer: New York, NY, USA, 2019; Volume 2006, pp. 247–267. ISBN 978-1-4939-9565-3.
69. Hyttel, P.; Pessôa, L.V.d.F.; Secher, J.B.-M.; Dittlau, K.S.; Freude, K.; Hall, V.J.; Fair, T.; Assey, R.J.; Laurincik, J.; Callesen, H.; et al. Oocytes, Embryos and Pluripotent Stem Cells from a Biomedical Perspective. *Anim. Reprod.* **2019**, *16*, 508–523. [CrossRef]
70. Flisikowski, K.; Flisikowska, T.; Sikorska, A.; Perkowska, A.; Kind, A.; Schnieke, A.; Switonski, M. Germline Gene Polymorphisms Predisposing Domestic Mammals to Carcinogenesis: Gene Polymorphisms Predisposing to Carcinogenesis. *Vet. Comp. Oncol.* **2017**, *15*, 289–298. [CrossRef]
71. Ostrander, E.A.; Dreger, D.L.; Evans, J.M. Canine Cancer Genomics: Lessons for Canine and Human Health. *Annu. Rev. Anim. Biosci.* **2019**, *7*, 449–472. [CrossRef] [PubMed]
72. Makino, S. Some epidemiologic aspects of venereal tumors of dogs as revealed by chromosome and DNA studies. *Ann. N. Y. Acad. Sci.* **2006**, *108*, 1106–1122. [CrossRef] [PubMed]
73. Reimann, N.; Nolte, I.; Bonk, U.; Werner, M.; Bullerdiek, J.; Bartnitzke, S. Trisomy 18 in a Canine Thyroid Adenoma. *Cancer Genet. Cytogenet.* **1996**, *90*, 154–156. [CrossRef]
74. Reimann, N.; Bartnitzke, S.; Bullerdiek, J.; Mischke, R.; Nolte, I. Trisomy 1 in a Canine Acute Leukemia Indicating the Pathogenetic Importance of Polysomy 1 in Leukemias of the Dog. *Cancer Genet. Cytogenet.* **1998**, *101*, 49–52. [CrossRef]
75. Winkler, S.; Reimann-Berg, N.; Escobar, H.M.; Loeschke, S.; Eberle, N.; Höinghaus, R.; Nolte, I.; Bullerdiek, J. Polysomy 13 in a Canine Prostate Carcinoma Underlining Its Significance in the Development of Prostate Cancer. *Cancer Genet. Cytogenet.* **2006**, *169*, 154–158. [CrossRef]
76. Reimann-Berg, N.; Willenbrock, S.; Murua Escobar, H.; Eberle, N.; Gerhauser, I.; Mischke, R.; Bullerdiek, J.; Nolte, I. Two New Cases of Polysomy 13 in Canine Prostate Cancer. *Cytogenet. Genome Res.* **2011**, *132*, 16–21. [CrossRef] [PubMed]
77. Thomas, R.; Bridge, W.; Benke, K.; Breen, M. Isolation and Chromosomal Assignment of Canine Genomic BAC Clones Representing 25 Cancer-Related Genes. *Cytogenet. Genome Res.* **2003**, *102*, 249–253. [CrossRef]
78. Breen, M.; Modiano, J.F. Evolutionarily Conserved Cytogenetic Changes in Hematological Malignancies of Dogs and Humans—Man and His Best Friend Share More than Companionship. *Chromosome Res.* **2008**, *16*, 145–154. [CrossRef] [PubMed]
79. Cruz Cardona, J.A.; Milner, R.; Alleman, A.R.; Williams, C.; Vernau, W.; Breen, M.; Tompkins, M. BCR-ABL Translocation in a Dog with Chronic Monocytic Leukemia: BCR-ABL Translocation in Canine CMoL. *Vet. Clin. Pathol.* **2011**, *40*, 40–47. [CrossRef] [PubMed]
80. Figueiredo, J.F.; Culver, S.; Behling-Kelly, E.; Breen, M.; Friedrichs, K.R. Acute Myeloblastic Leukemia with Associated BCR-ABL Translocation in a Dog. *Vet. Clin. Pathol.* **2012**, *41*, 362–368. [CrossRef] [PubMed]
81. Vozdova, M.; Kubickova, S.; Cernohorska, H.; Fröhlich, J.; Fictum, P.; Rubes, J. Structural and Copy Number Chromosome Abnormalities in Canine Cutaneous Mast Cell Tumours. *J. Appl. Genet.* **2019**, *60*, 63–70. [CrossRef] [PubMed]
82. Dunn, K.A.; Thomas, R.; Binns, M.M.; Breen, M. Comparative Genomic Hybridization (CGH) in Dogs—Application to the Study of a Canine Glial Tumour Cell Line. *Vet. J.* **2000**, *160*, 77–82. [CrossRef]
83. Thomas, R.; Fiegler, H.; Ostrander, E.A.; Galibert, F.; Carter, N.P.; Breen, M. A Canine Cancer-Gene Microarray for CGH Analysis of Tumors. *Cytogenet. Genome Res.* **2003**, *102*, 254–260. [CrossRef]
84. Thomas, R. Construction of a 2-Mb Resolution BAC Microarray for CGH Analysis of Canine Tumors. *Genome Res.* **2005**, *15*, 1831–1837. [CrossRef]
85. Thomas, R.; Duke, S.E.; Bloom, S.K.; Breen, T.E.; Young, A.C.; Feiste, E.; Seiser, E.L.; Tsai, P.-C.; Langford, C.F.; Ellis, P.; et al. A Cytogenetically Characterized, Genome-Anchored 10-Mb BAC Set and CGH Array for the Domestic Dog. *J. Hered.* **2007**, *98*, 474–484. [CrossRef] [PubMed]
86. Huber, D.; Voith von Voithenberg, L.; Kaigala, G.V. Fluorescence in Situ Hybridization (FISH): History, Limitations and What to Expect from Micro-Scale FISH? *Micro Nano Eng.* **2018**, *1*, 15–24. [CrossRef]

87. Beliveau, B.J.; Kishi, J.Y.; Nir, G.; Sasaki, H.M.; Saka, S.K.; Nguyen, S.C.; Wu, C.; Yin, P. OligoMiner Provides a Rapid, Flexible Environment for the Design of Genome-Scale Oligonucleotide in Situ Hybridization Probes. *Proc. Natl. Acad. Sci. USA* **2018**, *115*, E2183–E2192. [CrossRef] [PubMed]
88. Beliveau, B.J.; Joyce, E.F.; Apostolopoulos, N.; Yilmaz, F.; Fonseka, C.Y.; McCole, R.B.; Chang, Y.; Li, J.B.; Senaratne, T.N.; Williams, B.R.; et al. Versatile Design and Synthesis Platform for Visualizing Genomes with Oligopaint FISH Probes. *Proc. Natl. Acad. Sci. USA* **2012**, *109*, 21301–21306. [CrossRef] [PubMed]
89. Fan, J.; Wang, L.; Wang, H.; Ma, M.; Wang, S.; Liu, Z.; Xu, G.; Zhang, J.; Cram, D.S.; Yao, Y. The Clinical Utility of Next-Generation Sequencing for Identifying Chromosome Disease Syndromes in Human Embryos. *Reprod. Biomed. Online* **2015**, *31*, 62–70. [CrossRef] [PubMed]
90. Hu, L.; Liang, F.; Cheng, D.; Zhang, Z.; Yu, G.; Zha, J.; Wang, Y.; Xia, Q.; Yuan, D.; Tan, Y.; et al. Location of Balanced Chromosome-Translocation Breakpoints by Long-Read Sequencing on the Oxford Nanopore Platform. *Front. Genet.* **2020**, *10*, 1313. [CrossRef] [PubMed]
91. Sahajpal, N.S.; Barseghyan, H.; Kolhe, R.; Hastie, A.; Chaubey, A. Optical Genome Mapping as a Next-Generation Cytogenomic Tool for Detection of Structural and Copy Number Variations for Prenatal Genomic Analyses. *Genes* **2021**, *12*, 398. [CrossRef]
92. Wang, C.; Wallerman, O.; Arendt, M.-L.; Sundström, E.; Karlsson, Å.; Nordin, J.; Mäkeläinen, S.; Pielberg, G.R.; Hanson, J.; Ohlsson, Å.; et al. A Novel Canine Reference Genome Resolves Genomic Architecture and Uncovers Transcript Complexity. *Commun. Biol.* **2021**, *4*, 185. [CrossRef] [PubMed]
93. Hochstenbach, R.; Liehr, T.; Hastings, R.J. Chromosomes in the Genomic Age. Preserving Cytogenomic Competence of Diagnostic Genome Laboratories. *Eur. J. Hum. Genet.* **2020**. [CrossRef]

Review

Horse Clinical Cytogenetics: Recurrent Themes and Novel Findings

Monika Bugno-Poniewierska [1,*] and Terje Raudsepp [2,*]

[1] Department of Animal Reproduction, Anatomy and Genomics, University of Agriculture in Krakow, 31-120 Krakow, Poland
[2] Department of Veterinary Integrative Biosciences, Texas A&M University, College Station, TX 77843-4458, USA
* Correspondence: Monika.Bugno-Poniewierska@urk.edu.pl (M.B.-P.); traudsepp@cvm.tamu.edu (T.R.)

Simple Summary: Horse chromosomes have been studied for veterinary diagnostic purposes for over half a century. The findings show that changes in the chromosome number or structure are among the most common non-infectious causes of decreased fertility, infertility, and developmental abnormalities. Based on large-scale surveys, almost 30% of horses with reproductive or developmental problems have abnormal chromosomes. For a comparison, only 2–5% of horses in the general population have abnormal chromosomes. Most chromosome abnormalities are rare and found in one or a few animals. However, two conditions are recurrent: sterile mares with only one X chromosome, instead of two, and sterile mares with XY male sex chromosomes where the Y has lost the 'maleness' gene *SRY*. The two are signature features of chromosome abnormalities in the horse, being rare or absent in other domestic animals. The progress in horse genome sequencing and the development of molecular tools have improved the depth and quality of diagnostic chromosome analysis, allowing for an understanding of the underlying molecular mechanisms. Nevertheless, cutting-edge genomics tools are not about to entirely replace traditional chromosome analysis, which still is the most straightforward, cost-effective, and fastest approach for the initial evaluation of potential breeding animals and horses with reproductive or developmental disorders.

Abstract: Clinical cytogenetic studies in horses have been ongoing for over half a century and clearly demonstrate that chromosomal disorders are among the most common non-infectious causes of decreased fertility, infertility, and congenital defects. Large-scale cytogenetic surveys show that almost 30% of horses with reproductive or developmental problems have chromosome aberrations, whereas abnormal karyotypes are found in only 2–5% of the general population. Among the many chromosome abnormalities reported in the horse, most are unique or rare. However, all surveys agree that there are two recurrent conditions: X-monosomy and *SRY*-negative XY male-to-female sex reversal, making up approximately 35% and 11% of all chromosome abnormalities, respectively. The two are signature conditions for the horse and rare or absent in other domestic species. The progress in equine genomics and the development of molecular tools, have qualitatively improved clinical cytogenetics today, allowing for refined characterization of aberrations and understanding the underlying molecular mechanisms. While cutting-edge genomics tools promise further improvements in chromosome analysis, they will not entirely replace traditional cytogenetics, which still is the most straightforward, cost-effective, and fastest approach for the initial evaluation of potential breeding animals and horses with reproductive or developmental disorders.

Keywords: horse; chromosome aberration; aneuploidy; translocation; structural rearrangements; sex reversal; chimerism; molecular cytogenetics; FISH; CGH

Citation: Bugno-Poniewierska, M.; Raudsepp, T. Horse Clinical Cytogenetics: Recurrent Themes and Novel Findings. *Animals* **2021**, *11*, 831. https://doi.org/10.3390/ani11030831

Academic Editor: Leopoldo Iannuzzi

Received: 1 March 2021
Accepted: 13 March 2021
Published: 16 March 2021

Publisher's Note: MDPI stays neutral with regard to jurisdictional claims in published maps and institutional affiliations.

Copyright: © 2021 by the authors. Licensee MDPI, Basel, Switzerland. This article is an open access article distributed under the terms and conditions of the Creative Commons Attribution (CC BY) license (https://creativecommons.org/licenses/by/4.0/).

1. Introduction

Clinical cytogenetic research in horses has been ongoing for over half a century and has clearly demonstrated that chromosome abnormalities are associated with congenital disorders, embryonic loss, reduced fertility, and infertility. Changes in the chromosome number or structure typically result in genomic imbalance and affect meiotic cell division, gametogenesis, and the viability of zygotes and embryos. Genetically balanced chromosomal changes, such as translocations, can be transmitted, causing fertility problems in subsequent generations. In cases where chromosomal aberrations do not show phenotypic or behavioral effects, the carriers can be included in breeding, resulting in significant economic loss due to veterinary fees and the costs related to maintaining a sterile or a subfertile horse over the years. Therefore, the cytogenetic screening of potential breeding animals and clinical cytogenetic evaluation of problem horses are of economic importance for the equine industry, as well as for the owners and breeders.

During the peak of equine clinical cytogenetics in the 1970s–1990s, many abnormal karyotypes were published and in the following years, and the findings have been well-reviewed in books [1,2], book chapters [3–7], and multiple review papers, some specifically focusing on equine cytogenetics [8–10], others on cytogenetics of domestic animals, including the horse [11–14]. Since the last comprehensive and horse-focused reviews about a decade ago [4,5], equine clinical cytogenetics has advanced qualitatively, mainly thanks to the progress in equine genomics and the availability of new powerful genomic tools (reviewed by [15]). At the same time, the quantity of clinical cytogenetic publications in the horse has reduced compared to the pinnacle times in 1970s–1990s. This, however, is not because there are less horses with karyotype aberrations but rather because not every cytogenetic case results in a publication. The majority of recent reports combine conventional cytogenetics with molecular methods which allow for the validation and refinement of the findings, but also start revealing the underlying molecular causes and mechanisms of chromosome abnormalities in the horse.

In this review, we appraise the cytogenetic findings of the past and combine those with recent reports to identify novel findings and highlight recurrent patterns of chromosome abnormalities in the horse. We also discuss how molecular tools and the availability of the horse reference genome have essentially advanced equine clinical cytogenetics today and what the perspectives for the future are.

2. The Horse Chromosomes

2.1. Chromosome Number

The first reports about horse chromosomes date back to the early 20th century, when, using spermatogonial and meiotic preparations, it was proposed that the horse diploid number is approximately 20–37 [16–18] with an XO sex chromosome system [19]. Thanks to improvements in the chromosome analysis methodology, these early findings were soon revised showing that like in other mammals, horse has XY sex chromosome system [20] and the correct diploid number for the domestic horse (*Equus caballus*, ECA) is 2n = 64 [21–23].

2.2. Application of Different Chromosome Banding Techniques

Horse cytogenetics has evolved in conjunction with human cytogenetics and adopted from the latter all main chromosome differential staining or banding techniques (reviewed by [3]). Of the many techniques developed in the 1970s, only a few have remained in active everyday use in equine clinical chromosome analysis. Among these, the most common are G-banding [24] and its fluorescent version with 4′,6-diamidino-2-phenylindole, known as DAPI-banding [25]. The latter produces G-band-like pattern and is an essential part of all molecular cytogenetic methods (see Section 4). C-banding [26] is an excellent method to visualize horse sex chromosomes and is still in use as an additional method in cases involving sex chromosome abnormalities [27,28]. Compared to these, R-banding [29] and NOR-banding [30] are predominantly used for research [31–33] and not for routine clinical karyotyping.

2.3. Karyotype Features and Chromosome Nomenclature

In order to properly characterize chromosome abnormalities and communicate the findings between cytogenetic laboratories, standard karyotypes, and chromosome nomenclatures have been developed. These are agreements among researchers worldwide about how to number the chromosomes and arrange them by size, centromere position, and specific banding patterns, and demarcate individual chromosomal regions and bands.

To date, three standard karyotypes have been developed for the horse, the first from 1980 [34] and the second from 1990 [35], mainly differed by chromosome arrangement and numbering, but provided detailed description of the horse karyotype and chromosome banding patterns and established common grounds for clinical cytogeneticists. The third and current standard karyotype from 1997 [36] maintained the second arrangement [35] but included an enumerated nomenclature of chromosome bands.

According to ISCNH 1997 [36], the autosomes are divided into two groups: in the first group there are 13 pairs of meta- and sub-metacentric chromosomes, the second group includes 18 pairs of acrocentric chromosomes. Within each group, the autosomes are ordered by length. The sex chromosomes—a sub-metacentric X chromosome and an acrocentric Y chromosome [37]—are located in the center of the karyogram, next to the row of the three smallest bi-armed chromosomes. Horse sex chromosomes show distinct C-banding patterns. The X chromosome has two C-bands, one corresponding to the pericentromeric heterochromatin, another to an ampliconic array of *ETSTY7* sequences [38] interstitially in Xq17. The *ETSTY7* sequences prevail in the Y chromosome, which is almost completely C-band positive. Heterochromatin-rich pericentromeric C-bands are also present in most of the autosomes, except chr11. The latter is devoid of centromeric satellite DNA and presents an example of a chromosome with a neo-centromere where centromere function precedes satellite repeat accumulation [39]. The standard also describes the location of the 18S-5.8S-28S ribosomal RNA gene clusters or nucleolus organizer regions (NORs), which are in the telomeric region of chr1 and in the secondary constriction in chr28 and chr31. Some studies also found NOR in chr27 [40,41], though a more recent study [33] did not confirm the presence of a fourth pair of an NOR-bearing chromosome.

3. Chromosome Aberrations

3.1. Emerging Patterns of Chromosome Abnormalities—Large-Scale Studies

Diagnostic cytogenetic research in horses dates back to the late 1960s, preceding the introduction of chromosome banding techniques [42]. The first karyotype abnormalities detected by banding methods were published by Chandley et al. in 1975 [43]: in 7 mares referred for research due to reproductive problems, aneuploidies 63,X and 63,XXX; mosaicism 63,X/64,XX and 64,XY sex reversal were identified. Over the following years, several large-scale cytogenetic surveys [8,44–46] started to reveal the most prevalent and specific patterns of chromosome abnormalities in the horse.

A study of 180 mares with gonadal dysgenesis [44] found chromosomal abnormalities in 54%. The most common abnormality was X-monosomy (63,X), followed by 64,XY male-to-female sex reversal syndrome. Two mares showed structural abnormalities of one X chromosome [64,X,del(Xp)]. Chromosomal abnormalities, such as 63,X; 63,X/64,XX; 64,X,del(Xp) and 64,XX,i(26q), were also found in 4 fillies that were tested due to their small size and poor thriving [44].

A survey by Power [8], recorded X-monosomy in 204 mares (51%) out of 401 tested horses with chromosomal abnormalities. Of these, 70% had non-mosaic X-monosomy. Like in the survey by Bowling et al. [44], the second most frequent karyotype aberration was XY sex reversal, which was diagnosed in 27% out of the 401 horses. Over 13% of horses had various non-mosaic and mosaic forms of sex chromosome aneuploidies, such as 65,XXX; 65,XXY; 66,XXXY; 64,XX/65XXX; 63,X/64XY; 63,X/65,XYY; 64,XX/65,XXY; 63,X/64,XX/64,XY; 63,X/64,XY/65,XXY; 63,X/64,XX/65,XXY; 64,XX/64,XY/65,XXY; 63,X/64,XX/64,XY/65,XXY and 63,X/64,XX/65,XXX/65,XXY/66,XXXY/66,XXYY). The remaining 6% had structural aberrations (translocations, deletions, isochromosomes) or autosomal trisomies.

A third survey by Parada et al. [45] examined 244 mares with reproductive problems. Chromosome aberrations were found in 10 of the studied mares, which accounted for 4% of the entire study population, and 12.8% of completely sterile mares [45]. Like in the two previous large-scale surveys [8,44], the most common aberration was X-monosomy with non-mosaic 63,X in 3 mares, and mosaic 63,X/64,XX in four mares, which in total accounted for 70% of all aberrations [45]. Other findings included XX/XY leukocyte chimerism and an elongation of the p-arm of chr 12 [45]. The prevalence of sex chromosome abnormalities was also reported by a smaller-scale cytogenetic analysis of 42 mares with reproductive problems [47,48] showing mosaicism 63,X/64,XX in five individuals and a three-cell line mosaicism 63,X/64,XX/65,XXX in one mare. In addition, the analysis of two fillies and one colt from two different-sex twin pregnancies revealed one pair of twins with lymphocyte chimerism 64,XX/64,XY.

In order to find out the prevalence of chromosome abnormalities in general horse populations, Bugno et al. [46] conducted cytogenetic analysis in 500 young horses—272 fillies and 228 colts of 10 diverse breeds and breed crosses. Karyotype abnormalities were found in 10 young mares, which accounted for 2% of the entire population and 3.7% of the female population [46]. Among the diagnosed aberrations, 8 were X chromosome aneuploidies (80%)—one pure 63,X and seven 63,X/64,XX mosaics, one case of XX/XY chimerism, and one case of mosaicism for trisomy 31 (64,XX/65,XX,+31).

The above described surveys of reproductively abnormal and general horse populations were all conducted using conventional cytogenetic techniques. However, the development of molecular cytogenetic methods (see Section 4 for details), has increased the accuracy and power of diagnosis. For example, two recent studies validated the results of prior cytogenetic findings in a population of 500 young (up to 2 years) horses using fluorescence in situ hybridization (FISH) with molecular probes specific for the horse sex chromosomes [49,50]. The preliminary karyotyping results of 238 horses showed normal female or male karyotype in 225 animals. In 13 horses (5.5%) the following aberrations were found: 63,X/64,XX (3 mares); 63,X/64,XX/65,XXX (1 mare); 64,XX/65,XXX (2 mares); 64,XX/64,XX,del(Xp) (1 mare); 63,X/64,XX del(X)?/64,XX (1 mare); 63,X/64,XX/65,XXX del(X)? (1 mare); 64,XY/65XYY (1 stallion); 64,XX/64,XY (1 stallion); 64,XY *SRY*-negative sex reversal (1 mare), and one mare with a reciprocal translocation between chromosome 1 and X: 64,X,t(1p;Xp)(1q;Xq).

Finally, over the past 20 years (2001–2021), the Texas A&M Molecular Cytogenetics Laboratory (TAMUMCL) has analyzed 766 horses with congenital abnormalities, disorders of sex development, and/or reproductive problems, using a combination of conventional and molecular cytogenetic approaches. The data (Table 1, Figure 1) show that 28% of problem horses have karyotype abnormalities and like in all previous large-scale surveys, the most prevalent chromosome abnormalities are X-monosomy (35% of all chromosome abnormalities; 10% of all problem horses; 18% of all problem females) and *SRY*-negative XY sex reversal (11% of all chromosome abnormalities; 3% of all problem horses; 6% of problem females).

Table 1. Summary of clinical cytogenetic findings of the Texas A&M Molecular Cytogenetics Laboratory (TAMUMCL) in the period of 2001–2021.

Problem Horses	Total Number of Individuals	% of All Horses Studied	% of All Chromosome Abnormalities	Reference
Subjected for karyotyping due to reproductive or developmental problems	766	-	-	
Males	244	31.9	-	
Females	427	55.7	-	
Ambiguous sex	95	12.4	-	

Table 1. Cont.

Problem Horses	Total Number of Individuals	% of All Horses Studied	% of All Chromosome Abnormalities	Reference
Horses with chromosome abnormalities	215	28.1	-	
Types of chromosome abnormalities				
X-monosomy	76	9.9; (17.8) *	35.3	[51]
X-trisomy	5	0.7; (1.2) *	2.3	[51]
Sex chromosome and ploidy mosaicism • 63,X/64,XY • 64,XX/128,XXXX • 63,X/64,X,i(Yq) (Figure 1A)	3	0.4	1.4	[37]
X chromosome structural rearrangements • 64,X,del(Xp) • 64,X,i(Xq); 2 cases • 65,XX,+Xq (Figure 1B)	5	0.7	2.3	
X-autosome complex rearrangement • 63,X,der(X),del(q22),dup(q21q11), t(X;16)(q21;q11),dic(X;16) (Figure 1C)	1	0.1	0.5	[27]
Y chromosome structural rearrangements • 64,XY,del(Y)(q11q13) (Figure 1E)	1	0.1	0.5	
Y-autosome structural rearrangement • 64,XY,t(Yq;13p)	1	0.1	0.5	[52]
XX/XY blood chimerism	4	0.5	1.9	[53]
Autosomal translocations • 64,XX,t(2;13); familial, 2 cases • 64,XX;t(4;10) • 64,X?;t(5;16),+mar; familial, 9 cases • 64,XY;t(12q;25q),der(12p) • 64,XY;t(4p;30q); familial, 6 cases	20	2.6	9.3	[4,54–56]
Autosomal aneuploidies • 65,XY+27 • 65,XY+30; 2 cases • 64,XY,i(26q) or 64,XY,rob(26q26q)	4	0.5	1.9	[57]
XY Sex reversal conditions • SRY-neg XY DSD females; 24 cases (Figure 1D) • SRY-pos XY DSDs female-like; 20 cases	44 24 20	5.7; (10.3) * 3.1; (5.6) * 2.6; (4.7) *	20.5 11.2 9.3	[28,58]
XX DSDs; ambiguous sex	41	5.4	19.1	[58]
SRY-pos XY DSDs; male-like	8	1.0	3.7	[58]

* numbers in parentheses show percent of all problem females; DSD—Disorder of Sex Development.

3.2. Sex Chromosome Aneuploidies

As shown by large-scale surveys and by many individual case reports (reviewed by [3,5,9,10,51]), the most common karyotype abnormality in horses worldwide is X-monosomy (63,X) and its mosaic forms 63,X/64,XX and 63,X/64,XY. Occasionally, X-monosomy has been found together with a second, also abnormal cell line, e.g., 63,X/65,XXX [59] or 63,X/65,XYY [60,61], or as a mosaic of several cell lines [62–67].

X-monosomy. Mares with X-monosomy are often characterized by a lower height than age- and breed-mates with normal karyotype. They usually have properly developed external genitalia but have often underdeveloped small hypoplastic ovaries with no palpable follicles, and a small and flaccid uterus. Mares with X-monosomy show decreased steroidogenic activity of the ovaries and have overall higher levels of the luteinizing hormone, and lower levels of estrogen, progesterone, testosterone and cortisol [45,68,69]. The consequence of these changes are disturbances in the development and functioning of the reproductive system, leading to the absence of the estrus cycle and sterility. While the typical consequence of non-mosaic X-monosomy is sterility, a few cases of foals born to mares with a mosaic karyotype 63,X/64,XX have been described [8,44,47,66,69–71].

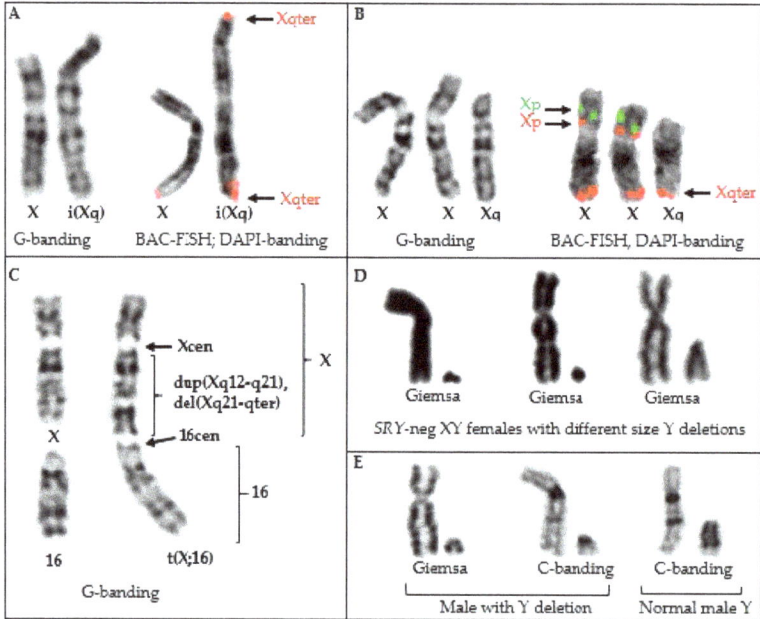

Figure 1. Examples of horse sex chromosome structural rearrangements (TAMUMCL archive). (**A**) Isochromosome Xq in a mare with short stature, no ovaries and 64,X,i(Xq) karyotype; (**B**) Trisomy Xq in a non-cycling mare with 65,XX,+Xq karyotype; (**C**) Complex dicentric X-autosome rearrangement in a mare with short stature but no other obvious problems, fertility unknown; (**D**) Sex chromosomes of three mares with *SRY*-neg XY male-to-female sex reversal syndrome; the first two have large deletions in the Y, the third one has a cytogenetically normal-looking Y, but a submicroscopic deletion around the *SRY* gene; (**E**) Partial Y chromosome deletion in a Shetland pony without penis (left, middle), comparison with the sex chromosomes of a normal male (right).

Sex chromosome trisomies—XXX, XXY and XYY. The second type of aneuploidy diagnosed in horses is sex chromosome trisomy—the presence of supernumerary X or Y chromosomes. These abnormalities are rare and like X-monosomy occur as non-mosaic 65,XXX [43,72–75] and 65,XXY [76–78], or as a mosaic of two [59,79–81], or more cell lines [46,63,82–84]. Mares with X-trisomy may look phenotypically normal. Some, especially those with mosaic X-trisomy, may show signs of estrus, but are rarely able to produce offspring because of hypoplastic gonads [59,74,85]. Likewise, stallions with XXY sex chromosomes may look normal and show normal male behavior but are sterile due to testicular hypoplasia and azoospermia [76–79,86]. Cases of male horses with an extra Y chromosome (65,XYY) are rare, show various forms of Disorders of Sex Development (DSDs), and have been described as pseudohermaphrodites [60,61,64]. In the 1970s, it was believed that the presence of two Y chromosomes could positively affect the performance of stallions. However, the cytogenetic research carried out among champions at that time did not confirm these expectations [87].

3.3. Autosomal Aneuploidies

Autosomal aneuploidies are rare in horses because the resulting genetic imbalance is typically lethal, and the few reported live-born cases are exclusively trisomies (Table 2). Among the large-scale cytogenetic surveys discussed in Section 3.1, autosomal aneuploidies were recorded only by two—the survey by M. Power [8] and TAMUMCL 20-year data.

In the latter, autosomal trisomies were found in just 4 animals out of the 766 abnormal horses studied, and account for less than 2% of all detected chromosome abnormalities (Table 1). In contrast, a recent whole genome analysis identified autosomal aneuploidies (both trisomies and monosomies) in over 20% of equine early pregnancy losses (EPLs) at 14–65 days of gestation [88]. This is in line with the data for humans where autosomal aneuploidies are well understood and described, and account for approximately 50% of all diagnosed chromosome disorders in miscarried fetuses [89,90].

Table 2. Summary data about all individual cases of autosomal aneuploidies reported for the horse.

Chr.	Karyotype/Type of Aneuploidy	Mosaicism	Phenotype	Methods	Breed	Maternal Age	Reference
1	n/a; trisomy	n/a	Early pregnancy loss fetus	SNP-CGH; WGS; ddPCR	WB	4	[88]
3	n/a; trisomy	n/a	Early pregnancy loss fetus	SNP-CGH; WGS; ddPCR	TB	6	[88]
15	n/a; trisomy	n/a	Early pregnancy loss fetus	SNP-CGH; WGS; ddPCR	TB	19	[88]
20	n/a; trisomy	n/a	Early pregnancy loss fetus	SNP-CGH; WGS; ddPCR	TB	13	[88]
20	n/a; trisomy	n/a	Early pregnancy loss fetus	SNP-CGH; WGS; ddPCR	TB	19	[88]
23	65,XY,+23	non-mosaic	Liveborn, congenital defects	G- and C-banding	STB	n/a	[84]
23, 24	n/a; double trisomy	n/a	Early pregnancy loss fetus	SNP-CGH; WGS; ddPCR	TB	3	[88]
26	64,XX,i(26q) or 64,XX,rob(26q26q)	non-mosaic	Liveborn, congenital defects, fertile	G-, R- and C-banding	TB	3	[92]
26	64,XX,i(26q) or 64,XX,rob(26q26q)	non-mosaic	Liveborn, congenital defects	G-banding; BAC-FISH	TB	5	TAMUMCL
27	65,XY,+27	non-mosaic	Liveborn, congenital defects	G-banding	QH	26	[93]
27	65,XY,+27	non-mosaic	Liveborn, congenital defects	G-banding	AR	25	[94]
27	65,XY,+27	non-mosaic	Liveborn, congenital defects	G-banding; BAC-FISH	STB	5	[57]
27	64,XX/65,XX,+27	mosaic	Liveborn, congenital defects	G-banding, BAC-FISH; SNP-CGH	FR	n/a	[95]
27	n/a; monosomy	n/a	Early pregnancy loss fetus	SNP-CGH; WGS; ddPCR	TB	10	[88]

51

Table 2. Cont.

Chr.	Karyotype/Type of Aneuploidy	Mosaicism	Phenotype	Methods	Breed	Maternal Age	Reference
27	n/a; monosomy	n/a	Early pregnancy loss fetus	SNP-CGH; WGS; ddPCR	TB	19	[88]
28	65,XY,+28	non-mosaic	Liveborn, congenital defects	G- and R-banding	TB	14	[91]
30	65,XX,+30	non-mosaic	Liveborn, congenital defects	G-, R- and C-banding	AR	23	[92]
30	n/a; trisomy	non-mosaic	Liveborn, congenital defects	SNP-CGH	WP	n/a	[95]
30	65,XX,+30	non-mosaic	Liveborn, congenital defects	G-banding; BAC-FISH	M	23	TAMUMCL
30	65,XY,+30	non-mosaic	Liveborn, congenital defects	G-banding; BAC-FISH	AR	9	TAMUMCL
30	64,XX/65,XX,+30	mosaic	Liveborn, fertile	G-banding	PK	n/a	[97]
30	n/a; trisomy	n/a	Early pregnancy loss fetus	SNP-CGH; WGS; ddPCR	TB	9	[88]
30	n/a; trisomy	n/a	Early pregnancy loss fetus	SNP-CGH; WGS; ddPCR	TB	19	[88]
31	65,XY,+31	non-mosaic	Liveborn, congenital defects	G-banding	TB	26	[96]
31	n/a; trisomy	n/a	Early pregnancy loss fetus	SNP-CGH; WGS; ddPCR	WB	10	[88]
31	64,XX/65,XX,+31	mosaic	Liveborn, normal	G-and C-banding	TB	n/a	[46]

CGH—comparative genomic hybridization; WGS—whole genome sequencing; ddPCR—digital droplet PCR; abbreviations of horse breeds: AR- Arabian; FR—Friesian; M—Morgan; PK—Polish Konik; QH—American Quarter Horse; STB—Standardbred; TB—Thoroughbred; WB—Warmblood; WP—Welsh Pony.

To date, 14 liveborn cases with trisomies involving 6 autosomes have been reported and in all, the extra chromosome is one of the smallest acrocentrics (Table 2). Phenotypes of the carriers vary but typically have numerous severe congenital malformations and primary infertility. The first diagnosed case was trisomy 28 in a Thoroughbred male with very short stature, cryptorchidism and azoospermia [91]. A foal with trisomy 23 had numerous defects of the skeletal system and sexual organs [84]. Trisomy 26 has been reported twice: in a filly with poor constitution, neurologic and behavioral issues ([92], and a colt with neurologic and gait defects and poor thriving (TAMUMCL). Curiously, in both cases, the chromosome number was normal 2n = 64, either due to the formation of isochromosome 26 or by Robertsonian fusion of the extra chr26. The dams of the abnormal foals in both cases were relatively young (3 years-old and 5 years-old, respectively), thus excluding advanced maternal age as a contributing factor. However, the most notable is that the filly with trisomy 26 turned into a fertile mare who gave birth to a healthy and chromosomally normal colt [92]. Trisomy 27 has been found in 4 cases and in all, the affected foals had multiple congenital malformations, including contracted tendon [93], arthrogryposis [94], skeletal malformations [95], and gait and behavioral abnormalities [57]. Four cases have also been diagnosed with trisomy 30, all having multiple developmental and behavioral abnormalities, such as poor thriving (TAMUMCL cases); abnormal gait and limb malformations [92]; facial deformities, scoliosis and heart and artery defects [95]. In contrast, trisomy for the smallest equine autosome, chr31, has been reported only once—a colt with underdevelopment of the limbs and reproductive organs [96].

Primary infertility, which is associated with the majority of non-mosaic autosomal trisomies, prevents propagating the aberrations in the population. However, this is not the case of mosaic forms. For example, Kubień and Tischner [97] described a phenotypically normal Polish Konik mare with 64, XX/65,XX 0 karyotype that gave birth to a normal foal. Likewise, Bugno et al. [46] diagnosed a 64,XX/65,XX,+31 karyotype in a few months-old filly with no developmental anomalies at this age. The lack of developmental abnormalities and normal fertility in mosaic forms of autosomal trisomy is likely due to the presence of a cell line with a normal karyotype.

It is well-established that the risk for autosomal trisomies in humans increases with advancing maternal age [98]. No such statistics is available for the horse, mainly because of the small number of reported cases. However, in some of the above-described studies, advanced age of the dam has been considered as a contributing factor [44,92]. In others, however, foals with an autosomal trisomy have been born to mares of average reproductive age (Table 2). Likewise, no clear correlation between maternal age and aneuploidies were detected in the single study of EPLs [88]. Because horses are used for breeding at all ages, continuing collection of cases with phenotypic and parental information is needed to shed more light into this matter.

3.4. Structural Rearrangements

Structural rearrangements change the constitution of one or more chromosomes and are typically caused by double-stranded DNA breaks and subsequent mistakes in repair during meiosis [99]. Depending on their effect on genome integrity, structural rearrangements are classified as genetically balanced and unbalanced. Balanced rearrangements include inversions and most translocations, and do not change the DNA content of a cell. Balanced structural changes typically do not have noticeable phenotypic manifestation and can easily remain unnoticed in carrier animals. In contrast, unbalanced rearrangements, such as deletions, duplications, and unbalanced translocations, cause a gain or loss of the genetic material and depending on the size and content, may have more or less severe effects on development, viability and reproduction (reviewed by [13,14]).

Translocations. Translocations involve nonhomologous chromosomes which exchange parts or fuse, giving rise to reciprocal or nonreciprocal translocations, respectively [100]. Carriers of genetically balanced translocations appear phenotypically normal but have reduced fertility because of producing both genetically balanced and unbalanced gametes. The former can pass the translocation between generations, whereas fertilization of unbalanced gametes typically results in embryonic or fetal death, and is noticed as reduced fertility [14,54]. Carriers of unbalanced translocations, on the other hand, show a range of developmental and reproductive disorders depending on the extent of genetic imbalance and the regions involved [100].

Translocations are rare in horses and to date, only 15 unique translocations have been reported (Table 3). Of these, 11 are autosomal and 4 involve an autosome and a sex chromosome.

Autosomal translocations. The majority of autosomal translocations found in horses are balanced, thus not affecting the performance or appearance of the carrier animal. They were discovered because the carrier mare or stallion was subjected for chromosome analysis due to recurrent early embryonic loss (REEL) and subfertility (reviewed by [4,5,9,14]). Therefore, the actual frequency of balanced autosomal translocations in equine populations may be higher, but due to no phenotypic effect and because only select individuals are used for breeding, they remain undetected [54]. In contrast, the single case of a live horse with unbalanced autosomal translocation, a Warmblood colt with 64,XY,t(4;30),+4p (Table 3), was euthanized due to poor thriving [54]. Overall, live animals with unbalanced autosomal translocations are extremely rare and typically, the condition is not viable beyond preimplantation [101].

Table 3. Summary data of all translocations found in horses.

Karyotype	Type	Genetic Balance	Evidence of Transmission	Reproductive Phenotype	Methods	Breed	Reference
64,XX,t(1q;3q)	Reciprocal	balanced	no	REEL	G- and R-banding	TB	[102]
64,XY,t(1;30)	Tandem	balanced	no	subfertility	G- and C-banding	TB	[103]
64,XX,t(1;16)	Reciprocal	balanced	no	subfertility	G-and C-banding; BAC-FISH; Zoo-FISH	TB	[104]
64,XX,t(1;21)	nonreciprocal	balanced	no	REEL	G-and C-banding; BAC-FISH	TB	[105]
64,XX,t(2;13)	nonreciprocal	balanced	yes	REEL	G-banding; BAC-FISH	TB	[56]
64,XX,t(4;13)	Reciprocal	balanced	no	REEL	G-and C-banding; BAC-FISH	TB	[105]
64,XX,t(4;10)	nonreciprocal	balanced	no	REEL	G-banding; BAC-FISH	AR	[55]
64,XY,t(4;30),der(4q) and 64,XY,t(4;30),+4p	nonreciprocal	balanced/unbalanced	yes	foals with congenital abnormalities	G-banding; BAC-FISH	WB	[58]
64,XY,t(5;16),+mar	nonreciprocal	balanced	yes	REEL	G-banding; BAC-FISH	TB	[4]
64,XY,t(12;25),der(12p)	nonreciprocal	balanced	no	azoospermia, small testes	G-banding; BAC-FISH	AR	[58]
64,XX,t(16;22),+mar	Reciprocal	balanced	no	REEL	G-and C-banding; BAC-FISH	TB	[105]
64,X,t(1p;Xp)(1q;Xq)	Reciprocal	balanced	no	n/a	G-banding; BAC-FISH	n/a	[49]
63,X,t(Xq;16),+ complex X rearrangements	nonreciprocal	unbalanced	no	n/a	G-banding; BAC-FISH	TB	[27]
64,X,t(15;X),-Xp,+15 *	nonreciprocal	unbalanced	no	infertility	G- and R-banding	TB	[91]
64,X,t(13;Y)	Reciprocal	balanced	no	azoospermia	G-and C-banding; BAC-FISH	FR	[52]

* According to ISCNH1997, the involved autosome is chr17; REEL—Recurrent Early Embryonic Loss; Abbreviations of horse breeds: AR—Arabian; FR-Friesian; TB—Thoroughbred; WB—Warmblood.

Balanced translocations are among the few hereditary chromosome abnormalities because the carriers can pass the condition to their offspring. If transmitted, the translocation will cause similar subfertility issues in the next generation [14]. In horse breeding where sires and dams are selected based on their athletic performance, appearance, and pedigrees, rather than reproductive performance, this can lead to propagating translocations over generations. Currently, there is cytogenetic evidence for three such 'translocation families' (Table 3): an elite Thoroughbred stallion with 64,XY,t(5;16),+mar passing the rearrangement to 8 offspring [4]; an elite Warmblood stallion with 64,XY,t(4;30),der(4q) passing the rearrangement to 5 offspring [54], and a Thoroughbred mare with 64,XX,t(2;13) passing the rearrangement to a single foal [54,56]. Therefore, systematic chromosome analysis of prospective breeding animals is needed for early detection of translocation carriers to prevent transmission and reduce economic loss due to subfertility.

Autosome and sex chromosome translocations. The phenotypic effects of translocations involving sex chromosomes differ from those of autosomes, as well as from each other. In each case, the genetic consequences depend on whether the horse is male or female and which X chromosome regions are involved. This is because random X inactivation (XCI) in mammalian females [106,107], balances X chromosome gene dosage between sexes, but also buffers deleterious effects of X chromosome mutations [108,109]. This is illustrated by phenotypic differences between the three reported cases of X-autosome translocations (Table 3). The mare with a balanced 64,X,t(1p;Xp)(1q;Xq) karyotype was phenotypically normal [49]. The Thoroughbred mare with unbalanced, dicentric complex X chromosome rearrangement and t(16;X) (Figure 1C), had only mild phenotype with short stature. This is because there was no autosomal imbalance and the rearranged X chromosome portion was subject for XCI, which probably did not spread over to chr16 [27]. In contrast, the Thoroughbred mare with unbalanced 64,X,t(15;X),-Xp,+15 karyotype had a short stature and was infertile [91], which is consistent with monosomy for Xp [14]. However, this mare also had trisomy 15, which most likely should not be viable, but because one copy of chr15 was translocated to Xq, it was functionally silenced by XCI [91].

A unique case is a balanced reciprocal Y-chr13 translocation in a Friesian stallion with complete azoospermia [52]. It is the first case of azoospermia in stallions with a cytogenetically detected Y chromosome abnormality. However, because balanced translocations typically cause only subfertility, the complete meiotic arrest and azoospermia in this stallion remained a puzzle [52]. The mystery was resolved by a recent hypothesis about Y-linked meiotic executioner genes which are necessary for successful meiosis but must also be subjected to meiotic sex chromosome inactivation (MSCI) [110]. If such genes are translocated to an autosome, ectopic expression of these genes during MSCI results in fatal meiotic arrest [110]. Thus, the Friesian stallion with Y-autosome translocation is a proof-of-principle to this hypothesis and another example illustrating different genetic consequences of translocations involving autosomes only compared to those involving an autosome and a sex chromosome.

Translocation-prone chromosomes. Even though none of equine translocations have been recurrent, i.e., have not independently occurred in unrelated individuals, some horse chromosomes tend to be engaged more often than others. For example, only 14 autosomes (out of 31) and the sex chromosomes have been involved in the 15 currently known translocations (Table 3) [54]. Of these, chr1 has been involved five times, chr16 four times, chr4, 13, and X three times each, chr30 twice, and chr2, 3, 5, 10, 12, 17, 21, 22, 25, and Y once each. Whether or not the involvement of particular chromosomes is random or associated with specific molecular features, remains a topic for future research. Studies in humans and pigs, where translocation frequency is high, indicate that translocation breakpoints are not random and occur preferentially in regions with open chromatin (G-negative bands), higher gene density and common fragile sites, and are demarcated by repetitive elements such as LINEs, SINEs, and endogenous retroviral elements [99,111]. Continuing the collection of additional clinical and cytogenetic data on translocations is the key for revealing their molecular patterns in horse chromosomes.

Deletions and duplications. Deletions and duplications decrease or increase the total amount of DNA in a cell, respectively, and cause genomic imbalance. Loss or gain of large chromosomal segments is usually lethal or accompanied by severe malformations and infertility. Smaller submicroscopic deletions and duplications, also known as DNA copy number variants (CNVs), may or may not have any evident phenotypic effect, and their contribution to equine health and fertility is, as of yet, poorly understood [112].

Chromosomal deletions and duplications are part of unbalanced translocations and were discussed in the previous section (Table 3). Otherwise, there are just two reports on cytogenetically detected autosomal deletions: deletion of chr13qter [64,XY,del(13)(qter)] in a Standardbred stallion [63] and an Arabian stallion with mosaicism for XX/XY and chr10 deletion [64,XY/63,XY,−10; 64,XX/63,XX,−10;] [113]. Both cases were identified due to fertility issues. However, it must be noted that these studies predated the availability of molecular cytogenetic tools to validate the findings.

Horse Y chromosome is particularly prone for deletions, most of which cause *SRY*-negative XY sex reversal syndrome and are discussed in Sections 3.6 and 5. In addition, TAMUMCL has studied a case of a male Shetland pony with no penis. The horse was

SRY-positive and had 64,XY karyotype with an unusually small Y chromosome that had lost the majority of the C-band positive heterochromatin (Table 1, Figure 1E).

Inversions. An inversion occurs when a piece of a chromosome breaks and reinserts within the same chromosome in inverted orientation [100]. Inversions are hard to detect both by conventional and molecular cytogenetic approaches, and to date there are no reports about equine clinical cases caused by inversions. The only known cytogenetically detectable inversion in the horse is the over 40 Mb-size inversion in chr3q causing the tobiano color pattern but is not associated with a disease or disorder [114].

Isochromosomes. Isochromosomes (i) are structurally abnormal chromosomes that are formed through a centric mis-division and result in chromosome arms which are mirror image of each other and genetically identical [115]. If one copy of the normal chromosome is also retained, the result is trisomy for that chromosome arm. Thus, isochromosome formation is both a structural and numerical rearrangement.

Isochromosomes have been reported for horse sex chromosomes –i(Xq) ([116]; TAMUMCL, Table 1, Figure 1A) and i(Yq) [37,117,118]. The two cases of 64,X,i(Xq) ([116]; TAMUMCL) were both described as having short stature and small inactive ovaries—a typical phenotype for X-monosomy or the deletion of Xp [14]. Isochromosome Y has been found only in mosaic form as 63,X/64,X,i(Yq) [37,117,118] (Figure 2). Two cases had similar DSD phenotypes—a male pseudohermaphrodite [117] and an intersex [37]. The third case was slightly different describing a female horse with abnormal external genitalia (Figure 2), but here the researchers also detected by PCR analysis Y chromosome microdeletions [118].

Among autosomes, there are two cases with putative isochromosome 26—a fertile Thoroughbred mare [92] and a Thoroughbred colt with neurologic issues, gait problems and poor growth (TAMUMCL, Table 1). Both horses had normal chromosome number with 64,XX or 64,XY karyotypes, respectively, but carried one normal chr26 and one metacentric marker chromosome which both arms corresponded to chr26. Further analysis by microsatellite genotyping (see Section 4.3) is needed to reveal whether the abnormal metacentric chromosome is an isochromosome or a result of Robertsonian fusion. In the first case, all chr26 markers should be bi-allelic, in the latter, three alleles can be detected. Regardless, recurrence of i(26q) or rob(26q26q) in unrelated horses is certainly a curious observation.

Fragile sites. Fragile sites are specific chromosomal loci that exhibit gaps and breaks on metaphase chromosomes following partial inhibition of DNA synthesis. Common fragile sites are found in all individuals, while rare fragile sites occur infrequently, are inherited in the Mendelian manner and can be associated with congenital disorders [119,120].

Chromosomal fragility has also been reported in horses in connection with sterility and reduced fertility [32], though the mechanism underlying this connection remains unclear [119]. Difficulty of interpretation is probably one of the reasons why fragile sites or chromosome breaks have not been included in any equine clinical cytogenetic case report, even though breaks and gaps have been observed and recorded in laboratory notes (TAMUMCL archive). Fragile sites certainly deserve further attention by clinical cytogenetics and basic genome research because in both humans and horses, they have been co-localized with interstitial telomeric sequences, known as genomic 'scars' marking DNA break/repair sites and possibly more unstable genomic regions [121].

3.5. Chimerism

The term *chimera* is defined as an individual that has two cell lines derived from two separate zygotes. This disorder may appear as a result of the early fusion of zygotes, which are the result of fertilization of the egg and polar body. In a situation where their genetic sex is different (XX and XY), chimerism involving all tissues leads to intersex with ovotestis. Chimerism may also be caused by early embryo fusion [122].

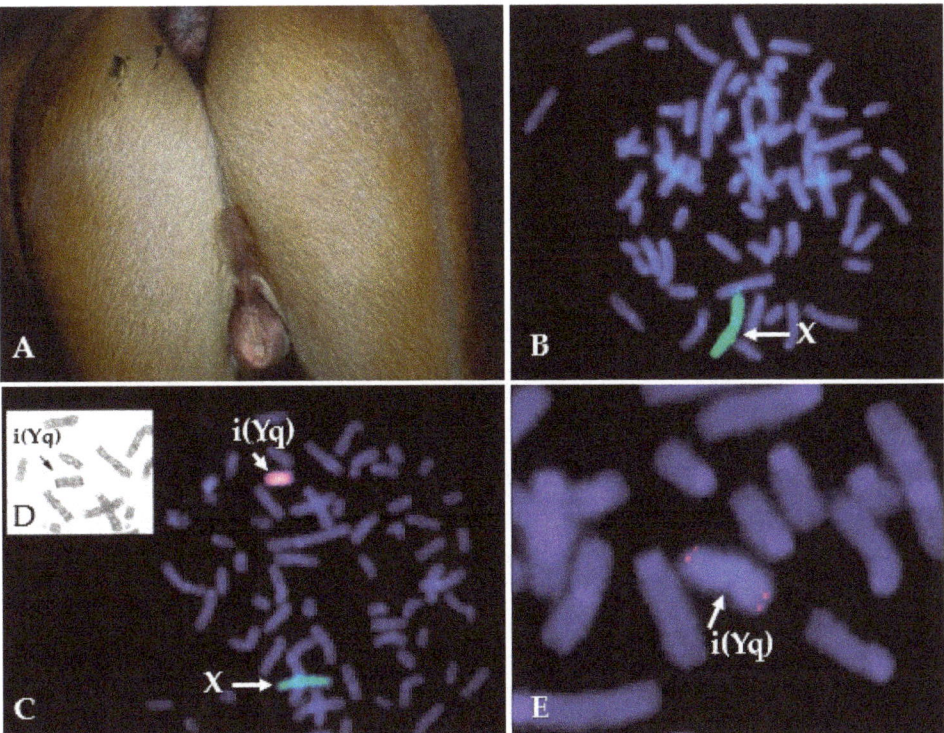

Figure 2. Example of a horse with isochromosome Y in a mosaic karyotype 63,X/64,X,i(Yq). (**A**) Abnormal external genitalia of a 6 month-old mare; (**B**) Cell line with X-monosomy 63,X (97%); the single X is labeled green with X-specific painting probe; (**C**) Cell line with isochromosome Y and 64,X,i(Yq) karyotype (3%); X is labeled green with X-specific painting probe and Y is labeled red with Y-specific painting probe; (**D**) Part of the same metaphase after G-banding showing i(Yq) (arrow); (**E**) Partial metaphase after FISH with a probe specific for *USP9Y* region in Y (note the red signal at both ends of the isochromosome Y).

The best-known form of chimerism in mammals is the presence of two cell populations in individuals born from twin or multiple pregnancy, both the same-sex and different-sex. In these cases, the occurrence of cellular chimerism is the result of the formation of a common bloodstream through anastomoses, i.e., vascular connections between the fetal membranes of co-twins [123,124]. The consequence is the exchange of hematopoietic cells and the interaction of the endocrine and immune systems between the twins. The consequence of the formation of anastomoses are changes in the female reproductive system, which most often lead to infertility (cattle, sheep) or reproductive problems (horses) [53,125–131].

The cases of fertile mares with XX/XY leukocyte chimerism may indicate a late contact of placental vessels between fetuses [129]. If the fusion takes place after sex determination and differentiation towards ovaries, they do not undergo masculinization, which occurs in freemartin heifers derived from different-sex twin pregnancies [131].

3.6. Cytogenetics of Sex Reversal Conditions

The term *sex reversal* has been used to describe situations where the genetic sex as determined by sex chromosomes does not agree with the gonadal and/or external phenotypic sex. In human medicine, due to social and ethical issues, gender-based diagnostic labels such as *sex reversal*, *intersex*, *hermaphroditism*, and *pseudohermaphroditism*, have been replaced by a more neutral term *disorders of sex development* (*DSDs*) [132]. The term has also been extended to veterinary medicine to denote congenital conditions in which development of chromosomal, gonadal, or anatomic sex is atypical [11]. Conditions with a normal karyotype but atypical or ambiguous sex development are classified according to the sex chromosomes into XY and XX DSDs.

64,XY DSDs. Many cases of male-to-female sex reversal or XY DSDs have been described in horses and after X-monosomy, they are the most common equine chromosome abnormalities (reviewed by [4,5,10–14,133,134]). It has been estimated that approximately 12% to 30% of all cytogenetically abnormal cases are XY DSDs [28,44,135] (Table 1).

In earlier studies, when XY DSDs were recognized solely by karyotyping, without testing for the *SRY* gene, the researchers observed that phenotypes of XY sex reversal horses can vary in a broad range—from a very feminine to a greatly masculinized mare [44,135–137]. With the inception of *SRY* testing in equine cytogenetics, first by Southern blotting [138] and thereafter by PCR [73,139,140], horse XY DSDs are categorized as *SRY*-negative and *SRY*-positive.

SRY-negative XY DSDs. This is the most prevalent form of XY DSDs and encompasses the majority of the feminine-type cases. The affected mares typically have normal female external genitalia and no somatic or behavioral abnormalities but are sterile due to ovarian and uterine dysgenesis [10,11,28,73,78,133,138,141–144]. The phenotype of *SRY*-negative XY DSD very closely resembles that of X-monosomy, indicating that while the absence of *SRY* blocks the male development pathway in these individuals, normal female development still requires the presence of two X chromosomes. At molecular level, *SRY*-negative XY DSD in horses is caused by Y chromosome deletions (Figure 1D) and discussed in detail in Section 6.

SRY-positive XY DSDs. This group of XY DSDs encompasses female-like horses showing various degrees of masculinization and virilization, as well as stallion-like behavior. These horses usually have abnormally developed genital tract, the gonads can range from ovotestes to testicular feminization, and the cases are described as male pseudohermaphrodites, intersex or ambiguous sex [28,44,133,134,136,137,145–149].

Despite female-like appearance, these horses are genetically male with an intact Y chromosome and a normal *SRY* gene [28]. Molecular causes for abnormal sex development are known or suggested only for a few cases. In three families of different breeds, *SRY*-positive XY DSD was associated with androgen insensitivity syndrome and with different mutations in the androgen receptor gene [150–152]. In two related male pseudohermaphrodite Standardbreds with *SRY*-positive XY DSD [146], a large (~200 kb) homozygous deletion in chr29 was found and proposed as a likely cause because the deletion removed a cluster of genes (*AKR1C* family) with known functions in steroid hormone biosynthesis, including androgens and estrogens [58,112]. However, *AR* mutations or the deletion in chr29 are not present in many other cases of *SRY*-positive XY DSDs [58], suggesting that the molecular causes of the condition are heterogeneous.

64,XX DSDs. Horses with XX DSDs are *SRY*-negative and cytogenetically indistinguishable from normal females. However, all equine XX DSDs cases have highly abnormal and ambiguous sex phenotypes (reviewed by [10,12–14]). In contrast to humans where multiple cases of *SRY*-positive XX males have been reported [153], true XX female-to-male sex reversal condition has not been found in horses. Over the years, tens of XX DSD cases have been described [133,139,154–160], including 41 unpublished cases from TAMUMCL (Table 1). While clinical details of individual cases may vary, they are typically reported as intersex, hermaphrodite or ambiguous sex because of difficulties to decide about the gonadal and/or phenotypic sex of the horse (reviewed by [134]). Molecular causes of equine XX DSD are unknown.

4. Molecular Cytogenetic Methods and Applications

During the past three decades, largely thanks to the progress in horse gene mapping and genome sequencing (reviewed by [15]), methodological advancements have also shaped equine cytogenetics, leading to improved resolution and accuracy in detecting various types of chromosome abnormalities. Clinical cytogenetics today is essentially a combination of conventional chromosome analysis by banding techniques and a variety of molecular approaches.

4.1. Fluorescence In Situ Hybridization (FISH)

The most widely used molecular approach in equine clinical cytogenetics is fluorescence in situ hybridization (FISH). The method was developed in the 1980s (reviewed by [161]), relies on the Watson–Crick DNA base-pairing complementarity principle and permits identification of the location of DNA sequences in their original place (in situ) in mitotic and meiotic chromosomes at different stages of the cell cycle [162,163]. The most commonly used probes for FISH in cytogenetic studies are clones from the horse genomic bacterial artificial chromosome (BAC) library CHORI-241 (https://bacpacresources.org/ accessed on 1 March 2021). This is because thanks to the whole-genome radiation hybrid and FISH map [164] and available end sequence data for approximately 315,000 BACs [165], precise chromosomal and sequence map locations are known for thousands of clones from this library. Therefore, if a BAC clone is needed for FISH to identify the chromosomes involved in numerical or structural rearrangements or for determining rearrangement breakpoints, it can be found from the Genomic Clones track of EquCab3 assembly in NCBI Genome (https://www.ncbi.nlm.nih.gov/genome/?term=horse accessed on 1 March 2021) and ordered from the CHORI BACPAC resources (https://bacpacresources.org/ accessed on 1 March 2021).

Other important FISH probes are horse chromosome-specific paints generated by chromosome flow sorting [72,166] or microdissection [167–170]. The latter method also allows for the preparation of probes specific for chromosomal segments and has been used to generate painting probes for the short- and long arm of the horse X chromosome [171].

In addition to BACs and chromosome painting probes, FISH probes are available for vertebrate telomeric repeats (Discovery®: https://www.discoverypeptides.com/pna/pna-telomere-fish-probes accessed on 1 March 2021), multicopy 18S-5.8S-28S ribosomal DNA (rDNA) sequences [33], also known as nucleolus organizer regions (NORs), and horse centromeres [172]. Alternatively, researchers have used *primed* in situ *DNA* synthesis (PRINS) for the detection of telomere, centromere, and rDNA repeat sequences in horse chromosomes [33,173,174].

4.2. Application of FISH in Horse Clinical Cytogenetics

The first application of FISH in equine clinical cytogenetics was the use of a flow sorted X chromosome paint to detect X chromosome aneuploidies [72]. Since then, X chromosome paints have been used in multiple cytogenetic cases for the detection of mosaic (see Figure 2B,C) and non-mosaic X chromosome aneuploidies [64,74,167,169], aneuploidies of X chromosome arms [171], sex chromosome mosaicism [175], and in one recent case, to show premature X chromosome centromere division in a Hucul mare [176]. Combination of both the X and the Y chromosome paints (Figure 2C) or the Y paint alone, have been used to confirm sex chromosome complement and large Y chromosome deletions in cases of XY male-to-female sex reversal [28,141].

In contrast to the wide use of sex chromosome paints, there are no reports about FISH with horse autosomal paints to analyze cases of autosomal aneuploidies or structural rearrangements. So far, all FISH experiments validating and refining various horse translocations (Table 3) have used BAC clones. Likewise, BAC-FISH has also been instrumental for the accurate identification of the small autosomes involved in trisomies [57,95] (Table 2), for confirming isochromosome formation [37], and for characterizing two cases with complex structural rearrangements. The first one involved 5;16 translocation and a de novo small marker chromosome [4], another had a dicentric X;16 translocation with partial Xq duplication and deletion [27]. The use of BACs instead of chromosome paints in these cases is probably because BAC-FISH provides better resolution for resolving rearrangement breakpoints, but also because horse chromosome painting probes are not commercially available.

Compared to BACs and sex chromosome paints, the use of FISH with centromeric, telomeric, or rDNA probes in equine clinical cytogenetics has been limited. The few examples include centromere-FISH to locate centromeric sequences in a dicentric derivative chromosome [27] and show the position of centromeres in isochromosome Y [37]. The latter study also determined that the horse Y chromosome is an acrocentric and not sub-metacentric as presented in ISCNH 1997 [36].

4.3. Cytogenetic Evaluation of Stallions by Sperm-FISH

Sperm-FISH is a state-of-art technique to analyze chromosomal constitution of mature spermatozoa, which have highly condensed chromatin, do not undergo cell division and cannot be studied by conventional cytogenetic approaches. Sperm-FISH is carried out on decondensed sperm nuclei using chromosome-specific paints or BAC clones. The method was initially developed for men [177] but has been optimized for domestic species, including the stallion [178]. While karyotyping evaluates chromosomes in diploid somatic cells, sperm-FISH allows determining the chromosomal constitution of mature haploid sperm and is potentially more informative for fertility evaluation. However, the ability to detect aneuploidies is limited to the availability of chromosome-specific probes and the number of fluorochromes that can be simultaneously visualized under the microscope.

Owing to these limitations, sperm-FISH studies in stallions have been restricted to analyzing sex chromosome aneuploidies in reproductively normal [179–181] and subfertile stallions [182]. These studies indicate that sex chromosome aneuploidy rate in normal stallions is in the range of 0.32–1.14% [179–181] with the highest frequency for sex chromosome nullisomy (0.47–1.22%) and the lowest for trisomy XXX or XXY (0.008–0.02%) [180,181]. Correlation has also been found between stallion age and the total number of aberrations in sperm [180,181]. Compared to normal stallions, subfertile Sorraia stallions have over five times more (5.83%) sex chromosome aneuploidies [182], whereas stallions in general show the highest rate of X and Y aneuploidies among domestic species (reviewed by [14]). Based on these data, it is tempting to speculate that the relatively high frequency of X-monosomy found in horses (see Section 3.2.) is partially caused by sperm aneuploidies, particularly the sex chromosome nullisomy.

4.4. Whole Genome Analysis by Comparative Genomic Hybridization and Sequencing

Comparative genomic hybridization (CGH) was originally designed for human cancer cytogenetics to overcome the difficulties to obtain high-quality metaphase spreads from various solid tumors [183]. The technique uses competitive hybridization of two differently labeled (red and green) DNA probes, one from a normal control, another from a cancer cell to normal metaphase chromosomes. The measurement of the ratios of red-to-green fluorescence along chromosomes will identify gains and losses in the cancer genome compared to the control. In horses, the CGH technique has been used to identify chromosome rearrangements involving large deletions and amplifications in equine sarcoid cells [184]. With the development of array technology, CGH has been adapted for SNP and oligonucleotide tiling arrays, known as array CGH (aCGH) [185,186].

The contribution of aCGH to horse clinical cytogenetics has been limited to just four studies. The first one used the Equine SNP50 BeadChip (Illumina) to confirm previously known cases of X monosomy and trisomy 31 and identified new cases with trisomy 27 and 31 [95]. The second study, used aCGH to investigate CNVs in normal horse genome, but coincidentally discovered a large, over 200 kb deletion in chr29 of two female-like horses with 64,XY *SRY*-positive DSD [112]. The third study applied aCGH to identify chromosome rearrangements in an intersex horse [187]. The most recent study used the high density 670K equine SNP array [188] and detected 12 different, mostly novel, autosomal aneuploidies in fetuses from early pregnancy loss [88] (Table 2). The findings were confirmed by whole genome sequencing (WGS) and digital droplet PCR, suggesting that advanced molecular methods will gradually become an integral part of equine clinical cytogenetics.

4.5. Immunolocalization of Chromosomal Proteins

The use of fluorescently labeled antibodies for chromosomal proteins such as centromere kinetochore proteins, synaptonemal complex proteins SCP1 and SCP3, proteins associated with double stranded break repair and recombination or meiotic silencing, has considerably improved the knowledge about chromosome function in normal cells, as well as in cells with chromosomal aberrations (reviewed by [189]). Immunostaining is often combined with FISH, which further increases the power and resolution of analysis.

In horse cytogenetics, immunostaining has been used to understand the organization, function, and evolution of centromeres [190,191] and for the study of synaptonemal complexes, recombination sites and the chiasmata in meiosis prophase and MI of normal stallions [192,193]. To date, immunostaining has not been used for the study of aberrant horse chromosomes.

4.6. STR Genotyping in Cytogenetics—Advantages and Limitations

Short tandem repeats (STR), also known as microsatellites, are widely used markers in parentage testing. The International Society of Animal Genetics (ISAG) recommends the use of a properly standardized panel of 17 microsatellite markers for horses, located on 12 different autosomes and the X chromosome. By adding to this set additional X and Y chromosome markers, whole-genome STR genotyping for parentage testing can simultaneously be used for the initial detection of chromosomal aberrations, such as monosomy, trisomy, XY sex reversal syndrome and chimerism [53,126,127,194–196]. Microsatellite genotyping has also been used to determine the parental origin of an aberrant chromosome [27,197] and in a study of two cloned horses, one with a de novo autosomal translocation, to confirm that the clones and their sire were genetically identical [54]. Other potential uses of STR genotyping are the identification of isochromosomes and determining the parental origin of the single X chromosome in X-monosomy. The latter will improve our currently limited understanding about the underlying mechanisms of this most common cytogenetic abnormality in horses.

The greatest advantages of this method are high sensitivity and specificity, speed of analysis, ease of interpretation of the results, and relatively low cost. As discussed in Section 3, chromosomal aberrations are most often associated with disorders of reproductive function, which are the most common reason for referral of horses for cytogenetic diagnostic testing. Therefore, in many cases, chromosome abnormalities are detected in adult individuals. Parentage tests, on the other hand, are usually performed in yearlings, which allows for the early diagnosis of any problems. It should be emphasized that DNA for STR analysis can be isolated from tissues other than blood and does not require lymphocyte culture for several days. Despite the many advantages, STR genotyping also has limitations in karyotype analysis. It cannot detect balanced structural aberrationsor aneuploidy in a mosaic form Though, in cases of X-monosomy, STR genotyping can be used to exclude mosaicism.

5. Molecular Underpinnings of the Unique Patterns of Horse Chromosome Abnormalities

All large-scale cytogenetic surveys (see Section 3.1) unanimously agree that the two most frequent chromosome abnormalities in the horse are X-monosomy and *SRY*-negative XY male-to-female sex reversal syndrome (*SRY*-negative XY DSD). The high prevalence of the two conditions is a signature feature of equine clinical cytogenetics, with no similar patterns found in other domestic species [14,28,51]. Recent advances in horse genomics (reviewed by [15]), particularly in the genomics of horse sex chromosomes [38,198], start to provide the first clues for these signatures.

X-monosomy. The high frequency of viable X-monosomy in horses, but not in other domestic species, has been associated with the molecular features of the horse pseudoautosomal region (PAR) [51]. The equine PAR is approximately 2 Mb in size [199], which is several magnitudes smaller than the 6–9 Mb-size PARs in other domestic species, such as cattle, sheep, goat, pig, camelids, dog and cat [14,51,198]. Since PAR genes escape X chromosome inactivation (XCI) in females and must be expressed bi-allelically, X-monosomy causes haploinsufficiency for these genes. It has been theorized that because the horse PAR is relatively small, less genes are affected by X-monosomy, resulting in viable live birth, whereas X-monosomy in species with larger PARs causes embryonic or fetal loss [14,51]. An alternative hypothesis, however, proposes that the rate of sex chromosome rearrangements, including aneuploidies, increases when the PAR shrinks because reduced X-Y synapsis in male meiosis causes more mistakes [200]. Both theories are consistent with the relatively high frequency of X-monosomy in humans (0.04% of live female births) [201], which is another species with a small PAR (2.7 Mb) [202]. However, the small PAR does not explain the dramatic differences between horses and humans regardings X chromosome and PAR overdose. Compared to X-monosomy, XXX or XXY aneuploidies are rare in the horse and the few reported cases show gonadal dysgenesis and infertility (see Section 3.2). In contrast, the XXY Klinefelter's syndrome and X-trisomy are the most common sex chromosome abnormalities in humans, affecting 0.1–0.2% of male births [203] and 0.1% of female births [204], respectively. Furthermore, many women with X-trisomy are fertile, thus potentially increasing the number of XXY male births. It has been proposed that the low number of identified 65,XXX horses is because the majority of mares with X-trisomy are normal fertile and escape detection [8]. However, in such cases, the incidence of 65,XXY male horses should be higher. A more plausible explanation is that despite similar PARs, the molecular regulation of the X chromosome in horses and humans is different, though more research is needed to confirm this.

SRY-negative 64,XY sex reversal. Thanks to recent sequencing of the horse Y chromosome [38], more is known about the molecular underpinnings of the relatively high frequency of mares with *SRY*-negative XY DSD. It appears that the single-copy horse *SRY* is located in a structurally unstable region in the Y chromosome, being embedded between ampliconic sequences and surrounded by direct and inverted repeats [38]. Such a location facilitates *SRY* involvement in ectopic inter-and intra-chromatid gene conversion and recombination within the Y chromosome [205]. These events may result in *SRY* deletion in one sperm and duplication in another [28,38]. Therefore, *SRY*-negative XY DSD females may have male siblings with two copies of *SRY*. The latter probably has no effect on the phenotype and remains undetected. Since the organization and content of mammalian Y chromosomes is different across species [38], this also explains why *SRY*-negative XY sex reversal is rare or absent in other species studied, including humans. For example, only 10–20% of human XY females (Swyer syndrome) have *SRY* mutations and the majority carry normal *SRY* [206].

6. Summary and Future Directions

Equine clinical cytogenetics has come a long and eventful way since the first description of karyotype abnormalities in horses in 1975 [43]. Starting with basic karyotyping of routinely Giemsa-stained chromosomes, it soon developed into an international, actively publishing, and methodologically sophisticated field of research to study the genetic causes of equine reproductive and congenital disorders. Despite the predictions that with the development of molecular methods, classical chromosome analysis will gradually disappear, horse clinical cytogenetics has remained. It successfully survived the golden days of equine gene mapping in the 1990s by adopting molecular cytogenetic methods, such as FISH and CGH. It remained active during the years of horse genome sequencing by applying genomic information to make chromosome analysis more refined and accurate. In the post-genome era today, cytogenetic research is blending with whole genome sequencing (WGS) and other cutting-edge technologies. Among the latter, perhaps the most promising for clinical cytogenetics is BioNano Genomics, which utilizes nanochannel technology and high-resolution imaging of ultra-high molecular weight DNA. The platform can detect almost any structural or numerical changes in the genome, including balanced translocations and inversions [207], and differently from WGS, the estimated price per sample of BioNano analysis is comparable to that of conventional clinical cytogenetics. The downsides, however, are that BioNano sets extremely high requirements for sample quality, and both WGS and BioNano are bioinformatically demanding. Therefore, despite the promises of new technologies, it is unlikely that they will entirely replace conventional and FISH-based chromosome analysis. Traditional clinical cytogenetics is still the most straightforward, cost-effective, and fastest approach to diagnose chromosome abnormalities, and will remain the gold standard for the initial evaluation of potential breeding animals and horses with reproductive or developmental disorders.

Author Contributions: The authors contributed equally to the preparation of the manuscript. Both authors have read and agreed to the published version of the manuscript.

Funding: M.B.-P. was supported by the University of Agriculture in Krakow SUB/020013-D015 and by National Centre for Research and Development, NCBiR: BIOSTRATEG2/297267/14/NCBR/2016; T.R. was supported by Texas A&M Molecular Cytogenetics service, Morris Animal Foundation grant D19EQ-051, and USDA-AFRI grant 2018-06521.

Institutional Review Board Statement: Not applicable.

Data Availability Statement: Not applicable.

Conflicts of Interest: The authors declare no conflict of interest.

References

1. Bowling, A.T. *Horse Genetics*; CAB International: Wallingford, UK, 1996; p. 200.
2. Bailey, E.; Brooks, S.A. *Horse Genetics*, 2nd ed.; CAB International: Wallingford, UK, 2013; p. 200.
3. Chowdhary, B.P.; Raudsepp, T. Cytogenetics and physical gene maps. In *The Genetics of the Horse*; Bowling, A.T., Ruvinsky, A., Eds.; CABI: Wallingford, UK, 2000; pp. 171–242.
4. Durkin, K.; Raudsepp, T.; Chowdhary, B. Cytogenetic Evaluation of the Stallion. In *Equine Reproduction*; Vaala, W.E., Varner, D.D., Eds.; Wiley Blackwell: Chichester, UK, 2011; pp. 1462–1468.
5. Lear, T.L.; Villagomez, D.A.F. Cytogenetic evaluation of mares and foals. In *Equine Reproduction*; Vaala, W.E., Varner, D.D., Eds.; Blackwell Publishing Ltd.: Oxford, UK, 2011; pp. 1951–1962.
6. Penedo, M.C.T.; Raudsepp, T. Molecular Genetic Testing and Karyotyping in the Horse. In *Equine Genomics*; Chowdhary, B.P., Ed.; Wiley-Blackwell: Oxford, UK, 2013; pp. 241–254.
7. Raudsepp, T.; Das, P.J. Genomics of reproduction and fertility. In *Equine Genomics*; Chowdhary, B.P., Ed.; Wiley-Blackwell: Oxford, UK, 2013; pp. 199–216.
8. Power, M.M. Chromosomes of the horse. *Adv. Vet. Sci. Comp. Med.* **1990**, *34*, 131–167.
9. Lear, T.L.; Bailey, E. Equine clinical cytogenetics: The past and future. *Cytogenet. Genome Res.* **2008**, *120*, 42–49. [CrossRef]
10. Lear, T.L.; McGee, R.B. Disorders of sexual development in the domestic horse, *Equus caballus*. *Sex. Dev.* **2012**, *6*, 61–71. [CrossRef] [PubMed]
11. Villagomez, D.A.; Iannuzzi, L.; King, W.A. Disorders of sex development in domestic animals. Preface. *Sex. Dev.* **2012**, *6*, 5–6.
12. Villagomez, D.A.; Parma, P.; Radi, O.; di Meo, G.; Pinton, A.; Iannuzzi, L.; King, W.A. Classical and molecular cytogenetics of disorders of sex development in domestic animals. *Cytogenet. Genome Res.* **2009**, *126*, 110–131. [CrossRef] [PubMed]
13. Szczerbal, I.; Switonski, M. Chromosome Abnormalities in Domestic Animals as Causes of Disorders of Sex Development or Impaired Fertility. *Insights Anim. Reprod.* **2016**, *9*, 207–225.
14. Raudsepp, T.; Chowdhary, B.P. Chromosome Aberrations and Fertility Disorders in Domestic Animals. *Annu. Rev. Anim. Biosci.* **2016**, *4*, 15–43. [CrossRef]
15. Raudsepp, T.; Finno, C.J.; Bellone, R.R.; Petersen, J.L. Ten years of the horse reference genome: Insights into equine biology, domestication and population dynamics in the post-genome era. *Anim. Genet.* **2019**, *50*, 569–597. [CrossRef] [PubMed]
16. Kirillow, S. Die Spermatogenese beim Pferde. *Arch. Fuer Mikrosk. Anat.* **1912**, *79*, A125–A127. [CrossRef]
17. Masui, K. A spermatogenesis of domestic mammals. I. The spermatogenesis of the horse (*Equus caballus*). *J. Coll. Agric. Tokyo Imp. Univ.* **1919**, *3*.
18. Painter, T.S. Studies in mammalian spermatogenesis. V. The chromosomes of the horse. *J. Exp. Zool.* **1924**, *39*, 229–247. [CrossRef]
19. Wodsedalek, J.E. Spermatogenesis of the horse with special reference to the accessory chromosome and the chromatoid body. *Biol. Bull.* **1914**, *27*, 295–325. [CrossRef]
20. Makino, S. The chromosomes of the horse (*Equus caballus*). *Cytologia* **1942**, *13*, 26–38. [CrossRef]
21. Rothfels, K.H.; Alexrad, A.A.; Siminovitch, L.; Parker, R.C.M. The origin of altered cell lines from mouse, monkey and man, as indicated by chromosome and transplantation studies. *Proc. Can. Cancer Res. Conf.* **1959**, *3*, 189–214.

22. Moorhead, P.S.; Nowell, P.C.; Mellman, W.J.; Battips, D.M.; Hungerford, D.A. Chromosome preparations of leucocytes cultured from human peripheral blood. *Exp. Cell Res.* **1960**, *20*, 613–616. [CrossRef]
23. Makino, S.; Sofuni, T.; Sasaki, M.S. A revised study of the chromosomes of the horse, the ass and the mule. *Proc. Jpn. Acad.* **1963**, *39*, 176–181. [CrossRef]
24. Seabright, M. A rapid banding technique for human chromosomes. *Lancet* **1971**, *2*, 971–972. [CrossRef]
25. Schweizer, D.; Ambros, P.; Andrle, M. Modification of DAPI banding on human chromosomes by prestaining with a DNA-binding oligopeptide antibiotic, distamycin A. *Exp. Cell Res.* **1978**, *111*, 327–332. [CrossRef]
26. Arrighi, F.E.; Hsu, T.C. Localization of heterochromatin in human chromosomes. *Cytogenetics* **1971**, *10*, 81–86. [CrossRef]
27. Mendoza, M.N.; Schalnus, S.A.; Thomson, B.; Bellone, R.R.; Juras, R.; Raudsepp, T. Novel complex unbalanced dicentric X-autosome rearrangement in a Thoroughbred mare with a mild effect on the phenotype. *Cytogenet. Genome Res.* **2020**, *160*, 597–609. [CrossRef]
28. Raudsepp, T.; Durkin, K.; Lear, T.L.; Das, P.J.; Avila, F.; Kachroo, P.; Chowdhary, B.P. Molecular heterogeneity of XY sex reversal in horses. *Anim. Genet.* **2010**, *41* (Suppl. 2), 41–52. [CrossRef]
29. Dutrillaux, B.; Lejeune, J. A new technic of analysis of the human karyotype. *C R Acad. Hebd. Seances Acad. Sci. D* **1971**, *272*, 2638–2640.
30. Goodpasture, C.; Bloom, S.E. Visualization of nucleolar organizer regions im mammalian chromosomes using silver staining. *Chromosoma* **1975**, *53*, 37–50. [CrossRef]
31. Iannuzzi, L.; di Meo, G.P.; Perucatti, A.; Incarnato, D.; Peretti, V.; Ciotola, F.; Barbieri, V. An improved characterization of horse (*Equus caballus*, 2n = 64) chromosomes by using replicating G and R banding patterns. *Caryologia* **2003**, *56*, 205–211. [CrossRef]
32. Rønne, M. Putative fragile sites in the horse karyotype. *Hereditas* **1992**, *117*, 127–136. [CrossRef] [PubMed]
33. Wnuk, M.; Villagomez, D.A.; Bugno-Poniewierska, M.; Tumidajewicz, P.; Carter, T.F.; Slota, E. Nucleolar organizer regions (NORs) distribution and behavior in spermatozoa and meiotic cells of the horse (*Equus caballus*). *Theriogenology* **2012**, *77*, 579–587. [CrossRef]
34. Ford, C.E.; Pollock, D.L.; Gustavsson, I. Proceedings of the First International Conference for the Standardisation of Banded Karyotypes of Domestic Animals. University of Reading Reading, England, 2nd–6th August 1976. *Hereditas* **1980**, *92*, 145–162. [CrossRef] [PubMed]
35. Richer, C.L.; Power, M.M.; Klunder, L.R.; McFeely, R.A.; Kent, M.G. Standard karyotype of the domestic horse (*Equus caballus*). Committee for standardized karyotype of *Equus caballus*. The Second International Conference for Standardization of Domestic Animal Karyotypes, INRA, Jouy-en Josas, France, 22nd–26th May 1989. *Hereditas* **1990**, *112*, 289–293. [CrossRef] [PubMed]
36. Bowling, A.T.; Breen, M.; Chowdhary, B.P.; Hirota, K.; Lear, T.; Millon, L.V.; Ponce, D.L.F.; Raudsepp, T.; Stranzinger, G. International system for cytogenetic nomenclature of the domestic horse. Report of the Third International Committee for the Standardization of the domestic horse karyotype, Davis, CA, USA 1996. *Chromosome Res.* **1997**, *5*, 433–443. [CrossRef]
37. Das, P.J.; Lyle, S.K.; Beehan, D.; Chowdhary, B.P.; Raudsepp, T. Cytogenetic and molecular characterization of Y chromosome in a 63XO/64Xi(Yq) mosaic karyotype of an intersex horse. *Sex. Dev.* **2012**, *6*, 117–127. [CrossRef]
38. Janečka, J.E.; Davis, B.W.; Ghosh, S.; Paria, N.; Das, P.J.; Orlando, L.; Schubert, M.; Nielsen, M.K.; Stout, T.A.E.; Brashear, W.; et al. Horse Y chromosome assembly displays unique evolutionary features and putative stallion fertility genes. *Nat. Commun.* **2018**, *9*, 2945. [CrossRef]
39. Wade, C.M.; Giulotto, E.; Sigurdsson, S.; Zoli, M.; Gnerre, S.; Imsland, F.; Lear, T.L.; Adelson, D.L.; Bailey, E.; Bellone, R.R.; et al. Genome sequence, comparative analysis, and population genetics of the domestic horse. *Science* **2009**, *326*, 865–867. [CrossRef]
40. Deryusheva, S.E.; Loginova, Y.A.; Chiryaeva, O.G.; Yaschak, K.; Smirnov, A.F. Distribution of ribosomal RNA genes on chromosomes of domestic horse (*Equus caballus*) revealed by fluorescence in situ hybridization. *Genetika* **1997**, *33*, 1281–1286. (In Russian)
41. Loginova, J.; Derjusheva, S.; Jasczak, K. Some cases of NOR instability in horse chromosomes. *Cytogenet. Cell Genet.* **1996**, *74*, 236.
42. Payne, H.W.; Ellsworth, K.; DeGroot, A. Aneuploidy in an infertile mare. *J. Am. Vet. Assoc.* **1968**, *153*, 1293–1299.
43. Chandley, A.C.; Fletcher, J.; Rossdale, P.D.; Peace, C.K.; Ricketts, S.W.; McEnery, R.J.; Thorne, J.P.; Short, R.V.; Allen, W.R. Chromosome abnormalities as a cause of infertility in mares. *J. Reprod. Fertil. Suppl.* **1975**, 377–383.
44. Bowling, A.T.; Millon, L.; Hughes, J.P. An update of chromosomal abnormalities in mares. *J. Reprod. Fertil. Suppl.* **1987**, *35*, 149–155.
45. Parada, R.; Jaszczak, K.; Sysa, P.; Jaszczak, J. Cytogenetic investigations of mares with fertility disturbances. *Pr. Mat. Zoot.* **1996**, *48*, 71–81.
46. Bugno, M.; Slota, E.; Koscielny, M. Karyotype evaluation among young horse populations in Poland. *Schweiz. Arch. Fur Tierheilkd.* **2007**, *149*, 227–232. [CrossRef] [PubMed]
47. Wieczorek, M.; Switonski, M.; Yang, F. A low-level X chromosome mosaicism in mares, detected by chromosome painting. *J. Appl. Genet.* **2001**, *42*, 205–209.
48. Pawlak, M.; Rogalska-Niżnik, N.; Cholewiński, G.; Świtoński, M. Study on the origin of 64,XX/63,X karyotype in four sterile mares. *Vet. Med. Czech.* **2000**, *45*, 01–06.
49. Bugno-Poniewierska, M.; Wojtaszek, M.; Pawlina-Tyszko, K.; Kowalska, K.; Witarski, W.; Raudsepp, T. Evaluation of the prevalence of sex chromosome aberrations in a population of young horses—Preliminary results. In Proceedings of the Doroty Russell Havemeyer 12th International Horse Genome Workshop, Pavia, Italy, 12–15 September 2018.

50. Wojtaszek, M.; Kowalska, K.; Witarski, W.; Bugno-Poniewierska, M. Diagnostics of horse karyotype aberrations—The initial results of screening. *Wiadomości Zootech. R. Lv.* **2017**, *5*, 49–55.
51. Raudsepp, T.; Das, P.J.; Avila, F.; Chowdhary, B.P. The pseudoautosomal region and sex chromosome aneuploidies in domestic species. *Sex. Dev.* **2012**, *6*, 72–83. [CrossRef]
52. Ruiz, A.J.; Castaneda, C.; Raudsepp, T.; Tibary, A. Azoospermia and Y chromosome-autosome translocation in a Friesian stallion. *J. Equine Vet. Sci.* **2019**, *82*, 102781. [CrossRef]
53. Juras, R.; Raudsepp, T.; Das, P.J.; Conant, E.; Cothran, E.G. XX/XY Blood Lymphocyte Chimerism in Heterosexual Dizygotic Twins from an American Bashkir Curly Horse. Case Report. *J. Equine Vet. Sci.* **2010**, *30*, 575–580. [CrossRef]
54. Ghosh, S.; Carden, C.F.; Juras, R.; Mendoza, M.N.; Jevit, M.J.; Castaneda, C.; Phelps, O.; Dube, J.; Kelley, D.E.; Varner, D.D.; et al. Two Novel Cases of Autosomal Translocations in the Horse: Warmblood Family Segregating t(4;30) and a Cloned Arabian with a de novo t(12;25). *Cytogenet. Genome Res.* **2020**, *16*, 1–10.
55. Ghosh, S.; Das, P.J.; Avila, F.; Thwaits, B.K.; Chowdhary, B.P.; Raudsepp, T. A Non-Reciprocal Autosomal Translocation 64,XX, t(4;10)(q21;p15) in an Arabian Mare with Repeated Early Embryonic Loss. *Reprod. Domest. Anim.* **2016**, *51*, 171–174. [CrossRef] [PubMed]
56. Lear, T.L.; Raudsepp, T.; Lundquist, J.M.; Brown, S.E. Repeated early embryonic loss in a thoroughbred mare with a chromosomal translocation [64, XX, t (2; 13)]. *J. Equine Vet. Sci.* **2014**, *34*, 805–809. [CrossRef]
57. Brito, L.F.C.; Sertich, P.L.; Durkin, K.; Chowdhary, B.P.; Turner, R.M.; Greene, L.M.; McDonnell, S. Autosomic 27 trisomy in a standardbred colt. *J. Equine Vet. Sci.* **2008**, *28*, 431–436. [CrossRef]
58. Ghosh, S.; Davis, B.W.; Rosengren, M.; Jevit, M.J.; Castaneda, C.; Arnold, C.; Jaxheimer, J.; Love, C.C.; Varner, D.D.; Lindgren, G.; et al. Characterization of A Homozygous Deletion of Steroid Hormone Biosynthesis Genes in Horse Chromosome 29 as A Risk Factor for Disorders of Sex Development and Reproduction. *Genes* **2020**, *11*, 251. [CrossRef]
59. Gill, J.J.; Kempski, H.M.; Hallows, B.J.; Warren, A.M. A 64XX/65XXX mosaic mare (*Equus caballus*) and associated infertility. *Equine Vet. J.* **1988**, *20*, 128–130. [CrossRef]
60. Höhn, H.; Klug, E.; Rieck, G.W. A 63,XO/65,XYY mosaic in a case of questionable equine male pseudohermaphroditism. In Proceedings of the 4 th European Colloquium on Cytogenetics of Domestic Animals, Uppsala, Sweden, 10–13 June 1980; pp. 82–92.
61. Paget, S.; Ducos, A.; Mignotte, F.; Raymond, I.; Pinton, A.; Seguela, A.; Berland, H.M.; Brun-Baronnat, C.; Darre, A.; Darre, R.; et al. 63,XO/65,XYY mosaicism in a case of equine male pseudohermaphroditism. *Vet. Rec.* **2001**, *148*, 24–25. [CrossRef]
62. Stewart-Scott, I.A. Infertile mares with chromosome abnormalities. *N. Z. Vet. J.* **1988**, *36*, 63–65. [CrossRef] [PubMed]
63. Halnan, C.R.E.; Watson, J.I.; Pryde, L.C. Detection by G-and C-band karyotyping of gonosome anomalies in horses of different breeds. *J. Reprod. Fert. Suppl.* **1982**, *32*, 627–662.
64. Bugno, M.; Zabek, T.; Golonka, P.; Pienkowska-Schelling, A.; Schelling, C.; Slota, E. A case of an intersex horse with 63,X/64,XX/65,XX,del(Y)(q?) karyotype. *Cytogenet. Genome Res.* **2008**, *120*, 123–126. [CrossRef]
65. Klunder, L.R.; McFeely, R.A.; Willard, J.P. Six separate sex chromosome anomalies in an Arabian mare. *Equine Vet. J.* **1990**, *22*, 218–220. [CrossRef]
66. Neuhauser, S.; Handler, J.; Schelling, C.; Pienkowska-Schelling, A. Fertility and 63,X Mosaicism in a Haflinger Sibship. *J. Equine Vet. Sci* **2019**, *78*, 127–133. [CrossRef] [PubMed]
67. Mushtag, A.; Memon, B.V. Sterility Associated with 64,XX/64,XY,63,X0 mosaic karyotype in a Belgian mare. *Equine Pract.* **1994**, *16*, 24–26.
68. Mäkinen, A.; Hasegawa, T.; Syriä, P.; Katila, T. Infertile mares with XO and XY sex chromosome deviations. *Equine Vet. Educ.* **2006**, *18*, 60–62. [CrossRef]
69. Jaszczak, K.; Sysa, P. Anomalie chromosomowe u koni i ich skutki w reprodukcji. *Przegląd. Hod.* **1992**, *12*, 13–15.
70. Halnan, C.R. Equine cytogenetics: Role in equine veterinary practice. *Equine Vet. J.* **1985**, *17*, 173–177. [CrossRef] [PubMed]
71. Bugno, M.; Słota, E.; Ząbek, T. Two cases of subfertile mares with 64,XX/63,X mosa-ic karyotype. *Ann. Anim. Sci.* **2001**, *1*, 7–11.
72. Breen, M.; Langford, C.F.; Carter, N.P.; Fischer, P.E.; Marti, E.; Gerstenberg, C.; Allen, W.R.; Lear, T.L.; Binns, M.M. Detection of equine X chromosome abnormalities in equids using a horse X whole chromosome paint probe (WCPP). *Vet. J.* **1997**, *153*, 235–238. [CrossRef]
73. Mäkinen, A.; Hasegawa, T.; Makila, M.; Katila, T. Infertility in two mares with XY and XXX sex chromosomes. *Equine Vet. J.* **1999**, *31*, 346–349. [CrossRef] [PubMed]
74. Bugno, M.; Slota, E.; Wieczorek, M.; Yang, F.; Buczynski, J.; Switonski, M. Nonmosaic X trisomy, detected by chromosome painting, in an infertile mare. *Equine Vet. J.* **2003**, *35*, 209–210. [CrossRef] [PubMed]
75. De Lorenzi, L.; Molteni, L.; Zannotti, M.; Galli, C.; Parma, P. X trisomy in a sterile mare. *Equine Vet. J.* **2010**, *42*, 469–470. [CrossRef]
76. Kubien, E.M.; Pozor, M.A.; Tischner, M. Clinical, cytogenetic and endocrine evaluation of a horse with a 65,XXY karyotype. *Equine Vet. J.* **1993**, *25*, 333–335. [CrossRef]
77. Mäkinen, A.; Katila, T.; Andersson, M.; Gustavsson, I. Two sterile stallions with XXY-syndrome. *Equine Vet. J.* **2000**, *32*, 358–360. [CrossRef]
78. Iannuzzi, L.; Di Meo, G.P.; Perucatti, A.; Spadetta, M.; Incarnato, D.; Parma, P.; Iannuzzi, A.; Ciotola, F.; Peretti, V.; Perrotta, G.; et al. Clinical, cytogenetic and molecular studies on sterile stallion and mare affected by XXY and sex reversal syndromes, respectively. *Caryologia* **2004**, *57*, 400–404.

79. Bouters, R.; Vandeplassche, M.; de Moor, A. An intersex (male pseudohermaphrodite) horse with 64 XX-65 XXY mosaicism. *Equine Vet. J.* **1972**, *4*, 150–153. [CrossRef]
80. Bielański, W.; Kleczkowska, A.; Tischner, M.; Jagiarz, M. Comparative cytogenetic examinations of parents and sibilding of colt with a false masculine hermaphroditism. *Med. Wet.* **1980**, *36*, 492–494.
81. Bugno, M.; Pieńkowska-Schelling, A.; Schelling, C.; Włodarczyk, N.; Słota, E. A probe generated by chromosome microdissection, useful for detection of equine X chromosome aneuploidy. *Ann. Anim. Sci.* **2006**, *6*, 205–210.
82. Basrur, P.K.; Kanagawa, H.; Gilman, J.P. An equine intersex with unilateral gonadal agenesis. *Can. J. Comp. Med.* **1969**, *33*, 297–306.
83. Dunn, H.O.; Vaughan, J.T.; McEntee, K. Bilaterally cryptorchid stallion with female karyotype. *Cornell Vet.* **1974**, *64*, 265–275.
84. Klunder, L.R.; McFeeley, R.A. Chromosome analysis of 130 equine clinical cases. In Proceedings of the 6th North American Colloquium on Cytogenetics of Domestic Animals, West Lafayette, IN, USA; 1989; p. 8.
85. Chandley, A.C. Infertility and chromosome abnormality. *Oxf. Rev. Rep. Biol.* **1984**, *6*, 1–46.
86. Nicolas, F.W. *Veterinary Genetics*; Oxford University Press: New York, NY, USA, 1987; p. 580.
87. Marx, M.B.; Melnyk, J.; Persinger, G.; Ono, S.; McGee, W.; Kaufman, W.; Pessin, A.; Gillespie, R. Cytogenetics of the superhorse. *J. Hered.* **1973**, *64*, 95–98. [CrossRef] [PubMed]
88. Shilton, C.A.; Kahler, A.; Davis, B.W.; Crabtree, J.R.; Crowhurst, J.; McGladdery, A.J.; Wathes, D.C.; Raudsepp, T.; de Mestre, A.M. Whole genome analysis reveals aneuploidies in early pregnancy loss in the horse. *Sci. Rep.* **2020**, *10*, 13314. [CrossRef] [PubMed]
89. Colley, E.; Hamilton, S.; Smith, P.; Morgan, N.V.; Coomarasamy, A.; Allen, S. Potential genetic causes of miscarriage in euploid pregnancies: A systematic review. *Hum. Reprod. Update* **2019**, *25*, 452–472. [CrossRef] [PubMed]
90. Hyde, K.J.; Schust, D.J. Genetic considerations in recurrent pregnancy loss. *Cold Spring Harb. Perspect. Med.* **2015**, *5*, a023119. [CrossRef] [PubMed]
91. Power, M.M. Equine half sibs with an unbalanced X;15 translocation or trisomy 28. *Cytogenet. Cell Genet.* **1987**, *45*, 163–168. [CrossRef]
92. Bowling, A.T.; Millon, L.V. Two autosomal trisomies in the horse: 64,XX,-26,+t(26q26q) and 65,XX,+30. *Genome* **1990**, *33*, 679–682. [CrossRef]
93. Zhang, T.Q.; Bellamy, J.; Buwn, L.C.; Weber, A.F.; Ruth, G.R. Autosomal trisomy in a foal with contracted tendon syndrome. In Proceedings of the 10th European Colloqium on Cytogenetics of Domestic Animals, Utrecht, The Netherlands, 18–21 August 1992; pp. 281–284.
94. Buoen, L.C.; Zhang, T.Q.; Weber, A.F.; Turner, T.; Bellamy, J.; Ruth, G.R. Arthrogrypsis in the foal and its possible relation to autosomal trisomy. *Equine Vet. J.* **1997**, *29*, 60–62. [CrossRef] [PubMed]
95. Holl, H.M.; Lear, T.L.; Nolen-Walston, R.D.; Slack, J.; Brooks, S.A. Detection of two equine trisomies using SNP-CGH. *Mamm. Genome* **2013**, *24*, 252–256. [CrossRef]
96. Lear, T.L.; Cox, J.H.; Kennedy, G.A. Autosomal trisomy in a Thoroughbred colt: 65,XY,+31. *Equine Vet. J.* **1999**, *31*, 85–88. [CrossRef] [PubMed]
97. Kubien, E.M.; Tischner, M. Reproductive success of a mare with a mosaic karyotype: 64,XX/65,XX,+30. *Equine Vet. J.* **2002**, *34*, 99–100. [CrossRef] [PubMed]
98. Cuckle, H.; Morris, J. Maternal age in the epidemiology of common autosomal trisomies. *Prenat. Diagn.* **2020**. online ahead of print. [CrossRef] [PubMed]
99. Weckselblatt, B.; Rudd, M.K. Human Structural Variation: Mechanisms of Chromosome Rearrangements. *Trends Genet.* **2015**, *31*, 587–599. [CrossRef]
100. Morin, S.J.; Eccles, J.; Iturriaga, A.; Zimmerman, R.S. Translocations, inversions and other chromosome rearrangements. *Fertil. Steril.* **2017**, *107*, 19–26. [CrossRef] [PubMed]
101. King, W.A. Chromosome variation in the embryos of domestic animals. *Cytogenet. Genome Res.* **2008**, *120*, 81–90. [CrossRef]
102. Power, M.M. The first description of a balanced reciprocal translocation [t(1q;3q)] and its clinical effects in a mare. *Equine Vet. J.* **1991**, *23*, 146–149. [CrossRef]
103. Long, S.E. Tandem 1;30 translocation: A new structural abnormality in the horse (*Equus caballus*). *Cytogenet. Cell Genet.* **1996**, *72*, 162–163. [CrossRef] [PubMed]
104. Lear, T.L.; Layton, G. Use of zoo-FISH to characterise a reciprocal translocation in a thoroughbred mare: T(1;1 6)(q16;q21.3). *Equine Vet. J.* **2002**, *34*, 207–209. [CrossRef] [PubMed]
105. Lear, T.L.; Lundquist, J.; Zent, W.W.; Fishback, W.D., Jr.; Clark, A. Three autosomal chromosome translocations associated with repeated early embryonic loss (REEL) in the domestic horse (*Equus caballus*). *Cytogenet. Genome Res.* **2008**, *120*, 117–122. [CrossRef]
106. Lyon, M.F. Gene action in the X-chromosome of the mouse (*Mus musculus* L.). *Nature* **1961**, *190*, 372–373. [CrossRef] [PubMed]
107. Lyon, M.F. Sex chromatin and gene action in the mammalian X-chromosome. *Am. J. Hum. Genet.* **1962**, *14*, 135–148. [PubMed]
108. Cantone, I.; Fisher, A.G. Human X chromosome inactivation and reactivation: Implications for cell reprogramming and disease. *Philos. Trans. R. Soc. Lond. B Biol. Sci.* **2017**, *372*. [CrossRef] [PubMed]
109. Carrel, L.; Brown, C.J. When the Lyon(ized) chromosome roars: Ongoing expression from an inactive X chromosome. *Philos Trans. R. Soc. Lond. B Biol. Sci.* **2017**, *372*. [CrossRef] [PubMed]
110. Waters, P.D.; Ruiz-Herrera, A. Meiotic Executioner Genes Protect the Y from Extinction. *Trends Genet.* **2020**, *36*, 728–738. [CrossRef]
111. Donaldson, B.; Villagomez, D.A.F.; Revay, T.; Rezaei, S.; King, W.A. Non-Random Distribution of Reciprocal Translocation Breakpoints in the Pig Genome. *Genes* **2019**, *10*, 769. [CrossRef]

112. Ghosh, S.; Qu, Z.; Das, P.J.; Fang, E.; Juras, R.; Cothran, E.G.; McDonell, S.; Kenney, D.G.; Lear, T.L.; Adelson, D.L.; et al. Copy number variation in the horse genome. *PLoS Genet.* **2014**, *10*, e1004712. [CrossRef]
113. McFeely, R.A.; Klunder, L.R.; Byars, D. Equine infertility associated with autosomal mixoploidy. In Proceedings of the 6th North American Colloquium on Domestic Animal Cytogenetics, West Lafayette, IN, USA; 1989; p. 7.
114. Brooks, S.A.; Lear, T.L.; Adelson, D.L.; Bailey, E. A chromosome inversion near the KIT gene and the Tobiano spotting pattern in horses. *Cytogenet. Genome Res.* **2007**, *119*, 225–230. [CrossRef]
115. ISCN. *An International System for Human Cytogenomic Nomenclature*; McGowan-Jordan, J., Hastings, R.J., Moore, S., Eds.; S. Karger AG: Basel, Switzerland, 2020; p. 163.
116. Mäkela, O.; Gustavsson, I.; Hollmen, T. A 64,X,i(Xq) karyotype in a standardbred filly. *Equine Vet. J.* **1994**, *26*, 251–254. [CrossRef]
117. Herzog, A.; Hohn, H.; Klug, E.; Hecht, W. A sex chromosome mosaic in male pseudohermaphroditism in a horse. *Tierarztl. Prax.* **1989**, *17*, 171–175.
118. Bugno-Poniewierska, M.; Ząbek, T.; Pawlina, K.; Klukowska-Rötzler, J.; Rojek, M.; Słota, E. Y isochromosome in mare with sex chromosome mosaicism 63,X/64,XYqi. In Proceedings of the 19th International Colloquium on Animal Cytogenetics and Gene Mapping, Balice/Kraków, Poland, 6–9 June 2010; p. 86.
119. Riggs, P.K.; Rønne, M. Fragile sites in domestic animal chromosomes: Molecular insights and challenges. *Cytogenet. Genome Res.* **2009**, *126*, 97–109. [CrossRef]
120. Durkin, S.G.; Glover, T.W. Chromosome fragile sites. *Annu. Rev. Genet.* **2007**, *41*, 169–192. [CrossRef] [PubMed]
121. Santagostino, M.; Piras, F.M.; Cappelletti, E.; del Giudice, S.; Semino, O.; Nergadze, S.G.; Giulotto, E. Insertion of Telomeric Repeats in the Human and Horse Genomes: An Evolutionary Perspective. *Int. J. Mol. Sci.* **2020**, *21*, 2838. [CrossRef] [PubMed]
122. Malan, V.; Vekemans, M.; Turleau, C. Chimera and other fertilization errors. *Clin. Genet.* **2006**, *70*, 363–373. [CrossRef]
123. Keszka, J.; Jaszczak, K.; Klewiec, J. High frequency of lymphocyte chimerism XX/XY and an analysis of hereditary occurrence of placental anastomoses in Booroola sheep. *J. Anim. Breed. Genet. Z. Fur Tierz. Und. Zucht.* **2001**, *118*, 135–140. [CrossRef]
124. Komisarek, J.; Dorynek, Z. Genetic aspects of twinning in cattle. *J. Appl. Genet.* **2002**, *43*, 55–68. [PubMed]
125. Albarella, S.; de Lorenzi, L.; Catone, G.; Magi, G.E.; Petrucci, L.; Vullo, C.; D'Anza, E.; Parma, P.; Raudsepp, T.; Ciotola, F.; et al. Diagnosis of XX/XY Blood Cell Chimerism at a Low Percentage in Horses. *J. Equine Vet. Sci.* **2018**, *69*, 129–135. [CrossRef]
126. Demyda-Peyras, S.; Anaya, G.; Bugno-Poniewierska, M.; Pawlina, K.; Membrillo, A.; Valera, M.; Moreno-Millan, M. The use of a novel combination of diagnostic molecular and cytogenetic approaches in horses with sexual karyotype abnormalities: A rare case with an abnormal cellular chimerism. *Theriogenology* **2014**, *81*, 1116–1122. [CrossRef] [PubMed]
127. Demyda-Peyras, S.; Membrillo, A.; Bugno-Poniewierska, M.; Pawlina, K.; Anaya, G.; Moreno-Millan, M. The use of molecular and cytogenetic methods as a valuable tool in the detection of chromosomal abnormalities in horses: A case of sex chromosome chimerism in a Spanish purebred colt. *Cytogenet. Genome Res.* **2013**, *141*, 277–283. [CrossRef]
128. Bowling, A.T.; Stott, M.L.; Bickel, L. Silent blood chimaerism in a mare confirmed by DNA marker analysis of hair bulbs. *Anim. Genet.* **1993**, *24*, 323–324. [CrossRef]
129. Bugno, M.; Slota, E.; Tischner, M.; Kozubska-Sobocinska, A. A case of 64,XX/64,XY leucocytic chimerism in a fertile mare of the Wielkopolska breed. *Ann. Anim. Sci.* **1999**, *26*, 9–16.
130. Gustavsson, I. Fertility of sires born as dizygotic twins and sex ratio in their progeny groups. *Ann. Genet. Sel Anim* **1977**, *9*, 531. [CrossRef]
131. Zhang, T.; Buoen, L.C.; Seguin, B.E.; Ruth, G.R.; Weber, A.F. Diagnosis of freemartinism in cattle: The need for clinical and cytogenic evaluation. *J. Am. Vet. Med. Assoc.* **1994**, *204*, 1672–1675. [PubMed]
132. Lee, P.A.; Houk, C.P.; Ahmed, S.F.; Hughes, I.A. International Consensus Conference on Intersex organized by the Lawson Wilkins Pediatric Endocrine, and E. the European Society for Paediatric. Consensus statement on management of intersex disorders. International Consensus Conference on Intersex. *Pediatrics* **2006**, *118*, e488–e500. [CrossRef]
133. Villagomez, D.A.; Lear, T.L.; Chenier, T.; Lee, S.; McGee, R.B.; Cahill, J.; Foster, R.A.; Reyes, E.; John, E.S.; King, W.A. Equine disorders of sexual development in 17 mares including XX, SRY-negative, XY, SRY-negative and XY, SRY-positive genotypes. *Sex. Dev.* **2011**, *5*, 16–25. [CrossRef] [PubMed]
134. Raudsepp, T. Genetics of Equine Reproductive Diseases. *Vet. Clin. N. Am. Equine Pract.* **2020**, *36*, 395–409. [CrossRef]
135. Power, M.M. XY sex reversal in a mare. *Equine Vet. J.* **1986**, *18*, 233–236. [CrossRef] [PubMed]
136. Kent, M.G.; Shoffner, R.N.; Buoen, L.; Weber, A.F. XY sex-reversal syndrome in the domestic horse. *Cytogenet. Cell Genet.* **1986**, *42*, 8–18. [CrossRef] [PubMed]
137. Kent, M.G.; Shoffner, R.N.; Hunter, A.; Elliston, K.O.; Schroder, W.; Tolley, E.; Wachtel, S.S. XY sex reversal syndrome in the mare: Clinical and behavioral studies, H-Y phenotype. *Hum. Genet.* **1988**, *79*, 321–328. [CrossRef] [PubMed]
138. Pailhoux, E.; Cribiu, E.P.; Parma, P.; Cotinot, C. Molecular analysis of an XY mare with gonadal dysgenesis. *Hereditas* **1995**, *122*, 109–112. [CrossRef] [PubMed]
139. Meyers-Wallen, V.N.; Hurtgen, J.; Schlafer, D.; Tulleners, E.; Cleland, W.R.; Ruth, G.R.; Acland, G.M. Sry-negative XX true hermaphroditism in a Pasa Fino horse. *Equine Vet. J.* **1997**, *29*, 404–408. [CrossRef]
140. Abe, S.; Miyake, Y.I.; Kageyama, S.I.; Watanabe, G.; Taya, K.; Kawakura, K. Deletion of the Sry region on the Y chromosome detected in a case of equine gonadal hypoplasia (XY female) with abnormal hormonal profiles. *Equine Vet. J.* **1999**, *31*, 336–338. [CrossRef]

141. Anaya, G.; Moreno-Millan, M.; Bugno-Poniewierska, M.; Pawlina, K.; Membrillo, A.; Molina, A.; Demyda-Peyras, S. Sex reversal syndrome in the horse: Four new cases of feminization in individuals carrying a 64,XY SRY negative chromosomal complement. *Anim. Reprod. Sci.* **2014**, *151*, 22–27. [CrossRef]
142. Pienkowska-Schelling, A.; Becker, D.; Bracher, V.; Pineroli, B.; Schelling, C. Cytogenetical and molecular analyses in a horse with SRY-negative sex reversal. *Schweiz. Arch. Tierheilkd.* **2014**, *156*, 341–344.
143. Martinez, M.M.; Costa, M.; Ratti, C. Molecular screening of XY SRY-negative sex reversal cases in horses revealed anomalies in amelogenin testing. *J. Vet. Diagn. Investig.* **2020**, *32*, 938–941. [CrossRef]
144. Bugno, M.; Klukowska, J.; Slota, E.; Tischner, M.; Switonski, M. A sporadic case of the sex-reversed mare (64,XY.; SRY-negative): Molecular and cytogenetic studies of the Y chromosome. *Theriogenology* **2003**, *59*, 1597–1603. [CrossRef]
145. Sant'Anna Monteiro da Silva, E.; Delfiol, D.J.Z.; Fabris, V.H.; Santos, B.M.; Nogueira, G.M.; Guimaraes, G.B.O.; Nogueira, P.P.d.; da Mota, L.S.L.S. Teratoma Associated With Testicular Tissue in a Female-Like Horse With 64,XY (SRY-Positive) Disorder of Sex Development. *J. Equine Vet. Sci.* **2020**, *92*, 103177. [CrossRef]
146. Knobbe, M.G.; Maenhoudt, C.; Turner, R.M.; McDonnell, S.M. Physical, behavioral, endocrinologic, and cytogenetic evaluation of two Standardbred racehorses competing as mares with an intersex condition and high postrace serum testosterone concentrations. *J. Am. Vet. Med. Assoc.* **2011**, *238*, 751–754. [CrossRef]
147. Howden, K.J. Androgen insensitivity syndrome in a thoroughbred mare (64, XY–testicular feminization). *Can. Vet. J.* **2004**, *45*, 501–503.
148. Crabbe, B.G.; Freeman, D.A.; Grant, B.D.; Kennedy, P.; Whitlatch, L.; MacRae, K. Testicular feminization syndrome in a mare. *J. Am. Vet. Med. Assoc.* **1992**, *200*, 1689–1691.
149. Switonski, M.; Chmurzynska, A.; Szczerbal, I.; Lipczynski, A.; Yang, F.; Nowicka-Posluszna, A. Sex reversal syndrome (64,XY.; SRY-positive) in a mare demonstrating masculine behaviour. *J. Anim. Breed. Genet.* **2005**, *122* (Suppl. 1), 60–63. [CrossRef] [PubMed]
150. Révay, T.; Villagomez, D.A.; Brewer, D.; Chenier, T.; King, W.A. GTG mutation in the start codon of the androgen receptor gene in a family of horses with 64,XY disorder of sex development. *Sex. Dev.* **2012**, *6*, 108–116. [CrossRef] [PubMed]
151. Bolzon, C.; Joone, C.J.; Schulman, M.L.; Harper, C.K.; Villagomez, D.A.; King, W.A.; Revay, T. Missense Mutation in the Ligand-Binding Domain of the Horse Androgen Receptor Gene in a Thoroughbred Family with Inherited 64,XY (SRY+) Disorder of Sex Development. *Sex. Dev.* **2016**, *10*, 37–44. [CrossRef] [PubMed]
152. Welsford, G.E.; Munk, R.; Villagomez, D.A.; Hyttel, P.; King, W.A.; Revay, T. Androgen Insensitivity Syndrome in a Family of Warmblood Horses Caused by a 25-bp Deletion of the DNA-Binding Domain of the Androgen Receptor Gene. *Sex. Dev.* **2017**, *11*, 40–45. [CrossRef] [PubMed]
153. Terribile, M.; Stizzo, M.; Manfredi, C.; Quattrone, C.; Bottone, F.; Giordano, D.R.; Bellastella, G.; Arcaniolo, D.; de Sio, M. 46,XX Testicular Disorder of Sex Development (DSD): A Case Report and Systematic Review. *Medicina (Kaunas)* **2019**, *55*, 371. [CrossRef] [PubMed]
154. Bannasch, D.; Rinaldo, C.; Millon, L.; Latson, K.; Spangler, T.; Hubberty, S.; Galuppo, L.; Lowenstine, L. SRY negative 64,XX intersex phenotype in an American saddlebred horse. *Vet. J.* **2007**, *173*, 437–439. [CrossRef] [PubMed]
155. Buoen, L.C.; Zhang, T.Q.; Weber, A.F.; Ruth, G.R. SRY-negative, XX intersex horses: The need for pedigree studies to examine the mode of inheritance of the condition. *Equine Vet. J.* **2000**, *32*, 78–81. [CrossRef] [PubMed]
156. Jaszczak, K.; Sysa, P.; Sacharczuk, M.; Parada, R.; Romanowicz, K.; Kawka, M.; Jarmuz, W. SRY-negative, 64,XX sex reversal in a Konik Polski horse: A case report. *Anim. Sci. Pap. Rep.* **2010**, *28*, 381–388.
157. Vaughan, L.; Schofield, W.; Ennis, S. SRY-negative XX sex reversal in a pony: A case report. *Theriogenology* **2001**, *55*, 1051–1057. [CrossRef]
158. Peretti, V.; Satue, K.; Ciotola, F.; Cristarella, S.; de Majo, M.; Biondi, V.; D'Anza, E.; Albarella, S.; Quartuccio, M. An Unusual Case of Testicular Disorder in Sex Development of Arabian Mare (64,XX SRY-Negative). *Animals* **2020**, *10*, 1963. [CrossRef] [PubMed]
159. Ciotola, F.; Albarella, S.; Pasolini, M.P.; Auletta, L.; Esposito, L.; Iannuzzi, L.; Peretti, V. Molecular and cytogenetic studies in a case of XX SRY-negative sex reversal in an Arabian horse. *Sex. Dev.* **2012**, *6*, 104–107. [CrossRef] [PubMed]
160. Torres, A.; Silva, J.F.; Bernardes, N.; Luis, J.S.; da Costa, L.L. 64, XX, SRY-negative, testicular DSD syndrome in a Lusitano horse. *Reprod. Domest. Anim.* **2013**, *48*, e33–e37. [CrossRef]
161. Huber, D.; von Voithenberg, L.V.; Kaigala, G.V. Fluorescence in situ hybridization (FISH): History, limitations and what to expect from micro-scale FISH? *Micro Nano Eng.* **2018**, *1*, 15–24. [CrossRef]
162. Raudsepp, T.; Chowdhary, B.P. FISH for mapping single copy genes. *Methods Mol. Biol.* **2008**, *422*, 31–49. [PubMed]
163. Rubes, J.; Pinton, A.; Bonnet-Garnier, A.; Fillon, V.; Musilova, P.; Michalova, K.; Kubickova, S.; Ducos, A.; Yerle, M. Fluorescence in situ hybridization applied to domestic animal cytogenetics. *Cytogenet. Genome Res.* **2009**, *126*, 34–48. [CrossRef]
164. Raudsepp, T.; Gustafson-Seabury, A.; Durkin, K.; Wagner, M.L.; Goh, G.; Seabury, C.M.; Brinkmeyer-Langford, C.; Lee, E.J.; Agarwala, R.; Stallknecht-Rice, E.; et al. A 4,103 marker integrated physical and comparative map of the horse genome. *Cytogenet. Genome Res.* **2008**, *122*, 28–36. [CrossRef]
165. Leeb, T.; Vogl, C.; Zhu, B.; de Jong, P.J.; Binns, M.M.; Chowdhary, B.P.; Scharfe, M.; Jarek, M.; Nordsiek, G.; Schrader, F.; et al. A human-horse comparative map based on equine BAC end sequences. *Genomics* **2006**, *87*, 772–776. [CrossRef]
166. Yang, F.; Fu, B.; O'Brien, P.C.; Nie, W.; Ryder, O.A.; Ferguson-Smith, M.A. Refined genome-wide comparative map of the domestic horse, donkey and human based on cross-species chromosome painting: Insight into the occasional fertility of mules. *Chromosome Res.* **2004**, *12*, 65–76. [CrossRef] [PubMed]

167. Bugno, M.; Slota, E.; Pienkowska-Schelling, A.; Schelling, C. Identification of chromosome abnormalities in the horse using a panel of chromosome-specific painting probes generated by microdissection. *Acta Vet. Hung.* **2009**, *57*, 369–381. [CrossRef]
168. Raudsepp, T.; Chowdhary, B.P. Construction of chromosome-specific paints for meta- and submetacentric autosomes and the sex chromosomes in the horse and their use to detect homologous chromosomal segments in the donkey. *Chromosome Res.* **1999**, *7*, 103–114. [CrossRef]
169. Bugno, M.; Slota, E.; Pienkowska-Schelling, A.; Schelling, C. Detection of equine X chromosome mosaicism in a mare using an equine X whole chromosome painting probe (WCPP)—A case report. *Acta Vet. Hung.* **2007**, *55*, 207–212. [CrossRef]
170. Pienkowska-Schelling, A.; Bugno, M.; Owczarek-Lipska, M.; Schelling, C.; Slota, E. Probe generated by Y chromosome microdissection is useful for analysing the sex chromosomes of the domestic horse. *J. Anim. Feed Sci.* **2006**, *15*, 173–178. [CrossRef]
171. Bugno, M.; Slota, E. Application of arm-specific painting probes of horse X chromosome for karyotype analysis in an infertile Hutsul mare with 64,XX/65,XX+Xp karyotype: Case report. *Acta Vet. Hung.* **2007**, *55*, 309–314. [CrossRef]
172. Alkan, C.; Cardone, M.F.; Catacchio, C.R.; Antonacci, F.; O'Brien, S.J.; Ryder, O.A.; Purgato, S.; Zoli, M.; della Valle, G.; Eichler, E.E.; et al. Genome-wide characterization of centromeric satellites from multiple mammalian genomes. *Genome Res.* **2011**, *21*, 137–145. [CrossRef] [PubMed]
173. Bugno-Poniewierska, M.; Wnuk, M.; Witarski, W.; Slota, E. The fluorescence in situ study of highly repeated DNA sequences in domestic horse (*Equus caballus*) and domestic donkey (*Equus asinus*)—Advantages and limits of usefulness in phylogenetic analyses. *J. Anim. Feed Sci.* **2009**, *18*, 723–732. [CrossRef]
174. Wnuk, M.; Bugno, M.; Slota, E. Application of primed in situ DNA synthesis (PRINS) with telomere human commercial kit in molecular cytogenetics of Equus caballus and Sus scrofa scrofa. *Folia Histochem. Cytobiol.* **2008**, *46*, 85–88. [CrossRef] [PubMed]
175. Bugno-Poniewierska, M.; Zabek, T.; Semik, E.; Pawlina, K.; Tischner, M. A case of sex chromosome mosaicism 64,XX/65,XXY/66,XXYY in mare. *Chromosome Res.* **2014**, *22*, 400–401.
176. Witarski, W.; Kij, B.; Nowak, A.; Bugno-Poniewierska, M. Premature centromere division (PCD) identified in a hucul mare with reproductive difficulties. *Reprod. Domest. Anim.* **2020**, *55*, 248–251. [CrossRef]
177. Wyrobek, A.J.; Alhborn, T.; Balhorn, R.; Stanker, L.; Pinkel, D. Fluorescence in situ hybridization to Y chromosomes in decondensed human sperm nuclei. *Mol. Reprod. Dev.* **1990**, *27*, 200–208. [CrossRef] [PubMed]
178. Bugno-Poniewierska, M.; Jablonska, Z.; Slota, E. Modification of equine sperm chromatin decondensation method to use fluorescence in situ hybridization (FISH). *Folia Histochem. Cytobiol.* **2009**, *47*, 663–666. [CrossRef] [PubMed]
179. Bugno, M.; Jablonska, Z.; Tischner, M.; Klukowska-Rotzler, J.; Pienkowska-Schelling, A.; Schelling, C.; Slota, E. Detection of sex chromosome aneuploidy in equine spermatozoa using fluorescence in situ hybridization. *Reprod. Domest. Anim.* **2010**, *45*, 1015–1019. [CrossRef]
180. Bugno-Poniewierska, M.; Kozub, D.; Pawlina, K.; Tischner, M.; Tischner, M., Jr.; Slota, E.; Wnuk, M. Determination of the correlation between stallion's age and number of sex chromosome aberrations in spermatozoa. *Reprod. Domest. Anim.* **2011**, *46*, 787–792. [CrossRef] [PubMed]
181. Bugno-Poniewierska, M.; Pawlina, K.; Tischner, M.; Tischner, M. Age-related effects on sex chromosome aberrations in equine spermatozoa. *J. Equine Vet. Sci.* **2014**, *34*, 34. [CrossRef]
182. Kjöllerström, H.J.; Oom, M.d.M.; Chowdhary, B.P.; Raudsepp, T. Fertility Assessment in Sorraia Stallions by Sperm-Fish and Fkbp6 Genotyping. *Reprod. Domest. Anim.* **2016**, *51*, 351–359. [CrossRef]
183. Kallioniemi, A.; Kallioniemi, O.P.; Sudar, D.; Rutovitz, D.; Gray, J.W.; Waldman, F.; Pinkel, D. Comparative genomic hybridization for molecular cytogenetic analysis of solid tumors. *Science* **1992**, *258*, 818–821. [CrossRef] [PubMed]
184. Bugno-Poniewierska, M.; Staron, B.; Potocki, L.; Gurgul, A.; Wnuk, M. Identification of Unbalanced Aberrations in the Genome of Equine Sarcoid Cells Using Cgh Technique. *Ann. Anim. Sci.* **2016**, *16*, 79–85. [CrossRef]
185. Pinkel, D.; Albertson, D.G. Comparative genomic hybridization. *Annu. Rev. Genom. Hum. Genet.* **2005**, *6*, 331–354. [CrossRef] [PubMed]
186. Pinkel, D.; Albertson, D.G. Array comparative genomic hybridization and its applications in cancer. *Nat. Genet.* **2005**, *37*, S11–S17. [CrossRef]
187. Bugno-Poniewierska, M.; Ząbek, T.; Gurgul, A.; Pawlina, K. Characteristics of chromosome rearrangements of an intersex horse using molecular cytogenetic techniques. In Proceedings of the 22nd International Colloquium on Animal Cytogenetics and Genomics, Tolouse, France, 2–5 July 2016; p. 38.
188. Schaefer, R.J.; Schubert, M.; Bailey, E.; Bannasch, D.L.; Barrey, E.; Bar-Gal, G.K.; Brem, G.; Brooks, S.A.; Distl, O.; Fries, R.; et al. Developing a 670k genotyping array to tag ~2M SNPs across 24 horse breeds. *BMC Genom.* **2017**, *18*, 565. [CrossRef]
189. Villagomez, D.A.; Pinton, A. Chromosomal abnormalities, meiotic behavior and fertility in domestic animals. *Cytogenet. Genome Res.* **2008**, *120*, 69–80. [PubMed]
190. Purgato, S.; Belloni, E.; Piras, F.M.; Zoli, M.; Badiale, C.; Cerutti, F.; Mazzagatti, A.; Perini, G.; della Valle, G.; Nergadze, S.G.; et al. Centromere sliding on a mammalian chromosome. *Chromosoma* **2015**, *124*, 277–287. [CrossRef]
191. Giulotto, E.; Raimondi, E.; Sullivan, K.F. The Unique DNA Sequences Underlying Equine Centromeres. *Cent. Kinetochores* **2017**, *56*, 337–354.
192. Al-Jaru, A.; Goodwin, W.; Skidmore, J.; Khazanehdari, K. Distribution of MLH1 foci in horse male synaptonemal complex. *Cytogenet. Genome Res.* **2014**, *142*, 87–94. [CrossRef]
193. Al-Jaru, A.; Goodwin, W.; Skidmore, J.; Raudsepp, T.; Khazanehdari, K. Male horse meiosis: Metaphase I chromosome configuration and chiasmata distribution. *Cytogenet. Genome Res.* **2014**, *143*, 225–231. [CrossRef] [PubMed]

194. Kakoi, H.; Hirota, K.; Gawahara, H.; Kurosawa, M.; Kuwajima, M. Genetic diagnosis of sex chromosome aberrations in horses based on parentage test by microsatellite DNA and analysis of X- and Y-linked markers. *Equine Vet. J.* **2005**, *37*, 143–147. [CrossRef]
195. Anaya, G.; Molina, A.; Valera, M.; Moreno-Millan, M.; Azor, P.; Peral-Garcia, P.; Demyda-Peyras, S. Sex chromosomal abnormalities associated with equine infertility: Validation of a simple molecular screening tool in the Purebred Spanish Horse. *Anim. Genet.* **2017**, *48*, 412–419. [CrossRef] [PubMed]
196. Gamo, S.; Tozaki, T.; Kakoi, H.; Hirota, K.I.; Nakamura, K.; Nishii, N.; Alumunia, J.; Takasu, M. X monosomy in the endangered Kiso horse breed detected by a parentage test using sex chromosome linked genes and microsatellites. *J. Vet. Med. Sci.* **2019**, *81*, 91–94. [CrossRef]
197. Bowling, A.T.; Millon, L.V.; Dileanis, S. Physical mapping of genetic markers to chromosome 30 using a trisomic horse and evidence for maternal origin of the extra chromosome. *Chromosome Res.* **1997**, *5*, 429–431. [CrossRef]
198. Raudsepp, T.; Chowdhary, B.P. The Eutherian Pseudoautosomal Region. *Cytogenet. Genome Res.* **2015**, *147*, 81–94. [CrossRef]
199. Chowdhary, B.P.; Raudsepp, T. The horse genome derby: Racing from map to whole genome sequence. *Chromosome Res.* **2008**, *16*, 109–127. [CrossRef]
200. Blackmon, H.; Brandvain, Y. Long-Term Fragility of Y Chromosomes Is Dominated by Short-Term Resolution of Sexual Antagonism. *Genetics* **2017**, *207*, 1621–1629. [CrossRef]
201. Bondy, C.A.; Cheng, C. Monosomy for the X chromosome. *Chromosome Res.* **2009**, *17*, 649–658. [CrossRef] [PubMed]
202. Lupski, J.R.; Stankiewicz, P. Genomic disorders: Molecular mechanisms for rearrangements and conveyed phenotypes. *PLoS Genet.* **2005**, *1*, e49. [CrossRef] [PubMed]
203. Bonomi, M.; Rochira, V.; Pasquali, D.; Balercia, G.; Jannini, E.A.; Ferlin, A.; Italia, N.G.K. Klinefelter syndrome (KS): Genetics, clinical phenotype and hypogonadism. *J. Endocrinol. Investig.* **2017**, *40*, 123–134. [CrossRef]
204. Skuse, D.; Printzlau, F.; Wolstencroft, J. Sex chromosome aneuploidies. *Handb. Clin. Neurol.* **2018**, *147*, 355–376. [PubMed]
205. Lange, J.; Skaletsky, H.; van Daalen, S.K.; Embry, S.L.; Korver, C.M.; Brown, L.G.; Oates, R.D.; Silber, S.; Repping, S.; Page, D.C. Isodicentric Y chromosomes and sex disorders as byproducts of homologous recombination that maintains palindromes. *Cell* **2009**, *138*, 855–869. [CrossRef]
206. Michala, L.; Goswami, D.; Creighton, S.M.; Conway, G.S. Swyer syndrome: Presentation and outcomes. *BJOG* **2008**, *115*, 737–741. [CrossRef]
207. Mantere, T.; Neveling, K.; Pebrel-Richard, C.; Benoist, M.; van der Zande, G.; Kater-Baats, E.; Baatout, I.; van Beek, R.; Yammine, T.; Oorsprong, M.; et al. Next generation cytogenetics: Genome-imaging enables comprehensive structural variant detection for 100 constitutional chromosomal aberrations in 85 samples. *bioRxiv* **2020**. [CrossRef]

Review

Classical, Molecular, and Genomic Cytogenetics of the Pig, a Clinical Perspective

Brendan Donaldson [1], Daniel A. F. Villagomez [2] and W. Allan King [1,3,*]

[1] Department of Biomedical Sciences, University of Guelph, Guelph, ON N1G 2W1, Canada; bdonalds@uoguelph.ca
[2] Departamento de Produccion Animal, Universidad de Guadalajara, Zapopan 44100, Mexico; dvilla@cucba.udg.mx
[3] Karyotekk Inc., Box 363 OVC, University of Guelph, Guelph, ON N1G 2W1, Canada
* Correspondence: waking@ovc.uoguelph.ca

Simple Summary: Chromosome rearrangements are one of the main etiological factors leading to impaired fertility in the domestic pig. The high prevalence of chromosome rearrangements in swine herds, coupled with the production of significantly lower litter sizes, has led to the implementation of cytogenetics techniques in screening prospective breeding boars for rearrangements. Beginning in the 1960s, classical cytogenetics techniques have been applied in laboratories, resulting in the identification of over 200 distinct chromosome rearrangements in the pig. More recently advances in technology, and the development of molecular cytogenetics and cytogenomics techniques, have enhanced the resolution of rearrangements and advanced diagnostic techniques, allowing for more precise and rapid diagnosis of rearrangements.

Abstract: The chromosomes of the domestic pig (*Sus scrofa domesticus*) are known to be prone to reciprocal chromosome translocations and other balanced chromosome rearrangements with concomitant fertility impairment of carriers. In response to the remarkable prevalence of chromosome rearrangements in swine herds, clinical cytogenetics laboratories have been established in several countries in order to screen young boars for chromosome rearrangements prior to service. At present, clinical cytogenetics laboratories typically apply classical cytogenetics techniques such as giemsa-trypsin (GTG)-banding to produce high-quality karyotypes and reveal large-scale chromosome ectopic exchanges. Further refinements to clinical cytogenetics practices have led to the implementation of molecular cytogenetics techniques such as fluorescent in-situ hybridization (FISH), allowing for rearrangements to be visualized and breakpoints refined using fluorescently labelled painting probes. The next-generation of clinical cytogenetics include the implementation of DNA microarrays, and next-generation sequencing (NGS) technologies such as DNA sequencing to better explore tentative genome architecture changes. The implementation of these cytogenomics techniques allow the genomes of rearrangement carriers to be deciphered at the highest resolution, allowing rearrangements to be detected; breakpoints to be delineated; and, most importantly, potential gene implications of those chromosome rearrangements to be interrogated. Clinical cytogenetics has become an integral tool in the livestock industry, identifying rearrangements and allowing breeders to make informed breeding decisions.

Keywords: clinical cytogenetics; genomics; chromosome abnormality; reciprocal translocation; domestic pig

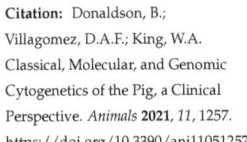

Citation: Donaldson, B.; Villagomez, D.A.F.; King, W.A. Classical, Molecular, and Genomic Cytogenetics of the Pig, a Clinical Perspective. *Animals* **2021**, *11*, 1257. https://doi.org/10.3390/ani11051257

Academic Editors: Leopoldo Iannuzzi and Pietro Parma

Received: 21 March 2021
Accepted: 23 April 2021
Published: 27 April 2021

Publisher's Note: MDPI stays neutral with regard to jurisdictional claims in published maps and institutional affiliations.

Copyright: © 2021 by the authors. Licensee MDPI, Basel, Switzerland. This article is an open access article distributed under the terms and conditions of the Creative Commons Attribution (CC BY) license (https://creativecommons.org/licenses/by/4.0/).

1. Introduction

The domestic pig (*Sus scrofa domesticus*) is known to have a high proportion of chromosomal rearrangements relative to other species [1]. Chromosome rearrangements are structural chromosome abnormalities that result from the breakage of one or more chromatids and a subsequent rearrangement or ectopic exchange of chromosome segments.

This results in the production of derivative chromosomes (e.g., translocation chromosomes), a gross change in the karyotype of the carrier that is often visible under light microscopy. Despite the large-scale rearrangement of genetic material within the genome, often on the scale of millions of base pairs, chromosome rearrangements typically occur with no associated observable signs of their presence [2]. Nevertheless, there are some reports describing chromosome abnormalities associated with physical malformations [3].

Most of the time, chromosome rearrangements appear to occur with minimal losses of genetic material, and thus produce no physical malformations typically associated with aneuploidy. Chromosome rearrangements are, however, known to cause predictable fertility loss in carriers. The litter size losses caused by chromosome rearrangements are variable among carriers and are dependent on a variety of factors, including the morphology of the rearrangement [4]. The derivative chromosomes must satisfy the need for homologous chromosomes to pair during meiosis and do so with a variety of complex formations between derivative chromosomes and their counterparts [5]. Asymmetric segregation of these chromosomes during meiosis leads to a subset of gametes being genetically unbalanced [6]. Resulting unbalanced embryos, or those with lethal mutation due to rearrangement, die early during post-fertilization development due to the presence of genetic imbalances—partial aneuploidies—from derivative chromosomes. Chromosome rearrangements, most notably reciprocal translocations (i.e., balanced exchanges between non-homologous chromosomes), are one of the leading causes of reproductive dysfunction in the domestic pig, with 50% of hypoprolific boars estimated to be carriers [2]. Balanced reciprocal translocations are known to cause the largest litter size losses, averaging 40% piglet loss relative to the herd average, while other chromosome rearrangements have a lesser yet significant impact on litter size [4,7].

Chromosome rearrangements are known to occur in various mammalian species, with reciprocal translocations being especially prevalent in the domestic pig relative to other species. Chromosome rearrangements, including balanced reciprocal translocations, are expected to occur frequently in swine herds throughout the world, being proposed to occur spontaneously in 1/200 live births [8,9]. Carriers of rearrangements, if permitted to breed, may then pass on the rearrangement to approximately 50% of their successful offspring, increasing the prevalence of chromosome rearrangements in swine herds over time [10]. In order to reduce the presence of chromosome rearrangements in swine herds, several labs operate cytogenetic screening programs in order to screen prospective breeding boars for chromosome rearrangements [11]. Although screening programs cannot totally eliminate rearrangements from herds, they can prevent carriers from breeding, resulting in the maintenance of litter size, eliminating the possibility of inheritance, and reducing the prevalence of rearrangements. In countries where such programs are available, many breeders will voluntarily submit their breeding boars for cytogenetic screening, seeing clear economic benefits to managing the presence of rearrangements [7]. Most large swine producing countries, however, fail to implement cytogenetic screening of their swine herds. Thus, the implementation of cytogenetic screening, or some other methods, to identify carriers or potential carriers have room to be widely implemented in the swine industry and greatly reduce the impact of chromosome rearrangements on swine herds.

The field of clinical cytogenetics seeks to apply various laboratory, molecular, and genomic techniques to the study of chromosome rearrangements in order to understand their possible implications on gene and genome functionality. Clinical cytogenetics laboratories, though relatively sparse and underutilized, serve an important purpose to identify chromosomal rearrangements and other chromosome abnormalities in the pig genome, to not only assist selection of boars for breeding services but also to help further the study and understanding of chromosome rearrangements in the domestic pig, including their breadth, diagnosis, origins, and the effect they have on meiosis and the genome itself.

2. Classical Cytogenetics

Conventional laboratory techniques applied to effectively view chromosomes under a light microscope are referred to as classical cytogenetics techniques. This encompasses the in vitro culture of cells such as peripheral blood lymphocytes and fibroblasts, and the subsequent preparation of chromosomes using a series of staining and banding techniques allowing metaphase chromosomes to be effectively viewed under a common light microscope. The first cytogenetics techniques employed in the pig used fixed testicular material and non-differential stains such as crystal violet, allowing the chromosomes to be differentiated from their surroundings, enabling descriptions of the general morphology and the determination of the diploid chromosome number [12]. The diploid chromosome number of the pig is $2n = 38$, consisting of 18 autosomal chromosome pairs, which vary in length and morphology (12-bi-armed and 6 one-armed pairs), and two sex chromosomes, XX or XY [12–14]. The employment of classical cytogenetics techniques to properly determine the diploid chromosome number in mammalian species, especially humans, was essential for the later development of clinical cytogenetics, which linked chromosomal aneuploidy such as trisomy of human chromosomes 13, 18, and 21 to known diseases and later linked sex chromosome aneuploidy and chromosome rearrangement to infertility [15–20].

Following the development of modern in vitro cell culture techniques, providing high-quality, well-spread metaphase chromosomes on glass slides [21,22], researchers began to employ banding techniques and differential staining in order to distinguish chromosomes from one another and observe those chromosomes at a higher resolution. One of the first banding techniques introduced was the hybridization of quinacrine mustard (QM) to chromosomes, which produced a distinct fluorescent pattern on each chromosome as a function of the relative density of guanine residues across each chromosome [23]. The resulting technique was referred to as quinacrine fluorescence (QFQ), or Q-banding. Soon after, other differential banding techniques were developed, including Giemsa-trypsin banding (GTG), or G-banding [24]; replication banding with Giemsa staining (RBG), or R-banding [25]; and reverse-banding with acridine orange staining (RBA), or R-banding [26].

GTG-banding technique employs the proteolytic enzyme trypsin to partially digest the chromosomes, and Giemsa stain, producing a distinct banding pattern on each chromosome where the condensed heterochromatic regions of chromosomes characterized as less transcriptionally active, late-replicating, and AT-rich are stained more intensely than the less condensed euchromatic regions characterized as more transcriptionally active, early-replicating, and GC-rich [24,27,28]. RBA-banding employs the use of the protease trypsin similar to GTG, and acridine orange fluorochrome to stain euchromatic regions more intensely, resulting in a banding pattern that is the inverse of GTG-banding [25,26,28]. RBG-banding employs bromodeoxyuridine (BrdU), which incorporates into DNA, substituting for thymidine residues, and Giemsa to stain regions of the chromosome where BrdU has incorporated, staining AT-rich heterochromatin deeply, producing R-bands [26]. Although these banding patterns were initially developed for the examination of human chromosomes, these methods have been adapted for use on porcine chromosomes and are still routinely applied to the study of porcine chromosomes [5,29].

Chromosomal banding techniques may also be complemented with a variety of staining protocols selective for specific chromosomal regions [30]. These staining techniques may be used to reveal chromosomal features such as constitutive heterochromatin blocks through the use of barium hydroxide (C-bands, CBG technique), nucleolar organizing regions by employing silver-staining techniques (Ag-NOR-bands, Ag-I technique), and telomeric regions through thermal denaturation (T-bands, THA technique) [31–33].

RBA-banding and especially GTG-banding are the most common banding techniques employed in porcine conventional cytogenetics. These banding methods produce approximately 300 bands across all chromosomes, thus the resolution provided by these methods is referred to as the 300 band-level [34]. More specifically, standard GTG-banding and RBA-banding produce a resolution of approximately 5–10 Mb, with chromosome features or rearrangements under 5 Mb in size typically t being indistinguishable [28,35].

Less condensed pro-metaphase chromosomes obtained through the use of replication or condensation blockers preventing cells from progressing to metaphase may be used to provide a higher resolution look at the structure and organization of chromosomes [36]. Pro-metaphase chromosomes may be banded using the GTG or RBA banding methods, producing more finely banded chromosomes with 600 total bands, increasing the resolution to 2–5 Mb [28,37,38]. These higher resolution banding methods enable more refined chromosome analysis in order to more accurately determine the sites of chromosome breakage and establish gene loci.

Classical cytogenetics banding techniques have been used to produce standard karyotypes of the pig, as the banding patterns allow for homologous chromosomes to be paired and their banded patterns converted into standard ideograms [30]. The first standard karyotype arranged for the pig used the QFQ-banding technique popular in the early years of porcine cytogenetics [39]. Previous to this, chromosomes were often arranged in an order that varied among laboratories and often simply ordered the chromosomes by length. Additional karyotypes were arranged using novel banding methods, including GTG-banding [30] and RBA-banding [40,41]. The standard application of the GTG and RBA banding techniques to porcine chromosomes, in tandem with the guidelines provided by the Reading conference [42], resulted in the establishment of a standard karyotype of the domestic pig and schematic representations of GTG and RBA-banded chromosomes [34]. Along with these standard karyotypes was the development of one of the first banding nomenclature systems in domestic species [34]. The development of a distinct nomenclature system allowed porcine cytogeneticists to characterize and report chromosome rearrangements and aberrations in a standardized way easily reported to and understood by other cytogeneticists, helping to further develop the field of clinical cytogenetics of the pig.

3. Chromosome Rearrangements in the Domestic Pig

Chromosome abnormalities, particularly structural chromosome rearrangements, are remarkably prevalent in the domestic pig relative to other species, with over 200 distinct structural rearrangements in the pig genome being identified [1,43]. The prevalence of structural chromosome rearrangements is variable between countries ranging from 0.47% to 3.3% and is largely influenced by access to screening laboratories that identify carriers and removal affected boars from breeding eligibility [8,11,44]. A variety of chromosome rearrangements are known to occur in the pig, including reciprocal translocations, Robertsonian translocations, tandem fusions, inversions, and deletions of chromosomes [1]. The tendency for pigs to experience balanced reciprocal translocations at the largest frequency results in a large variety of rearrangements reported, with all chromosomes of the pig known to be susceptible to ectopic rearrangements [45].

4. Clinical Cytogenetics

The recognized association between chromosome rearrangements and lower fertility has led to the development of routine cytogenetic screening programs in several countries [11]. Worldwide, cytogenetics laboratories primarily apply conventional cytogenetics techniques such as GTG-banding to screen young boars for chromosome rearrangements prior to entering artificial insemination (AI) centres. The National Sow Herd Management Program in France was the first of such cytogenetic screening programs and mandated that boars siring litters of eight piglets or less on average are to be cytogenetically examined prior to servicing additional sows [46,47]. This program has greatly increased the number of boars subject to cytogenetic examination in France over the years and resulted in 20 reported reciprocal translocations in French boars by 1999 [10,48]. This program was expanded in 1999 to include mandatory cytogenetic screening of all boars born of small litters prior to approval for A.I [7]. The success of this program is shown by the reduction in the prevalence of chromosome rearrangements in French boars, with France reporting the lowest prevalence of rearrangements, 0.47%, amongst countries reporting cytogenetic screening results [8]. Many French breeders now voluntarily submit boars for cytogenetic investigation regardless of whether the criteria for screening is met [7,49].

Clinical cytogenetics laboratories have been established in many countries, and at least seventeen countries have laboratories reporting rearrangements in the domestic pig. The countries with the largest cytogenetic screening programs include France, Poland, the Netherlands, Canada, and Spain, with several other countries, including Finland and the United Kingdom, also screening boars for rearrangements. As of 2017, the largest cytogenetic screening program for pigs is conducted at the National Veterinary School of France in Toulouse, with 31,000 boars having passed through this lab as of 2017 [50]. Other large cytogenetic screening programs take place in Spain, the Netherlands, Poland, and Canada, screening 800, 1000, 2000, and 7000 boars, respectively [4,11,43,44]. Clinical cytogenetics programs, although time-consuming, are effective at reducing the prevalence of rearrangements in herds and are cost-effective for breeders, with a cost-benefit ratio of 5.3/1 due to reduced losses from breeding carriers [9]. As of 2021, there are a limited number of cytogenetics laboratories that report the effects of pig screening programs on the prevalence of rearrangement within herds and describe individual chromosome rearrangements [51]. Despite these labs performing cytogenetic screening for breeding boars, just a small percentage of breeding boars worldwide are subject to cytogenetic screening. Further advances in technologies applied in cytogenetics laboratories that reduce the cost of screening, or allow for more rapid results, alongside awareness of the efficacy of cytogenetic screening may increase adoption amongst large swine breeders.

5. Reciprocal Translocations

Reciprocal translocations are the most prevalent chromosome rearrangement known to occur in the pig, representing 84% of all observed structural rearrangements [8]. Reciprocal translocations result from an exchange of chromatid segments between two non-homologous chromosomes following simultaneous chromatid breaks producing two novel derivative chromosomes (Figure 1). All chromosomes of the pig are susceptible to reciprocal translocation, although chromosomes 7, 10, 12, 14, and 17 appear more susceptible to rearrangement than chromosomes 2, 8, 9, 18, and Y [43,45]. Reciprocal translocations typically have unique breakpoints with just one rearrangement, t(12;14)(q13;q21), being observed in two unrelated boars [8,43].

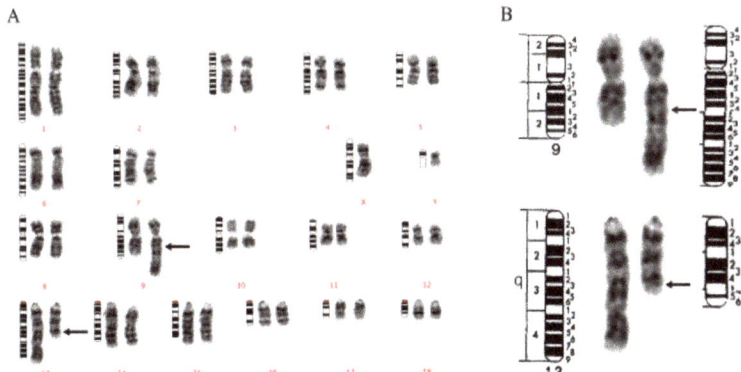

Figure 1. (A). GTG-banded karyotype of a Duroc boar carrying a t(9;13)(q24;q31). Derivative chromosomes are placed to the right. Arrows indicate presumptive breakpoints. (B). GTG-banded chromosomes of chromosomes 9, 13, and the derivative chromosomes formed by the reciprocal translocation event. The ideogram to the left of each chromosome pair indicates the normal chromosome structure. The ideogram to the right of each chromosome pair indicates the derivative chromosome structure. Arrows indicate presumptive breakpoints. Chromosome ideograms adapted from ref. [34].

Balanced structural chromosome rearrangements are a leading cause of fertility losses in pigs, particularly reciprocal translocations, with carriers experiencing average litter size losses of 40% (ranging between 10–100%) relative to the herd average [7,43]. In the case of reciprocal translocations, during meiosis the normal and derivative chromosomes form a quadrivalent shape as a result of full homologous pairing, which progressively may be segregated in a variety of ways, including alternate, adjacent-I, adjacent-II, 3:1 or 4:0, which allow for a high proportion (approximately 50%) of unbalanced gametes to be formed. The exact litter size losses are difficult to predict, with the morphology of chromosomes, size of the rearranged fragments, and involvement of the Y chromosome known to result in complete infertility, proposed to influence piglet loss [52]. The significant loss of litter size and high prevalence of reciprocal translocation in swine herds is one of the principle reasons for the adoption of routine cytogenetic screening programs for young boars prior to entering A.I centres [11].

6. Robertsonian Translocations and Tandem Fusions

Robertsonian translocations are a ubiquitous feature of the wild boar, known to have diploid chromosome number of $2n = 36$, or $2n = 37$ in the case hybridization with domestic hybrids, primarily due to the presence of rob(13;17) wild boar rearrangement in the homozygotic or heterozygotic state, respectively [53,54]. Robertsonian translocations, and especially tandem fusions, are by contrast relatively rare chromosomal events in the domestic pig. Just seven cases of Robertsonian translocation have been described in the pig, primarily cases of the same rob(13;17) rearrangement endemic in wild pigs [4,8,55–57]. Robertsonian translocation is known to occur in the acrocentric chromosomes 13, 14, 15, 16, 17, and 18. In these cases, the chromosomes fuse at the centromeric region, resulting in the production of two derivative chromosomes, a large bi-armed chromosome, and a secondary short chromosome often lost in subsequent cell divisions. Thus, the karyotype of a domestic pig carrying Robertsonian translocation has a distinct $2n = 37$ diploid chromosome number, with the noticeable addition of a novel large bi-armed chromosome. Tandem fusion occurs similarly to Robertsonian translocations; however, it instead involves the fusion of the telomeric region of one chromosome to the centromeric region of another. Tandem fusion is a rare event in the pig, with just one such rearrangement reported, 37, XY, der(14;17)(q29;q10) [58].

As with other chromosome rearrangements, carriers of Robertsonian rearrangements experience subfertility; however, the effect is less severe than in carriers of reciprocal translocations [4,57]. The trivalent formed by the Robertsonian rearrangement during meiosis segregates asymmetrically but results in a smaller percentage of unbalanced gametes that is sex-dependent, with males having a lower proportion of unbalanced gametes [3.2%] relative to female carriers [28.9%] [59]. The higher proportion of balanced gametes in Robertsonian translocations results in less severe litter size losses (5–22%) relative to carriers of reciprocal translocations [2,4,60,61].

7. Paracentric and Pericentric Inversions

Cases of chromosome inversion are relatively rare in the domestic pig with only 12 cases reported in the pig [8]. Chromosomal inversions may be divided into pericentric, involving the centromere, and paracentric, not involving the centromere. In cases of inversion there is no exchange of chromatid segments. Instead, inversions occur as a result of a broken chromatid segment that rotates 180 degrees and reattaches to the original chromosome. Carriers of chromosomal inversions experience minimum litter size loss amongst chromosome rearrangement carriers. Although inversions may result in a proportion of unbalanced gametes during meiosis, the overall proportion is estimated to be small, with average litter size losses of approximately 4% relative to the herd average [2,4,60,61].

8. Chromosomal Aneuploidy

Numerical chromosome aneuploidy in the pig is relatively rare and largely confined to observations in embryos and, in the case of live-birth, the sex chromosomes [62–66].

Cases of whole chromosome aneuploidy 37,X, X-chromosome monosomy [63], and 39,XXY Klinefelter Syndrome have been reported [64–66]. Cases of aneuploidy involving autosomal chromosomes are rare in the pig, as even partial chromosome aneuploidy, such is the case in the embryos of rearrangement carriers, is not tolerated by the pig. A handful of cases of live boars carrying a partial autosomal aneuploidy have been observed in the pig, the result of inheritance of an unbalanced rearrangement involving short segments in the telomeric regions [3,67,68]. In these cases, aneuploidy is accompanied by physical malformation such as cleft palate [3,68].

9. Mosaicism

Chromosome rearrangements are also known to occur in somatic cell lines. Mosaic chromosome rearrangements are estimated to occur frequently in pigs, with limited screening of karyotypes revealing mosaic rearrangements in the karyotypes of 1/300 boars [69]. Mosaic rearrangements arise in somatic cells, rather than the germ line, often appearing confined to certain cell types, such as peripheral blood leukocytes, and thus are not heritable. Indeed, carrier pigs experimentally bred were shown to have offspring with a normal karyotype composition [69]. In addition, mosaic rearrangements have a tendency for recurrence, with three mosaic rearrangements, t(7;9); t(7;18), and t(9;18), being shown to occur recurrently in swine herds [69,70]. Mosaic rearrangements share the same tendency for constitutional rearrangements to experience reciprocal translocation at the highest rate, with no current cases of Robertsonian translocation or inversion being observed in somatic cells. Mosaic rearrangements interestingly appear to tolerate conformations not seen in constitutional rearrangements, including a case of rearrangement between homologous chromosomes, t(7;7) [69]. Mosaic aneuploidies of the sex chromosomes have also been observed in a handful of cases [65,71]. Chimerism, the presence of two distinct sets of DNA in blood leukocytes, has also been described in the form of XX/XY individuals [8,69,72–76]. Somatic or mosaic rearrangements are well known in humans, and are often associated with cancers, especially leukemias and lymphomas that result from the aberrant rearrangement of genes [77]. Although there is no concrete evidence of a relationship between mosaic rearrangements and cancer in the pig, recurrent somatic rearrangements described above are homologous for recurrent somatic rearrangements associated with leukemias in humans [69,70,78].

10. Fragile Sites

Fragile sites are heritable chromosome regions known to break under exposure to distinct chemical stressors such as aphidicolin, bromodeoxyuridine (BrdU), and folate [79–81]. Sixty of these fragile sites are considered common amongst pigs and are expected to occur in most individuals [81]. Analysis of fragile sites in pigs has shown that cytogenetic bands harboring common fragile sites often overlap with known reciprocal translocation breakpoints, and the presence of fragile sites may be associated with higher frequency of rearrangement in those chromosome regions [4,43,81]. As such, it has been suggested that a subset of chromosome rearrangements in the pig are the result of the breakage of fragile chromosome regions as a response to chemical toxins present in farm environments [4,79,80,82].

11. Molecular Cytogenetics

Classical cytogenetics techniques have been instrumental in the development of clinical cytogenetics in the pig and the identification of structural chromosomes rearrangements; however, the limited resolution (>5 Mb) provided by classical cytogenetics impaired the ability to resolve rearrangements at the highest level. Molecular cytogenetics techniques developed in the 1980s were initially applied for physical gene mapping to chromosomes and were later implemented in clinical cytogenetics for the detection of chromosome rearrangements [83]. Molecular cytogenetics as a whole operates around two principles: the target and the probe. Molecular probes are developed to target regions as large as a whole chromosome, or more specific chromosome regions such as the centromere or a specific gene locus [84]. Probes are fluorescently labelled directly with fluorochromes or indirectly with molecules that bind to the probe via fluorochrome-conjugated antibodies [85]. The specificity of the probe for the target is based around the principle of DNA complementary base-pairing, whereby the nucleic acids of fluorescently labelled probes hybridize to the complementary DNA of the target, producing a specific fluorescent signal on the chromosome regions bound by the probe. The specificity of probes for targets enables molecular cytogenetics to achieve a much higher resolution (0.5–10 Mb) than can be obtained through the classical banding techniques [5,28,83].

In the pig, the most common technique for molecular cytogenetics analysis is fluorescent in situ hybridization (FISH) and primed in-situ labelling (PRINS) [5]. The DNA-probes used for FISH include whole chromosome painting probes and probes obtained by cloning genomic DNA inserts from genomic libraries. The probes obtained from genomic libraries may vary in size depending on the origin and the purpose, and may include cosmid probes with DNA insert sizes of <20–40 kb, bacterial probes, or bacterial artificial chromosome (BAC), with DNA insert sizes of 100–300 kb [86]. Chromosome-specific painting probes may be obtained through flow sorting chromosomes, a process that applies dyes to metaphase chromosome suspensions then runs the suspension through a flow cytometer with a laser exciting the chromosomes, and sorting the chromosomes according to relative amounts of genetic material present, roughly corresponding to the length of chromosomes [87–89]. Other techniques for the generation of chromosome painting probes include needle microdissection, which dissects a whole chromosome out of the nucleus or part of a chromosome such as an arm or band using a glass needle [90–92], and laser microdissection, which uses a laser to cut out a chromosome or chromosome arm from the metaphase cell [93,94]. Chromosome painting probes for one, two, or the whole chromosome set may then be applied to metaphase spreads. Different colored fluorescent probes may be applied such as single-color (one chromosome), dual-color (two chromosomes), or multi-color (three or more chromosomes) in order to visualize chromosome rearrangement within the genome [95] (Figure 2).

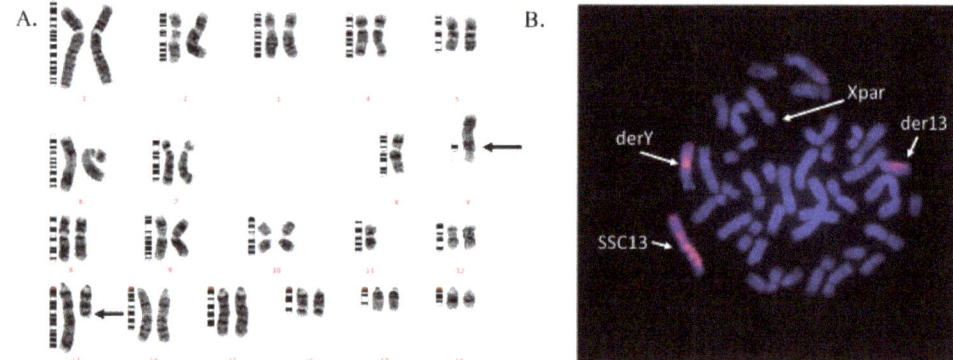

Figure 2. (**A**). GTG-banded karyotype of the t(Y:13) translocation carrier boar. (**B**). FISH chromosome painting of a metaphase plate from the t(Y:13) carrier, green signal pains the Y-chromosome segments, and redish signal pains the SSC13 chromosome segments. The presumptive psuedoautosomal region (green dots) on the X-chromosome is depicted.

PRINS is a technique that anneals short unlabeled oligonucleotide probes to complementary DNA sequences, which are subsequently extended by Taq DNA polymerase [96]. The PRINS technique is most useful for identifying repetitive DNA sequences such as telomeric and centromeric sequences [97,98]. In the pig, oligonucleotide probes for use of the PRINS technique are available, for telomeric (TTAGG)n repeats, centromeric sequences, and a subset of autosomal chromosomes (1, 9, 11, 14) and sex (Y) chromosomes [83,98–101]. Most often, PRINS is used as an alternative technique to FISH, for similar applications in the observation of chromosome rearrangements, and gene loci, with the focus on rearrangements located near repetitive sequences in the genome [102].

Another molecular cytogenetics technique that has been applied in the pig is interspecies in-situ hybridization (Zoo-FISH), which applies human genomic probes that hybridize to homologous sequences in animal genomes [103,104]. The use of Zoo-FISH has been used to study the evolution of mammalian karyotypes by analyzing regions of chromosome synteny between species, and for identifying chromosome segments shared by a common ancestor [104]. However, the expanded availability of flow sorted and microdissected chromosome probes available that are specific for the pig have resulted in Zoo-FISH being rarely implemented in the pig [89]. Another molecular cytogenetics technique, sperm-FISH, applies fluorescently labelled probes to decondensed sperm heads, allowing for the analysis of the sperm chromosome constitution [105]. The FISH technique may also be applied to the chromosomes of oocytes and embryos to analyze chromosome composition.

12. Implementation in Clinical Cytogenetics

Molecular cytogenetics techniques such as FISH have become essential diagnostic tools for the study of pig chromosome rearrangements, allowing for high-resolution viewing of chromosome rearrangements, their meiotic products, and the more accurate diagnosis of rearrangement breaks (e.g., delineation of chromosomal structural changes) [83]. Molecular cytogenetics is typically performed in a manner complementary to that of the classical cytogenetics techniques, and is most often applied in order to refine and/or verify the breakpoints of chromosome rearrangements originally discovered using banding techniques. The use of chromosome painting probes for FISH, or probes for centromeric and telomeric sequences for PRINS, have been used to examine and refine over 20 chromosome rearrangements [5]. The first instance of FISH being used in this way to study porcine rearrangements was by Konfortova et al. [106], who utilized single-colored painting probes in order to visualize the reciprocal exchange of a t(7;15)(q24;p12). Additional experiments applied flow-sorted probes for dual-color chromosome painting to demonstrate the exchange of small terminal chromosome segments not clearly visible via the classical banding techniques, and to verify several reciprocal translocations originally identified through GTG and RBA banding [7,10,48,49,107]. In these cases, the use of FISH was able to detect small exchanges of chromosome material and pinpoint breakpoints with greater accuracy than is available using banding techniques alone. Additionally, flow sorted dual color probes were used in this instance to correctly identify the breakpoints of a rearrangement, originally delineated as t(11;16)(p14;q14), to t(11;16)(p12;q12) [108]. The use of probes for centromeric sequences and the PRINS technique has also been incorporated into cytogenetic analysis to identify breakpoints and subsequent repositioning of the centromere in a rearrangement involving two pig chromosomes [108].

Other instances of cytomolecular analysis using DNA-probes generated by laser microdissection have been used in the analysis of mosaic rearrangements in pigs. Here, laser-microdissected probes specific to chromosomes 7, 9, and 18 were employed to identify three mosaic rearrangements, t(7;9), t(7;18), and t(9;18), amongst thousands of metaphase spreads [77]. Microdissected probes have also been used for the PRINS technique, with telomeric probes labelling the (TTAGG)n telomeric repeat sequence used to confirm the diagnosis of a reciprocal translocation previously identified through banding techniques t(7;13)(q13;q46) [102]. Notably, inversions may be difficult to discern in a banded karyotype due to there being no inter-chromosomal exchange, making it harder to compare chromosome banding patterns. The use of two painting probes, each specific to a chromosome arm obtained through glass-needle microdissection, have been employed to verify the presence of a peri-centric inversion inv(4)(p14;q23) [92]. Another instance of an inversion being re-examined using cytomolecular techniques was a paracentric inversion that employed BAC probes corresponding to microsatellite markers [109].

Although the application of molecular cytogenetics techniques to porcine chromosomes is typically done complimentary to GTG-banding, recent developments have been made to produce a FISH screening assay of multiple BAC probes specific for the subtelomeric regions of each chromosome arm to rapidly identify any chromosome rearrangement without the need to arrange karyotypes [110]. Preliminary research has shown that this method is useful for diagnosing reciprocal chromosomal rearrangements, and may enable the accurate detection of sub-microscopic rearrangements involving small telomeric exchanges of chromosome material near impossible to detect using classical banding techniques. The application of this assay confirmed the presence of four rearrangements originally identified through GTG banding while identifying a fifth rearrangement, involving small telomeric regions of pig chromosomes 5 and 6 not originally detected through GTG banding [110]. New research into the application of molecular cytogenetics into clinical cytogenetic screening laboratories may thus enable the rapid identification of chromosome rearrangements, helping to reduce the time and labor necessary for the production of GTG-banded karyotypes of each animal.

FISH may also be employed to analyze the interaction of chromosome rearrangements in germ cells, through analysis of synaptonemal complexes and meiotic segregation patterns in spermatocytes. Analysis of the meiotic segregation patterns of chromosome rearrangements provides an estimate of the prevalence of unbalanced gametes, facilitating estimates of fertility loss in carriers [111,112]. The application of FISH to synaptonemal complexes has revealed a complete loss of fertility in three carriers of reciprocal translocations involving the Y-chromosome, t(Y;1) [111], t(Y;14)(q11;q11) [113], and t(Y;13)(p13;q33) [52], and has been applied to other reciprocal translocations such as a t(3;15)(q27;q13) and t(12;14)(q13;q21), revealing expected losses of fertility of 47.83% and 24.33% relative to herd averages, respectively [114]. This technique has also been applied to non-reciprocal rearrangements such as a Robertsonian translocation rob(13;17) and paracentric inversions such as inv(2)(q13;q25), revealing less significant fertility losses of 2.96–3.83% and 4.12% respectively [59,115,116]. These studies have been essential in quantifying expected losses of fertility characteristic of each rearrangement, including analysis of the meiotic segregation profiles of male and female carriers, and comparing fertility losses between types of rearrangements and between the sexes [59,107,116].

Zoo-FISH is rarely used in porcine cytogenetic analyses except in rare instances such as to confirm the diagnosis of a tandem fusion-translocation der(14;17)(q29;q10) previously identified using GTG-banding [117]. A similar approach was used for Robertsonian translocation 15;17 in a European wild boar with a karyotype 37,XY,rob(15;17) in which human painting probes for the homologous chromosomes in the pig were used to demonstrate the fusion of the chromosomes and complement the initial diagnosis based upon GTG-banding [118]. In pigs, sperm-FISH has been used to validate the purity of flow cytometrically sorted boar sperm [119,120], to estimate the rate of aneuploidies in normal individuals [121], and to analyze meiotic segregation patterns in translocation and inversion carriers [114,115,122,123].

13. Cytogenomics

In recent years, cytogenomics, which refers to the use of DNA microarrays and whole genome sequencing (WGS) tools to visualize the genome at a high resolution, has increased in prominence. Cytogenomics tools allow the genome and structural variants within to be visualized at a higher resolution than molecular cytogenetics techniques, with resolutions of 100 kb in case of DNA microarrays and nucleotide level resolution in case of WGS [124,125]. The development of a high-quality annotated reference pig genome built upon the foundation laid by the development of autosomal radiation hybrid, recombination, cytogenetic, and BAC maps has been key to the implementation of cytogenomics technologies [126–131]. Genome sequences of the pig have been integral to the development and implementation of single nucleotide polymorphism (SNP) chips, and the WGS efforts in the pig providing a framework for genetic and structural variant discovery, and linkage with gene function, have helped to unravel the genetic and genomic factors associated with complex and disease traits [132,133].

DNA microarrays are tools used to analyze genomes consisting of a series of DNA probes attached to a solid surface (chip). SNP-arrays such as Illumina SNP-array genotyping use single-stranded DNA hybridized to fluorescently labelled DNA probes, producing a fluorescent signal that can be read and interpreted, providing an indication of the relative amount of genetic material present corresponding to a nucleotide base at each probe [134]. SNP array genotyping is most often employed to perform association studies between SNP genotypes and disease traits; however, it may also identify copy number variants (CNV), deletions, and duplications of genetic material, and unbalanced rearrangements (partial monosomy or trisomy) at probes expressing higher or lower signal intensity corresponding to proportional changes in the amount of genetic material. Genomic selection, which analyzes associations between tens of thousands of SNPs and specific trait variations in a phenotyped population, is one of the most widely adopted applications of SNP array genotyping in pigs. Genomic selection incorporates phenotypic and genotypic data from pigs and applies regression analysis to estimate the effect of a genotype on a phenotype, resulting in the estimated breeding value (EBV) used to select candidate breeding animals, resulting in genomics-enabled genetic improvement [135–137]. Initially introduced for application in dairy cattle breeding, genomic selection is now being used in many sectors within animal and plant breeding, including leading pig breeding companies [138,139]. Genomic selection in the pig has been continually improved through novel genome-wide association studies analyzing novel and refining established phenotype-genotype associations in the pig. These studies primarily focus on economically important traits such as back-fat thickness [140] and meat quality [141], and reproductive traits such as farrowing interval [142].

Comparative genomic hybridization (CGH) similarly uses competitive hybridization between normal and cancer cells to fluorescently labelled probes. The ratio of red-green fluorescence along a chromosome would then indicate the presence of gains or losses of genetic material in the chromosomes of cancer cells. This technique has been applied in porcine cytogenetics in order to detect small chromosomal losses and detect aneuploidy in porcine embryos [143,144]. Array-CGH refines this technology using bacterial artificial chromosome (BAC) clone inserts, or short oligonucleotide sequences spaced over the entire chromosome or a region of interest to enhance resolution [145]. The primary application of array-CGH is the detection of disease-associated complex chromosome rearrangements, such as rearrangements associated with tumors. Array-CGH may also have applications in detecting large copy number variants (deletions and duplications) that may be associated with a specific phenotype or disease [146,147].

WGS methods such as Illumina next-generation sequencing provide a higher resolution look at the genome than DNA microarrays by fragmenting DNA into short DNA segments several bases to hundreds of bases long. Adapters are ligated to these DNA segments and then amplified via polymerase chain reaction (PCR), producing several copies of each DNA segment. The DNA segments are then exposed to fluorescently labelled nucleotides and DNA polymerase, binding to one base at a time, and taking an image that is interpreted by computer software. This process is repeated several times, allowing for each segment to be sequenced several times over and aligned, producing an accurate sequence of the genomic region of interest. WGS has been applied by cytogeneticists to delineate chromosome rearrangement breakpoints, and the breakpoint signatures coinciding with the repair mechanism, small copy number variants and indels (CNV less than 1 kb in length) not visible via SNP array, and novel single nucleotide variants within genes that may be associated with disease [148]. Although a relatively new technology, the cost of WGS in the pig has dropped dramatically in recent years, enabling methods such as Illumina short-read sequencing to be more widely implemented in research, facilitating the sequencing of several hundred pig genomes for variant discovery [149–151].

The application of WGS results in reads that provide the base pair sequence of regions of chromosomes, or whole chromosomes. WGS may detect balanced rearrangements using split alignments that map to two different locations in the reference genome, and discordant read pairs, paired ends that do not align within an expected distance or orientation [152]. These methods indicate regions of chromosomes where breakpoint junctions may occur and are the most successful and precise methods of identifying balanced rearrangements [153]. The most used sequencing methods for the detection of balanced reciprocal translocation in humans include long-read, short-read, and linked-read sequencing. Long-read sequencing is an expensive technique capable of producing long continuous sequences reads of DNA > 10 kb [154]. Long-read sequencing may be used to detect any type of structural variation; however, it is most successful at the detection of complex rearrangements, and breakpoints present in repetitive elements relative to other methods of WGS [153,155,156]. Short-read sequencing in contrast produces many overlapping short reads (150–300 bases) of the DNA [157]. Short-read sequencing is more established and less expensive than long-read sequencing, and is most useful for detecting simple rearrangements such as reciprocal translocations [158–160]. Linked-read sequencing is a method that uses small amounts of high-molecular-weight genomic DNA, spread across 100,000 droplets, each of which is tagged with a barcode [161]. The barcode-tagged droplets undergo short-read sequencing, with a computer algorithm using the barcode to link the sequenced reads to the original molecules and construct continuous segments of DNA. From here, structural variants may be determined from reads belonging to disparate regions but sharing the same barcode. Although less developed than long-read and short-read sequencing, linked-read sequencing has been successfully implemented in the detection of rearrangements spanning repetitive elements, overcoming one of the deficiencies of short-read sequencing [162].

In research, WGS has been used to detect a wide range of mutations, including copy number variations [163–165] as well as balanced chromosomal rearrangements such as translocations [148,166] and inversions [167]. Approximately 90% of breakpoints of balanced rearrangements can be identified using WGS [159]. Notably, low-level mosaicism and Robertsonian rearrangements, as well as supernumerary chromosomes and a subset of reciprocal translocations, especially those with breakpoint in repetitive sequences, cannot be routinely detected through short-read sequencing [159,168]. The cost of WGS, and the fact that not all rearrangements may be detected by WGS, indicates that cytogenomics are unlikely to overtake the classical and molecular cytogenetics techniques as a standard test for the detection of chromosome rearrangements. Even in the case of WGS being used to identify a rearrangement, it is recommended that follow-up studies consist of karyotyping and/or FISH in order to visualize the rearrangement and determine the structural rearrangement underlying the imbalance [169,170].

14. Implementation of Cytogenomics in Clinical Cytogenetics

The use of DNA microarrays such as SNP-array genotyping is widespread in the pig, and is primarily used for the purposes of genomic selection [136]. SNP-array genotyping and array-CGH are comparatively seldom applied in the field of clinical cytogenetics as both techniques are ineffective at identifying the balanced chromosome rearrangements characteristic of the pig and instead are only capable of detecting rare unbalanced rearrangements [171]. Instead, DNA microarrays are more often applied in order to identify copy number variants in pigs, deletions and duplications of genomic material that may be identified as they are characterized by genomic imbalance. Analysis of CNV rarely falls under the purview of cytogenetics laboratories as these structural variants are often too small to be detected through classical or molecular cytogenetics techniques. The importance of CNV has increased in recent years, with these structural variants being linked to much of the genomic variation observed in mammalian species and associated with diseases seen in human and animal genomes [172]. Currently, studies of CNV have sought to link these variants to traits relevant for breeding such as meat quality [173] and fertility [174], as well as diseases such as porcine reproductive and respiratory syndrome [175]. The expansion of

WGS technologies will allow for more in-depth exploration of smaller structural variants in the genomes of pigs such as CNV and indels, along with an increased understanding of those variants with both economically desirable traits as well as disease. The linkage of CNV to fertility may indicate that one-day it may be routine to include screening for specific CNV alongside chromosome rearrangements in pigs prior to breeding [174].

SNP array genotyping has been used to identify the sire of unbalanced rearrangement carriers through identifying an imbalance in the inheritance of paternal alleles, resulting from a partial monosomy of chromosome 8, and a partial trisomy of chromosome 14 [3]. Although SNP genotypes may be used to identify unbalanced rearrangements and aneuploidies, such chromosomal events are rare in the pig, occurring in less than 10% of cases of clinically diagnosed chromosome abnormality [1]. The use of SNP array genotyping for the identification of balanced chromosome rearrangements may be applicable in a small subset of cases using karyomapping. This approach determines the linkage phase of SNPs, and has seen application as a pre-implantation genetic test for known rearrangement carriers, allowing those embryos with the same linkage phase as the rearrangement carrier to be identified [176]. This technique is unlikely to see widespread application in clinical cytogenetics programs as the carrier must be identified prior to karyomapping, and is only useful in cases where the goal is the identification of the offspring of carriers or of the parents of carriers.

WGS technologies provide the clearest avenue for implementation of cytogenomics techniques into clinical cytogenetics laboratories. Classical chromosome banding techniques as well as molecular technique such as FISH, although effective at identifying large rearrangements, may miss smaller terminal rearrangements and are incapable of identifying the precise breakpoints of the rearrangements [110]. The application of sequencing technologies will allow for further refinements to the identification and study of chromosome rearrangements, allowing for breakpoint junctions to be delineated. WGS has been successfully applied in humans to identify the precise breakpoint junctions of hundreds of balanced rearrangements [148,177]. Currently, WGS has been applied to just a handful of chromosome rearrangements in the pig, with the high cost of genome sequencing only recently reduced serving as a barrier to the more widespread implementation of WGS in the pig [126]. In one case, short-read sequencing was conducted on boars carrying unbalanced rearrangements [3]. The rearrangement was identified through reduced sequence coverage on chromosome 8, and increased sequence coverage on chromosome 14, corresponding to a partial monosomy and a partial trisomy, along with discordant paired-end sequence reads aligning on chromosomes 8 and 14, confirming the presence of a rearrangement [3].

The largest study of the applications of WGS into the study of porcine chromosome rearrangements performed short-read sequencing to seven carriers of karyotypically balanced rearrangements alongside 15 non-carriers [178]. Here, it was found that short-read sequencing was capable of accurately detecting the breakpoint junctions of six of seven carriers, with no false-positives detected. The main deficiency of short-read sequencing noted was that it was not capable of identifying breakpoint junctions occurring in repetitive sequences [178]. This study for the first time described the breakpoints of chromosome rearrangements in the pig genome, identifying several varieties of breakpoint signatures including microhomology, microinsertions, and blunt-end ligations also characteristic of human rearrangement breakpoint junctions [156,178]. Genes disrupted by the breakpoint interrupting the gene sequence were also found at the sites of breakpoint junctions, with the heterozygous nature of the disruption suggested to be protective of any phenotypic effect associated with the rearrangement. The sequencing of these breakpoints therefore indicates that so-called balanced rearrangements may not be as balanced as once thought, with the presence of small deletions, insertions, and gene disruptions noted as occurring in the pig [178].

The factors influencing the formation of chromosome rearrangements in pigs are still largely unknown, despite the sequencing of a handful of rearrangement breakpoint junctions. No clear pattern presents itself, with breakpoints appearing with different breakpoint signatures and in a variety of chromosome regions and landscapes [148,156,178]. A study of mosaic rearrangement carriers indicated that relatives of mosaic carriers themselves carried mosaic rearrangements at 2.5× the frequency as control animals, indicating a possible genetic component to the formation of chromosome rearrangements [74]. A preliminary genome-wide association study (GWAS) was performed in our laboratory on 15 carriers of reciprocal translocations, and 11 control boars with normal karyotypes, revealing the presence of five SNPs on three chromosomes associated with reciprocal translocation [179]. Functional analysis of these SNPs revealed that each was in close proximity (<2 Mb) to genes playing roles in the maintenance of DNA, detection of DNA damage, and the initiation of the DNA damage response [179]. These results indicate that genetic factors may play a large role in the susceptibility of pigs to produce de-novo chromosome rearrangements during meiotic events that are then passed on to offspring. The identification of SNPs closely linked to chromosome rearrangement could be incorporated alongside other cytogenomics analyses, as a control effort to identify boars at risk of producing carrier offspring.

15. Future Perspectives and Conclusions

The factors influencing the formation of chromosome rearrangements in mammalian genomes are still poorly understood. Various chromosomal characteristics such as relative chromosome density and the presence of common fragile sites have been associated with breakage hotspots in the pig genome [43]. The precise breakpoints, and the genomic landscape surrounding those breaks, is largely unknown, making it difficult to determine why a given chromosome region may experience more or less chromosome breaks and how this may subsequently lead to permanent rearrangement of chromosomes. The application of classical banding, and molecular cytogenetics techniques (although they have limited applications), will still be useful for primary identification of gross chromosome rearrangements. Therefore, it is necessary to incorporate WGS technologies into clinical cytogenetics programs in order to visualize rearrangements at the highest level, allowing for the delineation of breakpoints and the best understanding of any genomic consequences associated with the rearrangement.

Although currently too expensive for widespread application into clinical cytogenetics programs, the cost of WGS in humans has dramatically dropped in recent years from >$10,000 USD in 2010 to under $1000 USD as of 2020 [180]. Given this trend, it is not too unreasonable to predict that the price of DNA sequencing may further reduce in the coming years, facilitating the widespread introduction of WGS into the livestock industry. Access to sequencing data could allow for the routine detection of various chromosome rearrangements and aneuploidies ranging from balanced and unbalanced rearrangements, and CNV, to aneuploidies. This could be a cost-effective strategy for both breeders and cytogeneticists allowing for a wide range of accurate tests to be conducted on a single genomic data set. Although unlikely to fully replace banding techniques and FISH for chromosome analysis as both techniques allow for the visualization of the rearrangement, cost-reductions in WGS could enable DNA sequencing to be a first-tier genetic test for livestock [165,167].

The development and implementation of laboratory and genomic techniques for use in clinical cytogenetics has played an important role in swine breeding for the last forty years. From the first chromosome rearrangements and abnormalities identified, it has been clear that the pig is susceptible to a number of chromosome abnormalities resulting in impaired fertility to total infertility [181]. The development of the classical cytogenetics techniques and their subsequent application in pigs led to the first clinical cytogenetics programs, and to this day continue to form the basis of clinical cytogenetics operations. With the development of molecular cytogenetics techniques such as FISH, further refinements were made to the visualization of chromosome rearrangements, allowing rearrangements to be viewed at the highest resolution yet, along with refinements made to the identification of breakpoints, and the identification of sub-telomeric breaks. Lastly, as WGS technologies continue to be developed and applied in the pig, along with associated cost-reductions, there is an opportunity to revolutionize livestock breeding and clinical cytogenetics, allowing for DNA sequencing data to be used in concert with banding techniques or FISH to identify and study a large range of chromosomal rearrangements and abnormalities furthering the study and understanding of chromosome rearrangements. It is hopeful that with the implementation of WGS technologies, our understanding of chromosome rearrangements will increase and the factors influencing rearrangements in the pig genome be fully understood.

Author Contributions: Writing—Original Draft Preparation, B.D.; Writing—Review and Editing, B.D., D.A.F.V., W.A.K. All authors have read and agreed to the published version of the manuscript.

Funding: This research received no external funding.

Institutional Review Board Statement: Not applicable for this work.

Conflicts of Interest: The authors declare no conflict of interest.

References

1. Raudsepp, T.; Chowdhary, B.P. Cytogenetics and chromosome maps. In *the Genetics of the Pig*, 2nd ed.; CABI: Wallingford, UK, 2011; pp. 134–178.
2. Gustavsson, I. Chromosomes of the pig. In *Advances in Veterinary Science and Comparative Medicine*; Academic Press: Cambridge, MA, USA, 1990; Volume 34, pp. 73–107.
3. Grahofer, A.; Letko, A.; Häfliger, I.M.; Jagannathan, V.; Ducos, A.; Richard, O.; Peter, V.; Nathues, H.; Drögemüller, C. Chromosomal imbalance in pigs showing a syndromic form of cleft palate. *BMC Genom.* **2019**, *20*, 1–11. [CrossRef] [PubMed]
4. Quach, A.T.; Revay, T.; Villagomez, D.A.F.; Macedo, M.P.; Sullivan, A.; Maignel, L.; Wyss, S.; Sullivan, B.; King, W.A. Prevalence and consequences of chromosomal abnormalities in Canadian commercial swine herds. *Genet. Sel. Evol.* **2016**, *48*, 1–7. [CrossRef] [PubMed]
5. Danielak-Czech, B.; Kozubska-Sobocińska, A.; Rejduch, B. Molecular Cytogenetics in the Diagnostics of Balanced Chromosome Mutations in the Pig (*Sus scrofa*)—A Review. *Ann. Anim. Sci.* **2016**, *16*, 679–699. [CrossRef]
6. King, W.A.; Gustavsson, I.; Popescu, C.P.; Linares, T. Gametic products transmitted by rcp (13q−; 14q+) translocation heterozygous pigs, and resulting embryonic loss. *Hereditas* **1981**, *95*, 239–246. [CrossRef] [PubMed]
7. Pinton, A.; Ducos, A.; Berland, H.; Seguela, A.; Brun-Baronnat, C.; Darré, A.; Darré, R.; Schmitz, A.; Yerle, M. Chromosomal Abnormalities in Hypoprolific Boars. *Hereditas* **2004**, *132*, 55–62. [CrossRef]
8. Ducos, A.; Berland, H.-M.; Bonnet, N.; Calgaro, A.; Billoux, S.; Mary, N.; Garnier-Bonnet, A.; Darré, R.; Pinton, A. Chromosomal control of pig populations in France: 2002–2006 survey. *Genet. Sel. Evol.* **2007**, *39*, 583. [CrossRef] [PubMed]
9. King, W.A.; Donaldson, B.; Rezaei, S.; Schmidt, C.; Revay, T.; Villagomez, D.A.; Kuschke, K. Chromosomal abnormalities in swine and their impact on production and profitability. In *Comprehensive Biotechnology*, 3rd ed.; Moo-Young, M., Ed.; Pergamon Press: Oxford, UK, 2019; pp. 508–518.
10. Ducos, A.; Berland, H.M.; Pinton, A.; Guillemot, E.; Seguela, A.; Blanc, M.F.; Darre, A.; Darre, R. Nine new cases of reciprocal translocation in the domestic pig (*Sus scrofa domestica* L.). *J. Hered.* **1998**, *89*, 136–142. [CrossRef]
11. Ducos, A.; Revay, T.; Kovacs, A.; Hidas, A.; Pinton, A.; Bonnet-Garnier, A.; Molteni, L.; Slota, E.; Switonski, M.; Arruga, M.V.; et al. Cytogenetic screening of livestock populations in Europe: An overview. *Cytogenet. Genome Res.* **2008**, *120*, 26–41. [CrossRef]
12. Bryden, W. The chromosomes of the pig. *Cytologia* **1993**, *5*, 149–153. [CrossRef]
13. Krallinger, H.F. *Cytologische Studien an Einigen Haussäugetieren [Cytological Studies on Some Domestic Animals]*; Springer: Berlin/Heidelberg, Germany, 1931.
14. McConnell, J.; Fechheimer, N.S.; Gilmore, L.O. Somatic Chromosomes of the Domestic Pig. *J. Anim. Sci.* **1963**, *22*, 374–379. [CrossRef]
15. Lejeune, J. Etude des Chromosomes Somatiques de Neuf Enfants Mongoliens [Study of the Somatic Chromosomes of Nine Mongoloid Children]. *CR Acad. Sci.* **1959**, *248*, 1721–1722.
16. Patau, K.; Smith, D.; Therman, E.; Inhorn, S.; Wagner, H. Multiple congenital anomaly caused by an extra autosome. *Lancet* **1960**, *275*, 790–793. [CrossRef]
17. Edwards, J.; Harnden, D.; Cameron, A.; Crosse, V.; Wolf, O. A new trisomic syndrome. *Lancet* **1960**, *275*, 787–790. [CrossRef]
18. McIlree, M.; Price, W.; Brown, W.; Tulloch, W.; Newsam, J.; MacLean, N. sChromosome studies on testicular cells from 50 subfertile men. *Lancet* **1966**, *288*, 69–71. [CrossRef]
19. Philip, J.; Skakkebæk, N.E.; Hammen, R.; Johnsen, S.G.; Rebbe, H. Cytogenetic investigations in male infertility. *Acta Obstet. Gynecol. Scand.* **1970**, *49*, 235–239. [CrossRef]
20. Chandley, A.C.; Edmond, P.; Christie, S.; Gowans, L.; Fletcher, J.; Frackiewicz, A.; Newton, M. Cytogenetics and infertility in man. I. Karyotype and seminal analysis: Results of a five-year survey of men attending a subfertility clinic. *Ann. Hum. Genet.* **1975**, *39*, 231–254. [CrossRef]

21. Moorhead, P.S.; Nowell, P.C.; Mellman, W.J.; Battips, D.T.; Hungerford, D.A. Chromosome preparations of leukocytes cultured from human peripheral blood. *Exp. Cell Res.* **1960**, *20*, 613–616. [CrossRef]
22. Arakaki, D.; Sparkes, R. Microtechnique for Culturing Leukocytes from Whole Blood. *Cytogenet. Genome Res.* **1963**, *2*, 57–60. [CrossRef]
23. Caspersson, T.; Zech, L.; Johansson, C.; Modest, E.J. Identification of human chromosomes by DNA-binding fluorescent agents. *Chromosoma* **1970**, *30*, 215–227. [CrossRef] [PubMed]
24. Seabright, M. A rapid banding technique for human chromosomes. *Lancet* **1971**, *298*, 971–972. [CrossRef]
25. Dutrillaux, B.; Lejeune, J. Sur une Nouvelle Technique D'analyse du Caryotype Humain [On a New Technique for Analyzing the Human Karyotype]. *CR Acad. Sci.* **1971**, *272*, 2638–2640.
26. Dutrillaux, B. Coloration des Chromosomes Humains par L'acridine Orange Après Traitement par le 5 Bromodéoxyuridine [Staining of Human Chromosomes with Acridine Orange after Treatment with 5 Bromodeoxyuridine]. *CR Acad. Sci.* **1973**, *276*, 3179–3181.
27. Wang, H.C.; Fedoroff, S. Banding in Human Chromosomes treated with Trypsin. *Nat. New Biol.* **1972**, *235*, 52–54. [CrossRef]
28. Bickmore, A.W. Karyotype analysis and chromosome banding. *eLS* **2001**. [CrossRef]
29. Hageltorn, M.; Gustavsson, I. Giemsa staining patterns for identification of the pig mitotic chromosomes. *Hereditas* **2009**, *75*, 144–146. [CrossRef]
30. Gustavsson, I. Banding techniques in chromosome analysis of domestic animals. *Adv. Vet. Sci. Comp. Med.* **1980**, *24*, 245–289. [PubMed]
31. Sumner, A. A simple technique for demonstrating centromeric heterochromatin. *Exp. Cell Res.* **1972**, *75*, 304–306. [CrossRef]
32. Bloom, S.E.; Goodpasture, C. An improved technique for selective silver staining of nucleolar organizer regions in human chromosomes. *Qual. Life Res.* **1976**, *34*, 199–206. [CrossRef]
33. Dutrillaux, B. Nouveau Système de Marquage Chromosomique: Les Bandes T [New Chromosome Labeling System: T Bands]. *Chromosoma* **1973**, *41*, 395–402. [CrossRef]
34. Gustavsson, I. Standard karyotype of the domestic pig: Committee for the Standardized Karyotype of the Domestic Pig. *Hereditas* **1988**, *109*, 151–157. [CrossRef]
35. Vorsanova, S.G.; Yurov, Y.B.; Iourov, I.Y. Human interphase chromosomes: A review of available molecular cytogenetic technologies. *Mol. Cytogenet.* **2010**, *3*, 1. [CrossRef] [PubMed]
36. Berardino, D.D.; Lannuzzi, L.; Lioi, M. The high-resolution RBA-banding pattern of bovine chromosomes. *Cytogenet. Genome Res.* **1985**, *39*, 136–139. [CrossRef] [PubMed]
37. Rønne, M. Chromosome preparation and high resolution banding. *In Vivo* **1990**, *4*, 337–365. [PubMed]
38. Yerle, M.; Galman, O.; Echard, G. The high-resolution GTG-banding pattern of pig chromosomes. *Cytogenet. Genome Res.* **1991**, *56*, 45–47. [CrossRef]
39. Gustavsson, I.; Hageltorn, M.; Johansson, C.; Zech, L. Identification of the pig chromosomes by the quinacrine mustard fluorescence technique. *Exp. Cell Res.* **1972**, *70*, 471–474. [CrossRef]
40. Lin, C.C.; Biederman, B.M.; Jamro, H.K.; Hawthorne, A.B.; Church, R.B. Porcine (*Sus scrofa domestica*) chromosome identification and suggested nomenclature. *Can. J. Genet. Cytol.* **1980**, *22*, 103–116. [CrossRef]
41. Rønne, M.; Stefanova, V.; Di Berardino, D.; Poulsen, B.S. The R-banded karyotype of the domestic pig (*Sus scrofa dornestica* L.). *Hereditas* **2008**, *106*, 219–231. [CrossRef]
42. Ford, C.E.; Pollock, D.L.; Gustavsson, I. Proceedings of the First International Conference for the Standardisation of Banded Karyotypes of Domestic Animals University of Reading Reading, England, 2–6 August 1976. *Hereditas* **1980**, *92*, 145–162. [CrossRef]
43. Donaldson, B.; Villagomez, D.A.; Revay, T.; Rezaei, S.; King, W.A. Non-Random distribution of reciprocal translocation breakpoints in the pig genome. *Genes* **2019**, *10*, 769. [CrossRef]
44. Sánchez-Sánchez, R.; Gómez-Fidalgo, E.; Pérez-Garnelo, S.; Martín-Lluch, M.; De La Cruz-Vigo, P. Prevalence of chromosomal aberrations in breeding pigs in Spain. *Reprod. Domest. Anim.* **2019**, *54*, 98–101. [CrossRef]
45. Basrur, P.; Stranzinger, G. Veterinary cytogenetics: Past and perspective. *Cytogenet. Genome Res.* **2008**, *120*, 11–25. [CrossRef]
46. Dagorn, R. *Note Aux Établissements Départementaux de l'Elevage*; Institut Technique du Porc: Paris, France, 1978.
47. Popescu, C.P.; Boscher, J.; Tixier, M. Une nouvelle translocation réciproque t, rcp (7q−; 15q+) chez un verrat «hypoprolifique» [A new reciprocal translocation, rcp (7q−; 15q+) translocation in a "hypoprolific" boar]. *Génétique Sél. Évol.* **1983**, *15*, 479–488. [CrossRef]
48. Ducos, A.; Pinton, A.; Berland, H.-M.; Seguela, A.; Blanc, M.-F.; Darre, A.; Darre, R. Five New Cases of Reciprocal Translocation in the Domestic Pig. *Hereditas* **2004**, *128*, 221–229. [CrossRef]
49. Ducos, A.; Pinton, A.; Yerle, M.; Séguéla, A.; Berland, H.-M.; Brun-Baronnat, C.; Bonnet, N.; Darré, R. Cytogenetic and molecular characterization of eight new reciprocal translocations in the pig species. Estimation of their incidence in French populations. *Genet. Sel. Evol.* **2002**, *34*, 1–406. [CrossRef] [PubMed]
50. Ducos, A.; Calgaro, A.; Mouney-Bonnet, N.; Loustau, A.M.; Revel, C.; Barasc, H.; Mary, N.; Pinton, A. Chromosomal control of pig populations in France: A 20-year perspective. *Journées Rech. Porc. Fr.* **2017**, *49*, 49–50.
51. Pinton, A.; Calgaro, A.; Bonnet, N.; Mary, N.; Dudez, A.M.; Barasc, H.; Plard, C.; Yerle, M.; Ducos, A. Chromosomal control of pig populations in France: 2007–2010 survey. *Journées Rech. Porc. Franc.* **2012**, *44*, 43–44.

52. Villagómez, D.A.; Revay, T.; Donaldson, B.; Rezaei, S.; Pinton, A.; Palomino, M.; Junaidi, A.; Honaramooz, A.; King, W.A. Azoospermia and Testicular Hypoplasia in a Boar Carrier of a Novel Y-Autosome Translocation. *Sex. Dev.* **2017**, *11*, 46–51. [CrossRef]
53. Tikhonov, V.N.; Troshina, A.I. Chromosome translocations in the karyotypes of wild boars *Sus scrofa* L. of the European and the Asian areas of USSR. *Theor. Appl. Genet.* **1975**, *45*, 304–308. [CrossRef]
54. Rejduch, B.; Slota, E.; Rozycki, M.; Koscielny, M. Chromosome number polymorphism in a litter of European wild boar (*Sus scrofa scrofa* L.). *Anim. Sci. Pap. Rep.* **2003**, *1*, 57–62.
55. Miyake, Y.-I.; Kawata, K.; Ishikawa, T.; Umezu, M. Translocation heterozygosity in a malformed piglet and its normal littermates. *Teratology* **1977**, *16*, 163–167. [CrossRef]
56. Alonso, R.A.; Cantu, J.M. A Robertsonian translocation in the domestic pig (*Sus scrofa*) 37,XX,-13,-17,t rob(13;17). *Ann. Génétique* **1982**, *25*, 50–52.
57. Schwerin, M.; Golisch, D.; Ritter, E. A Robertsonian translocation in swine. *Genet. Sel. Evol.* **1986**, *18*, 1–7. [CrossRef] [PubMed]
58. Danielak-Czech, B.; Słota, E. A new case of reciprocal translocation t (10;13) (q16;q21) diagnosed in an AI boar. *J. Appl. Genet.* **2007**, *48*, 379–382. [CrossRef]
59. Pinton, A.; Calgaro, A.; Bonnet, N.; Ferchaud, S.; Billoux, S.; Dudez, A.; Mary, N.; Massip, K.; Bonnet-Garnier, A.; Yerle, M.; et al. Influence of sex on the meiotic segregation of a t (13;17) Robertsonian translocation: A case study in the pig. *Hum. Reprod.* **2009**, *24*, 2034–2043. [CrossRef]
60. Switonski, M.; Stranzinger, G. Studies of synaptonemal complexes in farm mammals—A review. *J. Hered.* **1998**, *89*, 473–480. [CrossRef] [PubMed]
61. Villagómez, D.; Pinton, A. Chromosomal abnormalities, meiotic behavior and fertility in domestic animals. *Cytogenet. Genome Res.* **2008**, *120*, 69–80.
62. McFeely, R.A. A direct method for the display of chromosomes from early pig embryos. *Reproduction* **1996**, *11*, 161–163. [CrossRef]
63. Lojda, L. The cytogenetic pattern in pigs with hereditary intersexuality similar to the syndrome of testicular feminization in man. *Acta Vet. Brno* **1975**, *8*, 71–82.
64. Breeuwsma, A.J. A case of XXY sex chromosome constitution in an intersex pig. *Reproduction* **1968**, *16*, 119–120. [CrossRef]
65. Hancock, J.L.; Daker, M.G. Testicular hypoplasia in a boar with abnormal sex chromosome constitution (39 XXY). *Reproduction* **1981**, *61*, 395–397. [CrossRef]
66. Mäkinen, A.; Andersson, M.; Nikunen, S. Detection of the X chromosomes in a Klinefelter boar using a whole human X chromosome painting probe. *Anim. Reprod. Sci.* **1998**, *52*, 317–323. [CrossRef]
67. Ducos, A.; Berland, H.M.; Pinton, A.; Calgaro, A.; Brun-Baronnat, C.; Bonnet, N.; Garnier-Bonnet, A.; Darré, R. Chromosome control of domestic animal populations in France. 16th European Colloquium on Animal Cytogenetics and Gene Mapping. *Cytogenet. Genome Res.* **2004**, *106*, 1–27.
68. Villagómez, D.A.F.; Gustavsson, I.; Jönsson, L.; Plöen, L. Reciprocal Chromosome Translocation, rcp(7;17)(q26;q11), in a Boar Giving Reduced Litter Size and Increased Rate of Piglets Dying in the Early Life. *Hereditas* **2004**, *122*, 257–267. [CrossRef]
69. Rezaei, S.; Donaldson, B.; Villagomez, D.A.F.; Revay, T.; Mary, N.; Grossi, D.A.; King, W.A. Routine Karyotyping Reveals Frequent Mosaic Reciprocal Chromosome Translocations in Swine: Prevalence, Pedigree, and Litter Size. *Sci. Rep.* **2020**, *10*, 1–9. [CrossRef]
70. Musilova, P.; Drbalova, J.; Kubickova, S.; Cernohorska, H.; Stepanova, H.; Rubes, J. Illegitimate recombination between T cell receptor genes in humans and pigs (*Sus scrofa domestica*). *Chromosom. Res.* **2014**, *22*, 483–493. [CrossRef]
71. Quilter, C.R.; Wood, D.; Southwood, O.I.; Griffin, D.K. X/XY/XYY mosaicism as a cause of subfertility in boars: A single case study. *Anim. Genet.* **2003**, *34*, 51–54. [CrossRef]
72. Bruère, A.; Fielden, E.; Hutchings, H. XX/XY mosaicism in lymphocyte cultures from a pig with freemartin characteristics. *N. Z. Vet. J.* **1968**, *16*, 31–38. [CrossRef]
73. Somlev, B.; Hansen-Melander, E.; Melander, Y.; Holm, L. XX/XY chimerism in leucocytes of two intersexual pigs. *Hereditas* **2009**, *64*, 203–210. [CrossRef]
74. Toyama, Y. Sex chromosome mosaicisms in five swine intersexes. *Jpn. J. Zootech. Sci.* **1974**, *45*, 551–557.
75. Christensen, K.; Nielsen, P.B. A case of blood chimerism (XX, XY) in pigs. *Anim. Blood Groups Biochem. Genet.* **1980**, *11*, 55–57. [CrossRef] [PubMed]
76. Clarkson, B.G.; Fisher, K.R.S.; Partlow, G.D. Agonadal presumptive XX/XY leukochimeric pig. *Anat. Rec. Adv. Integr. Anat. Evol. Biol.* **1995**, *242*, 195–199. [CrossRef]
77. Tsai, A.G.; Lieber, M.R. Mechanisms of chromosomal rearrangement in the human genome. *BMC Genom.* **2010**, *11*, 1–9. [CrossRef] [PubMed]
78. Hiraiwa, H.; Uenishi, H.; Kiuchi, S.; Watanabe, M.; Takagaki, Y.; Yasue, H. Assignment of T cell receptor (TCR) alpha-chain gene (A), beta-chain gene (B), gamma-chain gene (G), and delta-chain gene (D) loci on swine chromosomes by in situ hybridization and radiation hybrid mapping. *Cytogenet. Cell Genet.* **2001**, *93*, 94–99. [CrossRef]
79. Riggs, P.; Kuczek, T.; Chrisman, C.; Bidwell, C. Analysis of aphidicolin-induced chromosome fragility in the domestic pig (*Sus scrofa*). *Cytogenet. Genome Res.* **1993**, *62*, 110–116. [CrossRef]
80. Yang, M.; Long, S. Folate sensitive common fragile sites in chromosomes of the domestic pig (*Sus scrofa*). *Res. Vet. Sci.* **1993**, *55*, 231–235. [CrossRef]

81. Rønne, M. Localization of Fragile Sites in the Karyotype of *Sus scrofa domestica*: Present Status. *Hereditas* **2004**, *122*, 153–162. [CrossRef]
82. Riggs, P.; Rønne, M. Fragile Sites in Domestic Animal Chromosomes: Molecular Insights and Challenges. *Cytogenet. Genome Res.* **2009**, *126*, 97–109. [CrossRef]
83. Rubeš, J.; Pinton, A.; Bonnet-Garnier, A.; Fillon, V.; Musilova, P.; Michalova, K.; Kubíčková, S.; Ducos, A.; Yerle, M. Fluorescence in situ Hybridization Applied to Domestic Animal Cytogenetics. *Cytogenet. Genome Res.* **2009**, *126*, 34–48. [CrossRef] [PubMed]
84. Pinkel, D.; Straume, T.; Gray, J.W. Cytogenetic analysis using quantitative, high-sensitivity, fluorescence hybridization. *Proc. Natl. Acad. Sci. USA* **1986**, *83*, 2934–2938. [CrossRef]
85. Mao, S.Y.; Mullins, J.M. Conjugation of fluorochromes to antibodies. In *Immunocytochemical Methods and Protocols*; Humana Press: Totowa, NJ, USA, 2010; pp. 43–48.
86. Fahrenkrug, S.C.; Rohrer, G.A.; Freking, B.A.; Smith, T.P.; Osoegawa, K.; Shu, C.L.; Catanese, J.J.; De Jong, P.J. A porcine BAC library with tenfold genome coverage: A resource for physical and genetic map integration. *Mamm. Genome* **2001**, *12*, 472–474. [CrossRef]
87. Schmitz, A.; Chaput, B.; Fouchet, P.; Guilly, M.N.; Frelat, G.; Vaiman, M. Swine chromosomal DNA quantification by bivariate flow karyotyping and karyotype interpretation. *Cytom. J. Int. Soc. Anal. Cytol.* **1992**, *13*, 703–710. [CrossRef] [PubMed]
88. Telenius, H.; Ponder, B.A.J.; Tunnacliffe, A.; Pelmear, A.H.; Carter, N.P.; Ferguson-Smith, M.A.; Behmel, A.; Nordenskjöld, M.; Pfragner, R. Cytogenetic analysis by chromosome painting using dop-pcr amplified flow-sorted chromosomes. *Genes Chromosom. Cancer* **1992**, *4*, 257–263. [CrossRef]
89. Langford, C.F.; Telenius, H.; Miller, N.G.A.; Thomsen, P.D.; Tucker, E.M. Preparation of chromosome-specific paints and complete assignment of chromosomes in the pig flow karyotype. *Anim. Genet.* **2009**, *24*, 261–267. [CrossRef] [PubMed]
90. Scalenghe, F.; Turco, E.; Edström, J.E.; Pirrotta, V.; Melli, M. Microdissection and cloning of DNA from a specific region of Drosophila melanogaster polytene chromosomes. *Chromosoma* **1981**, *82*, 205–216. [CrossRef] [PubMed]
91. Chaudhary, R.; Kijas, J.; Raudsepp, T.; Guan, X.Y.; Zhang, H.; Chowdhary, B.P. Microdissection of pig chromosomes: Dissection of whole chromosomes, arms and bands for construction of paints and libraries. *Hereditas* **1998**, *128*, 265–271. [CrossRef]
92. Pinton, A.; Ducos, A.; Yerle, M. Chromosomal rearrangements in cattle and pigs revealed by chromosome microdissection and chromosome painting. *Genet. Sel. Evol.* **2003**, *35*, 1–12. [CrossRef] [PubMed]
93. Schermelleh, L.; Thalhammer, S.; Heckl, W.; Pösl, H.; Cremer, T.; Schütze, K.; Cremer, M. Laser Microdissection and Laser Pressure Catapulting for the Generation of Chromosome-Specific Paint Probes. *Biotechniques* **1999**, *27*, 362–367. [CrossRef] [PubMed]
94. Kubickova, S.; Cernohorska, H.; Musilova, P.; Rubes, J. The use of laser microdissection for the preparation of chromosome-specific painting probes in farm animals. *Chromosom. Res.* **2002**, *10*, 571–577. [CrossRef]
95. Ried, T.; Schröck, E.; Ning, Y.; Wienberg, J. Chromosome painting: A useful art. *Hum. Mol. Genet.* **1998**, *7*, 1619–1626. [CrossRef] [PubMed]
96. Koch, J.; Hindkjaer, J.; Mogensen, J.; Kølvraa, S.; Bolund, L. An improved method for chromosome-specific labeling of α satellite DNA in situ by using denatured double-stranded DNA probes as primers in a primed in situ labeling (PRINS) procedure. *Genet. Anal. Biomol. Eng.* **1991**, *8*, 171–178. [CrossRef]
97. Seña, C.D.L.; Chowdhary, B.P.; Gustavsson, I. Localization of the telomeric (TTAGGG) n sequences in chromosomes of some domestic animals by fluorescence in situ hybridization. *Hereditas* **1995**, *123*, 269–274. [CrossRef] [PubMed]
98. Gu, F.; Hindkjaer, J.; Gustavsson, I.; Bolund, L. A signal of telomeric sequences on porcine chromosome 6q21–q22 detected by primed in situ labelling. *Chromosom. Res.* **1996**, *4*, 251–252. [CrossRef]
99. Pellestor, F.; Girardet, A.; Lefort, G.; Andréo, B.; Charlieu, J.P. Use of the primed in situ labelling (PRINS) technique for a rapid detection of chromosomes 13, 16, 18, 21, X and Y. *Hum. Genet.* **1995**, *95*, 12–17. [CrossRef] [PubMed]
100. Miller, J.R.; Hindkjaer, J.; Thomsen, P.D. A chromosomal basis for the differential organization of a porcine centromere-specific repeat. *Cytogenet. Genome Res.* **1993**, *62*, 37–41. [CrossRef] [PubMed]
101. Rogel-Gaillard, C.; Bourgeaux, N.; Save, J.C.; Renard, C.; Coullin, P.; Pinton, P.; Yerle, M.; Vaiman, M.; Chardon, P. Construction of a swine YAC library allowing an efficient recovery of unique and centromeric repeated sequences. *Mamm. Genome* **1997**, *8*, 186–192. [CrossRef]
102. Danielak-Czech, B.; Rejduch, B.; Kozubska-Sobocińska, A. Identification of telomeric sequences in pigs with rearranged karyotype using PRINS technique. *Ann. Anim. Sci.* **2013**, *13*, 495–502. [CrossRef]
103. Goureau, A.; Yerle, M.; Schmitz, A.; Riquet, J.; Milan, D.; Pinton, P.; Frelat, G.; Gellin, J. Human and porcine correspondence of chromosome segments using bidirectional chromosome painting. *Genomics* **1996**, *36*, 252–262. [CrossRef] [PubMed]
104. Chowdhary, B.P.; Raudsepp, T.; Frönicke, L.; Scherthan, H. Emerging patterns of comparative genome organization in some mammalian species as revealed by Zoo-FISH. *Genome Res.* **1998**, *8*, 577–589. [CrossRef]
105. Sarrate, Z.; Anton, E. Fluorescence in situ hybridization (FISH) protocol in human sperm. *J. Vis. Exp. JoVE* **2009**, *31*, 1405. [CrossRef]
106. Konfortova, G.; Miller, N.; Tucker, E. A new reciprocal translocation (7q+;15q−) in the domestic pig. *Cytogenet. Genome Res.* **1995**, *71*, 285–288. [CrossRef]
107. Pinton, A.; Faraut, T.; Yerle, M.; Gruand, J.; Pellestor, F.; Ducos, A. Comparison of male and female meiotic segregation patterns in translocation heterozygotes: A case study in an animal model (*Sus scrofa domestica* L.). *Hum. Reprod.* **2005**, *20*, 2476–2482. [CrossRef]

108. Pinton, A.; Ducos, A.; Séguéla, A.; Berland, H.M.; Darré, R.; Darré, A.; Pinton, P.; Schmitz, A.; Cribiu, E.P.; Yerle, M. Characterization of reciprocal translocations in pigs using dual-colour chromosome painting and primed in situ DNA labelling. *Chromosome Res.* **1998**, *6*, 361–366. [CrossRef]
109. Pinton, A.; Pailhoux, E.; Piumi, F.; Rogel-Gaillard, C.; Darré, R.; Yerle, M.; Ducos, A.; Cotinot, C. A case of intersexuality in pigs associated with a de novo paracentric inversion 9 (p1. 2; p2. 2). *Anim. Genet.* **2002**, *33*, 69–71. [CrossRef]
110. O'Connor, R.E.; Fonseka, G.; Frodsham, R.; Archibald, A.L.; Lawrie, M.; Walling, G.A.; Griffin, D.K. Isolation of subtelomeric sequences of porcine chromosomes for translocation screening reveals errors in the pig genome assembly. *Anim. Genet.* **2017**, *48*, 395–403. [CrossRef]
111. Barasc, H.; Mary, N.; Letron, R.; Calgaro, A.; Dudez, A.; Bonnet, N.; Lahbib-Mansais, Y.; Yerle, M.; Ducos, A.; Pinton, A. Y-Autosome Translocation Interferes with Meiotic Sex Inactivation and Expression of Autosomal Genes: A Case Study in the Pig. *Sex. Dev.* **2012**, *6*, 143–150. [CrossRef] [PubMed]
112. Mary, N.; Barasc, H.; Ferchaud, S.; Billon, Y.; Meslier, F.; Robelin, D.; Calgaro, A.; Loustau-Dudez, A.M.; Bonnet, N.; Yerle, M.; et al. Meiotic recombination analyses of individual chromosomes in male domestic pigs (*Sus scrofa domestica*). *PLoS ONE* **2014**, *9*, e99123. [CrossRef] [PubMed]
113. Pinton, A.; Letron, I.R.; Berland, H.; Bonnet, N.; Calgaro, A.; Garnier-Bonnet, A.; Yerle, M.; Ducos, A. Meiotic studies in an azoospermic boar carrying a Y;14 translocation. *Cytogenet. Genome Res.* **2008**, *120*, 106–111. [CrossRef] [PubMed]
114. Bonnet-Garnier, A.; Guardia, S.; Pinton, A.; Ducos, A.; Yerle, M. Analysis using sperm-FISH of a putative interchromosomal effect in boars carrying reciprocal translocations. *Cytogenet. Genome Res.* **2009**, *126*, 194–201. [CrossRef]
115. Massip, K.; Bonnet, N.; Calgaro, A.; Billoux, S.; Baquié, V.; Mary, N.; Bonnet-Garnier, A.; Ducos, A.; Yerle, M.; Pinton, A. Male Meiotic Segregation Analyses of Peri- and Paracentric Inversions in the Pig Species. *Cytogenet. Genome Res.* **2009**, *125*, 117–124. [CrossRef]
116. Massip, K.; Yerle, M.; Billon, Y.; Ferchaud, S.; Bonnet, N.; Calgaro, A.; Mary, N.; Dudez, A.-M.; Sentenac, C.; Plard, C.; et al. Studies of male and female meiosis in inv (4) (p1.4; q2.3) pig carriers. *Chromosom. Res.* **2010**, *18*, 925–938. [CrossRef] [PubMed]
117. Danielak-Czech, B.; Kozubska-Sobocińska, A.; Rejduch, B. Diagnosis of tandem fusion translocation in the boar using FISH technique with human painting probes. *Ann. Anim. Sci.* **2010**, *10*, 361–366.
118. Rejduch, B.; Slota, E.; Sysa, P.; Koscielny, M.; Wrzeska, M.; Babicz, M. Synaptonemal complexes analysis of the European wild boars û carriers of the 15; 17 Robertsonian translocation. *Rocz. Nauk. Zootech.* **2003**, *3*, 255–262.
119. Kawarasaki, T.; Matsumoto, K.; ChiKyu, M.; Itagaki, Y.; Horiuchi, A.; Murofushi, J. Sexing of porcine embryo by in situ hybridization using chromosome Y- and 1-specific DNA probes. *Theriogenology* **2000**, *53*, 1501–1509. [CrossRef]
120. Parrilla, I.; Vázquez, J.M.; Oliver-Bonet, M.; Navarro, J.; Yelamos, J.; Roca, J.; Martínez, E.A. Fluorescence in situ hybridization in diluted and flow cytometrically sorted boar spermatozoa using specific DNA direct probes labelled by nick translation. *Reprod. Camb.* **2003**, *126*, 317–325. [CrossRef]
121. Rubeš, J.; Vozdova, M.; Kubíčková, S. Aneuploidy in pig sperm: Multicolor fluorescence in situ hybridization using probes for chromosomes 1, 10, and Y. *Cytogenet. Genome Res.* **1999**, *85*, 200–204. [CrossRef]
122. Pinton, A.; Ducos, A.; Yerle, M. Estimation of the proportion of genetically unbalanced spermatozoa in the semen of boars carrying chromosomal rearrangements using FISH on sperm nuclei. *Genet. Sel. Evol.* **2004**, *36*, 1–15. [CrossRef] [PubMed]
123. Massip, K.; Berland, H.; Bonnet, N.; Calgaro, A.; Billoux, S.; Baquié, V.; Mary, N.; Bonnet-Garnier, A.; Ducos, A.; Yerle, M.; et al. Study of inter- and intra-individual variation of meiotic segregation patterns in t (3;15) (q27;q13) boars. *Theriogenology* **2008**, *70*, 655–661. [CrossRef] [PubMed]
124. Russo, C.D.; Di Giacomo, G.; Cignini, P.; Padula, F.; Mangiafico, L.; Mesoraca, A.; D'Emidio, L.; McCluskey, M.R.; Paganelli, A.; Giorlandino, C. Comparative study of aCGH and Next Generation Sequencing (NGS) for chromosomal microdeletion and microduplication screening. *J. Prenat. Med.* **2014**, *8*, 57.
125. Crosetto, N.; Mitra, A.; Silva, M.J.; Bienko, M.; Dojer, N.; Wang, Q.; Karaca, E.; Chiarle, R.; Skrzypczak, M.; Ginalski, K.; et al. Nucleotide-resolution DNA double-strand break mapping by next-generation sequencing. *Nat. Methods* **2013**, *10*, 361–365. [CrossRef] [PubMed]
126. Warr, A.; Affara, N.; Aken, B.; Beiki, H.; Bickhart, D.M.; Billis, K.; Chow, W.; Eory, L.; Finlayson, H.A.; Flicek, P.; et al. An improved pig reference genome sequence to enable pig genetics and genomics research. *GigaScience* **2020**, *9*, giaa051. [CrossRef] [PubMed]
127. Servin, B.; Faraut, T.; Iannuccelli, N.; Zelenika, D.; Milan, D. High-Resolution autosomal radiation hybrid maps of the pig genome and their contribution to the genome sequence assembly. *BMC Genom.* **2012**, *13*, 1–12. [CrossRef] [PubMed]
128. Tortereau, F.; Servin, B.; Frantz, L.; Megens, H.J.; Milan, D.; Rohrer, G.; Wiedmann, R.; Beever, J.; Archibald, A.L.; Schook, L.B.; et al. A high density recombination map of the pig reveals a correlation between sex-specific recombination and GC content. *BMC Genom.* **2012**, *13*, 1–12. [CrossRef] [PubMed]
129. Yerle, M.; Lahbib-Mansais, Y.; Mellink, C.; Goureau, A.; Pinton, P.; Echard, G.; Gellin, J.; Zijlstra, C.; De Haan, N.; Bosma, A.A.; et al. The PiGMaP consortium cytogenetic map of the domestic pig (*Sus scrofa domestica*). *Mamm. Genome* **1995**, *6*, 176–186. [CrossRef]
130. Humphray, S.J.; Scott, C.E.; Clark, R.; Marron, B.; Bender, C.; Camm, N.; Davis, J.; Jenks, A.; Noon, A.; Patel, M.; et al. A high utility integrated map of the pig genome. *Genome Biol.* **2007**, *8*, 1–11. [CrossRef]
131. Groenen, M.A.; Archibald, A.L.; Uenishi, H.; Tuggle, C.K.; Takeuchi, Y.; Rothschild, M.F.; Schook, L.B. Analyses of pig genomes provide insight into porcine demography and evolution. *Nature* **2012**, *491*, 393–398. [CrossRef]

132. Ramos, A.M.; Crooijmans, R.P.M.A.; Affara, N.A.; Amaral, A.J.; Archibald, A.L.; Beever, J.E.; Bendixen, C.; Churcher, C.; Clark, R.; Dehais, P.; et al. Design of a High Density SNP Genotyping Assay in the Pig Using SNPs Identified and Characterized by Next Generation Sequencing Technology. *PLoS ONE* **2009**, *4*, e06524. [CrossRef]
133. Hu, Z.-L.; Park, C.A.; Reecy, J.M. Developmental progress and current status of the Animal QTLdb. *Nucleic Acids Res.* **2016**, *44*, D827–D833. [CrossRef]
134. Bumgarner, R. Overview of DNA microarrays: Types, applications, and their future. *Curr. Protoc. Mol. Biol.* **2013**, *101*, 22.
135. Meuwissen, T.; Hayes, B.; Goddard, M. Accelerating Improvement of Livestock with Genomic Selection. *Annu. Rev. Anim. Biosci.* **2013**, *1*, 221–237. [CrossRef]
136. Ibáñez-Escriche, N.; Forni, S.; Noguera, J.L.; Varona, L. Genomic information in pig breeding: Science meets industry needs. *Livest. Sci.* **2014**, *166*, 94–100. [CrossRef]
137. Meuwissen, T.H.; Hayes, B.J.; Goddard, M.E. Prediction of total genetic value using genome-wide dense marker maps. *Genetics* **2001**, *157*, 1819–1829.
138. Robinson, J.; Buhr, M. Impact of genetic selection on management of boar replacement. *Theriogenology* **2005**, *63*, 668–678. [CrossRef]
139. Christensen, O.; Madsen, P.; Nielsen, B.; Ostersen, T.; Su, G. Single-step methods for genomic evaluation in pigs. *Animal* **2012**, *6*, 1565–1571. [CrossRef] [PubMed]
140. Lee, Y.-S.; Shin, D. Genome-Wide Association Studies Associated with Backfat Thickness in Landrace and Yorkshire Pigs. *Genom. Inform.* **2018**, *16*, 59–64. [CrossRef]
141. Verardo, L.L.; Sevón-Aimonen, M.-L.; Serenius, T.; Hietakangas, V.; Uimari, P. Whole-genome association analysis of pork meat pH revealed three significant regions and several potential genes in Finnish Yorkshire pigs. *BMC Genet.* **2017**, *18*, 13. [CrossRef] [PubMed]
142. Wang, Y.; Ding, X.; Tan, Z.; Ning, C.; Xing, K.; Yang, T.; Pan, Y.; Sun, D.; Wang, C. Genome-Wide Association Study of Piglet Uniformity and Farrowing Interval. *Front. Genet.* **2017**, *8*, 194. [CrossRef]
143. Apiou, F.; Vincent-Naulleau, S.; Spatz, A.; Vielh, P.; Geffrotin, C.; Frelat, G.; Dutrillaux, B.; Le Chalony, C. Comparative genomic hybridization analysis of hereditary swine cutaneous melanoma revealed loss of the swine 13q36-49 chromosomal region in the nodular melanoma subtype. *Int. J. Cancer* **2004**, *110*, 232–238. [CrossRef] [PubMed]
144. Horňák, M.; Hulinska, P.; Musilova, P.; Kubíčková, S.; Rubeš, J. Investigation of Chromosome Aneuploidies in Early Porcine Embryos Using Comparative Genomic Hybridization. *Cytogenet. Genome Res.* **2009**, *126*, 210–216. [CrossRef] [PubMed]
145. Shinawi, M.; Cheung, S.W. The array CGH and its clinical applications. *Drug Discov. Today* **2008**, *13*, 760–770. [CrossRef] [PubMed]
146. Redon, R.; Ishikawa, S.; Fitch, K.R.; Feuk, L.; Perry, G.H.; Andrews, T.D.; Fiegler, H.; Shapero, M.H.; Carson, A.R.; Chen, W.; et al. Global variation in copy number in the human genome. *Nature* **2006**, *444*, 444–454. [CrossRef]
147. Fadista, J.; Nygaard, M.; Holm, L.-E.; Thomsen, B.; Bendixen, C. A Snapshot of CNVs in the Pig Genome. *PLoS ONE* **2008**, *3*, e03916. [CrossRef] [PubMed]
148. Nilsson, D.; Pettersson, M.; Gustavsson, P.; Förster, A.; Hofmeister, W.; Wincent, J.; Zachariadis, V.; Anderlid, B.-M.; Nordgren, A.; Mäkitie, O.; et al. Whole-Genome Sequencing of Cytogenetically Balanced Chromosome Translocations Identifies Potentially Pathological Gene Disruptions and Highlights the Importance of Microhomology in the Mechanism of Formation. *Hum. Mutat.* **2017**, *38*, 180–192. [CrossRef] [PubMed]
149. Li, M.; Chen, L.; Tian, S.; Lin, Y.; Tang, Q.; Zhou, X.; Li, D.; Yeung, C.K.L.; Che, T.; Li, X.; et al. Comprehensive variation discovery and recovery of missing sequence in the pig genome using multiple de novo assemblies. *Genome Res.* **2017**, *27*, 865–874. [CrossRef] [PubMed]
150. Frantz, L.A.F.; Schraiber, J.G.; Madsen, O.D.; Megens, H.-J.; Cagan, A.; Bosse, M.; Paudel, Y.; Crooijmans, R.P.M.A.; Larson, G.; Groenen, M.A.M. Evidence of long-term gene flow and selection during domestication from analyses of Eurasian wild and domestic pig genomes. *Nat. Genet.* **2015**, *47*, 1141–1148. [CrossRef]
151. Groenen, M.A.M. A decade of pig genome sequencing: A window on pig domestication and evolution. *Genet. Sel. Evol.* **2016**, *48*, 1–9. [CrossRef]
152. Liu, B.; Conroy, J.M.; Morrison, C.D.; Odunsi, A.O.; Qin, M.; Wei, L.; Trump, D.L.; Johnson, C.S.; Liu, S.; Wang, J. Structural variation discovery in the cancer genome using next generation sequencing: Computational solutions and perspectives. *Oncotarget* **2015**, *6*, 5477–5489. [CrossRef]
153. Ho, S.S.; Urban, A.E.; Mills, R.E. Structural variation in the sequencing era. *Nat. Rev. Genet.* **2019**, *21*, 171–189. [CrossRef] [PubMed]
154. Pollard, M.O.; Gurdasani, D.; Mentzer, A.J.; Porter, T.; Sandhu, M.S. Long reads: Their purpose and place. *Hum. Mol. Genet.* **2018**, *27*, R234–R241. [CrossRef]
155. Chow, J.F.; Cheng, H.H.; Lau, E.Y.; Yeung, W.S.; Ng, E.H. Distinguishing between carrier and noncarrier embryos with the use of long-read sequencing in preimplantation genetic testing for reciprocal translocations. *Genomics* **2020**, *112*, 494–500. [CrossRef] [PubMed]
156. Hu, L.; Liang, F.; Cheng, D.; Zhang, Z.; Yu, G.; Zha, J.; Wang, Y.; Xia, Q.; Yuan, D.; Tan, Y.; et al. Location of Balanced Chromosome-Translocation Breakpoints by Long-Read Sequencing on the Oxford Nanopore Platform. *Front. Genet.* **2020**, *10*, 1313. [CrossRef]
157. Goodwin, S.; McPherson, J.D.; McCombie, W.R. Coming of age: Ten years of next-generation sequencing technologies. *Nat. Rev. Genet.* **2016**, *17*, 333. [CrossRef]

158. Dong, Z.; Jiang, L.; Yang, C.; Hu, H.; Wang, X.; Chen, H.; Choy, K.W.; Hu, H.; Dong, Y.; Hu, B.; et al. A Robust Approach for Blind Detection of Balanced Chromosomal Rearrangements with Whole-Genome Low-Coverage Sequencing. *Hum. Mutat.* **2014**, *35*, 625–636. [CrossRef] [PubMed]
159. Redin, C.; Brand, H.; Collins, R.L.; Kammin, T.; Mitchell, E.; Hodge, J.C.; Hanscom, C.; Pillamarri, V.; Seabra, C.M.; Abbott, M.-A.; et al. The genomic landscape of balanced cytogenetic abnormalities associated with human congenital anomalies. *Nat. Genet.* **2017**, *49*, 36–45. [CrossRef]
160. Talkowski, M.E.; Ernst, C.; Heilbut, A.; Chiang, C.; Hanscom, C.; Lindgren, A.; Kirby, A.; Liu, S.; Muddukrishna, B.; Ohsumi, T.K.; et al. Next-Generation Sequencing Strategies Enable Routine Detection of Balanced Chromosome Rearrangements for Clinical Diagnostics and Genetic Research. *Am. J. Hum. Genet.* **2011**, *88*, 469–481. [CrossRef] [PubMed]
161. Zheng, G.X.Y.; Lau, B.T.; Schnall-Levin, M.; Jarosz, M.; Bell, J.M.; Hindson, C.M.; Kyriazopoulou-Panagiotopoulou, S.; Masquelier, D.A.; Merrill, L.; Terry, J.M.; et al. Haplotyping germline and cancer genomes with high-throughput linked-read sequencing. *Nat. Biotechnol.* **2016**, *34*, 303–311. [CrossRef] [PubMed]
162. Uguen, K.; Jubin, C.; Duffourd, Y.; Bardel, C.; Malan, V.; Dupont, J.M.; Khattabi, L.E.; Chatron, N.; Vitobello, A.; Sanlaville, D.; et al. Genome sequencing in cytogenetics: Comparison of short-read and linked-read approaches for germline structural variant detection and characterization. *Mol. Genet. Genom. Med.* **2020**, *8*, e1114. [CrossRef]
163. Gross, A.M.; Ajay, S.S.; Rajan, V.; Brown, C.; Bluske, K.; Burns, N.J.; Chawla, A.; Coffey, A.J.; Malhotra, A.; Scocchia, A.; et al. Copy-Number variants in clinical genome sequencing: Deployment and interpretation for rare and undiagnosed disease. *Genet. Med.* **2019**, *21*, 1121–1130. [CrossRef] [PubMed]
164. Trost, B.; Walker, S.; Wang, Z.; Thiruvahindrapuram, B.; MacDonald, J.R.; Sung, W.W.; Pereira, S.L.; Whitney, J.; Chan, A.J.S.; Scherer, S.W.; et al. A comprehensive workflow for read depth-based identification of copy-number variation from whole-genome sequence data. *Am. J. Hum. Genet.* **2018**, *102*, 142–155. [CrossRef]
165. Ellingford, J.M.; Campbell, C.; Barton, S.; Bhaskar, S.; Gupta, S.; Taylor, R.L.; Sergouniotis, P.I.; Horn, B.; Lamb, J.A.; Michaelides, M.; et al. Validation of copy number variation analysis for next-generation sequencing diagnostics. *Eur. J. Hum. Genet.* **2017**, *25*, 719–724. [CrossRef]
166. Bramswig, N.C.; Lüdecke, H.J.; Pettersson, M.; Albrecht, B.; Bernier, R.A.; Cremer, K.; Eichler, E.E.; Falkenstein, D.; Gerdts, J.; Wieczorek, D.; et al. Identification of new TRIP12 variants and detailed clinical evaluation of individuals with non-syndromic intellectual disability with or without autism. *Hum. Genet.* **2017**, *136*, 179–192. [CrossRef]
167. Grigelioniene, G.; Nevalainen, P.I.; Reyes, M.; Thiele, S.; Tafaj, O.; Molinaro, A.; Takatani, R.; Ala-Houhala, M.; Nilsson, D.; Jüppner, H.; et al. A large inversion involving GNAS exon A/B and all exons encoding Gsα is associated with autosomal dominant pseudohypoparathyroidism type Ib (PHP1B). *J. Bone Miner. Res.* **2017**, *32*, 776–783. [CrossRef]
168. Hochstenbach, R.; Liehr, T.; Hastings, R.J. Chromosomes in the genomic age. Preserving cytogenomic competence of diagnostic genome laboratories. *Eur. J. Hum. Genet.* **2021**, *29*, 541–552. [CrossRef] [PubMed]
169. Nowakowska, B.A.; De Leeuw, N.; Al Ruivenkamp, C.; Sikkema-Raddatz, B.; Crolla, J.A.; Thoelen, R.; Koopmans, M.; Hollander, N.D.; Van Haeringen, A.; Van Der Kevie-Kersemaekers, A.-M.; et al. Parental insertional balanced translocations are an important cause of apparently de novo CNVs in patients with developmental anomalies. *Eur. J. Hum. Genet.* **2011**, *20*, 166–170. [CrossRef] [PubMed]
170. Silva, M.; De Leeuw, N.; Mann, K.; Schuring-Blom, H.; Morgan, S.; Giardino, D.; Rack, K.; Hastings, R. European guidelines for constitutional cytogenomic analysis. *Eur. J. Hum. Genet.* **2018**, *27*, 1–16. [CrossRef] [PubMed]
171. Treff, N.R.; Tao, X.; Schillings, W.J.; Bergh, P.A.; Scott, R.T., Jr.; Levy, B. Use of single nucleotide polymorphism microarrays to distinguish between balanced and normal chromosomes in embryos from a translocation carrier. *Fertil. Steril.* **2011**, *96*, e58–e65. [CrossRef]
172. Zhang, F.; Gu, W.; Hurles, M.E.; Lupski, J.R. Copy Number Variation in Human Health, Disease, and Evolution. *Annu. Rev. Genom. Hum. Genet.* **2009**, *10*, 451–481. [CrossRef]
173. Wang, L.; Xu, L.; Liu, X.; Zhang, T.; Li, N.; Hay, E.H.; Zhang, Y.; Yan, H.; Zhao, K.; Liu, G.E.; et al. Copy number variation-based genome wide association study reveals additional variants contributing to meat quality in Swine. *Sci. Rep.* **2015**, *5*, 12535. [CrossRef]
174. Revay, T.; Quach, A.T.; Maignel, L.; Sullivan, B.; King, W.A. Copy number variations in high and low fertility breeding boars. *BMC Genom.* **2015**, *16*, 280. [CrossRef]
175. Hay, E.H.A.; Choi, I.; Xu, L.; Zhou, Y.; Rowland, R.R.R.; Lunney, J.K.; Liu, G.E. CNV Analysis of Host Responses to Porcine Reproductive and Respiratory Syndrome Virus Infection. *J. Genom.* **2017**, *5*, 58–63. [CrossRef]
176. Handyside, A.H.; Harton, G.L.; Mariani, B.; Thornhill, A.R.; Affara, N.; Shaw, M.-A.; Griffin, D.K. Karyomapping: A universal method for genome wide analysis of genetic disease based on mapping crossovers between parental haplotypes. *J. Med. Genet.* **2009**, *47*, 651–658. [CrossRef]
177. Luukkonen, T.M.; Mehrjouy, M.M.; Pöyhönen, M.; Anttonen, A.K.; Lahermo, P.; Ellonen, P.; Paulin, L.; Tommerup, N.; Palotie, A.; Varilo, T. Breakpoint mapping and haplotype analysis of translocation t (1; 12) (q43; q21. 1) in two apparently independent families with vascular phenotypes. *Mol. Genet. Genom. Med.* **2018**, *6*, 56–68. [CrossRef] [PubMed]
178. Bouwman, A.C.; Derks, M.F.L.; Broekhuijse, M.L.W.J.; Harlizius, B.; Veerkamp, R.F. Using short read sequencing to characterise balanced reciprocal translocations in pigs. *BMC Genom.* **2020**, *21*, 1–14. [CrossRef] [PubMed]

179. Donaldson, B. Reciprocal Chromosome Translocations in the Domestic Pig, the Prevalence, Genetic and Genomic Factors Associated with Breakpoint Formation. Ph.D. Thesis, University of Guelph, Guelph, ON, Canada, 2020.
180. Wetterstrand, K.A. DNA Sequencing Costs: Data from the NHGRI Genome Sequencing Program (GSP). 2020. Available online: www.genome.gov/sequencingcostsdata (accessed on 25 February 2021).
181. Madan, K.; Ford, C.E.; Polge, C. A reciprocal translocation, t (6p+; 14q−), in the pig. *Reproduction* **1978**, *53*, 395–398. [CrossRef] [PubMed]

Article

Anchoring the CerEla1.0 Genome Assembly to Red Deer (*Cervus elaphus*) and Cattle (*Bos taurus*) Chromosomes and Specification of Evolutionary Chromosome Rearrangements in Cervidae

Miluse Vozdova *, Svatava Kubickova, Halina Cernohorska, Jan Fröhlich and Jiri Rubes

Department of Genetics and Reproductive Biotechnologies, Central European Institute of Technology—Veterinary Research Institute, 62100 Brno, Czech Republic; kubickova@vri.cz (S.K.); cernohorska@vri.cz (H.C.); frohlich@vri.cz (J.F.); rubes@vri.cz (J.R.)
* Correspondence: vozdova@vri.cz; Tel.: +42-05-3333-1422

Citation: Vozdova, M.; Kubickova, S.; Cernohorska, H.; Fröhlich, J.; Rubes, J. Anchoring the CerEla1.0 Genome Assembly to Red Deer (*Cervus elaphus*) and Cattle (*Bos taurus*) Chromosomes and Specification of Evolutionary Chromosome Rearrangements in Cervidae. *Animals* **2021**, *11*, 2614. https://doi.org/10.3390/ani11092614

Academic Editors: Leopoldo Iannuzzi, Pietro Parma and Cristina Sartori

Received: 30 March 2021
Accepted: 2 September 2021
Published: 6 September 2021

Publisher's Note: MDPI stays neutral with regard to jurisdictional claims in published maps and institutional affiliations.

Copyright: © 2021 by the authors. Licensee MDPI, Basel, Switzerland. This article is an open access article distributed under the terms and conditions of the Creative Commons Attribution (CC BY) license (https:// creativecommons.org/licenses/by/ 4.0/).

Simple Summary: The red deer (*Cervus elaphus*) de novo genome assembly (CerEla1.0) has provided a great resource for genetic studies in various deer species. In this study, we used gene order comparisons between *C. elaphus* CerEla1.0 and *B. taurus* ARS-UCD1.2 genome assemblies and fluorescence in situ hybridization (FISH) with bovine BAC probes to verify the red deer-bovine chromosome relationships and anchor the CerEla1.0 C-scaffolds to karyotypes of both species. We showed the homology between bovine and deer chromosomes and determined the centromere-telomere orientation of the CerEla1.0 C-scaffolds. Using a set of BAC probes, we were able to narrow the positions of evolutionary chromosome breakpoints defining the family Cervidae. In addition, we revealed several errors in the current CerEla1.0 genome assembly. Finally, we expanded our analysis to other Cervidae and confirmed the locations of the cervid evolutionary fissions and orientation of the fused chromosomes in eight cervid species. Our results can serve as a basis for necessary improvements of the red deer genome assembly and provide support to other genetic studies in Cervidae.

Abstract: The family Cervidae groups a range of species with an increasing economic significance. Their karyotypes share 35 evolutionary conserved chromosomal segments with cattle (*Bos taurus*). Recent publication of the annotated red deer (*Cervus elaphus*) whole genome assembly (CerEla1.0) has provided a basis for advanced genetic studies. In this study, we compared the red deer CerEla1.0 and bovine ARS-UCD1.2 genome assembly and used fluorescence in situ hybridization with bovine BAC probes to verify the homology between bovine and deer chromosomes, determined the centromere-telomere orientation of the CerEla1.0 C-scaffolds and specified positions of the cervid evolutionary chromosome breakpoints. In addition, we revealed several incongruences between the current deer and bovine genome assemblies that were shown to be caused by errors in the CerEla1.0 assembly. Finally, we verified the centromere-to-centromere orientation of evolutionarily fused chromosomes in seven additional deer species, giving a support to previous studies on their chromosome evolution.

Keywords: BAC mapping; comparative cytogenetics; chromosome fission; chromosome fusion; FISH; genome assembly; karyotype

1. Introduction

The family Cervidae (Ruminantia) groups more than fifty extant deer species, including species with growing economic importance. Deer species can be divided into three subfamilies: Cervinae, Capreolinae and Hydropotinae [1] and show a great karyotype diversity reflecting chromosome evolution of the taxon. The diploid chromosome numbers range from 2n = 6 in the female Indian muntjac (*Muntiacus vaginalis*) to 2n = 70 in several species of Caprolinae [2–4]. The 2n = 70 karyotypes of *Hydropotes inermis* and *Mazama gouzoubira*, involving 68 acrocentric autosomes, an acrocentric X and a small submetacentric Y, most probably represent an ancestral cervid karyotype [4] which evolved from the hypothetical ancestral pecoran karyotype (2n = 58) by six chromosome fissions [5].

Comparative cytogenetic studies revealing interspecies chromosome homologies and tracking of evolutionary karyotype rearrangements have been still scarce in Cervidae, with the exception of Muntjacini. The published studies were based mostly on standard banding methods [6–8] or on fluorescence in situ hybridisation (FISH) using whole chromosome painting probes [5,9–13]. The known data show that the most common mechanism of karyotype evolution in Cervidae is represented by Robertsonian (centric) fusions [4,7], whereas tandem fusions were described as the major evolutionary karyotype shaping factor in Muntiacini [9,11]. On the other hand, fissions of several ancestral pecoran chromosomes conserved in *Bos taurus* (BTA, 2n = 60) as BTA1, 2, 5, 6, 8, 9 and intrachromosomal rearrangements of the BTA1 orthologue and the X chromosome were also detected in Cervidae using bovine BAC (Bacterial Artificial Chromosome) probes [13,14].

However, the recent rapid development of high throughput molecular methods, namely whole genome sequencing, has brought new resources for comparative phylogenetic studies. At the level of chromosomes and their parts, an analysis of the next generation sequencing data can enable a precise determination of evolutionary chromosome breakpoints and allow a detection of small or intrachromosomal rearrangements that cannot be visualized by conventional cytogenetics or FISH with whole chromosome painting probes. In ruminants, cryptic interspecies chromosome differences as small as 3.3 Mb were identified in cattle and sheep using an in silico comparative bioinformatic approach [15]. This indicates that the use of sensitive methods can bring interesting discoveries even in seemingly well-described taxa.

Unfortunately, this approach is only limited to species with completely sequenced and well-assembled genomes. Regarding Cervidae, whole genome assembly divided to chromosome-scale scaffolds (C-scaffolds) and including basic gene annotation is available only for the red deer (*Cervus elaphus*, CEL, 2n = 68) [16]. The CerEla 1.0 assembly available in the NCBI database has a total length of 3438.62 Mb and a total ungapped length 1960.83 Mb. It includes 406,637 contigs, 11,479 scaffolds and 35 chromosome-scale scaffolds (C-scaffolds) (https://www.ncbi.nlm.nih.gov/assembly/GCA_002197005.1/#/st accessed on 5 November 2020). The C-scaffolds in the CerEla1.0 genome assembly currently available in the NCBI database are arranged in accordance with the red deer genetic linkage map [17]. As a result, their order does not comply with the physical chromosome length and the chromosome order and centromere-telomere orientation in the red-deer karyotype.

Generally, the use of other methods, i.e. BAC FISH mapping, is recommended to verify the newly established genome assemblies and physically anchor them to to chromosomes, thus upgrading them to a chromosome level [18–21]. In this study, we used comparisons with cattle (*B. taurus*), a closely related species used as model for comparative studies among Cetartiodactyla, with a range of available BACs and, above all, a well established whole genome sequence that served as a reference sequence for the CerEla1.0 assembly establishment [16]. We paired the 34 deer chromosome-scale scaffolds of the *C. elaphus* (CerEla1.0) genome assembly with bovine chromosomes by comparison of the gene annotation of the *C. elaphus* (CerEla1.0) and *B. taurus* (ARS-UCD1.2) assemblies available in the NCBI database. We selected bovine BACs for a construction of FISH probes that we used to anchor the CerEla1.0 C-scaffolds to *C. elaphus* karyotype, to compare the centromere-telomere orientation of the deer and bovine chromosomes and to analyse cervid evolutionary chromosome rearrangements. Using this approach, we revealed and corrected several incongruences between the CerEla1.0 and ARS-UCD1.2 genome assemblies, specified the orientation of the *C. elaphus* C-scaffolds and adjusted the predicted positions of evolutionary breakpoints characteristic for the cervid lineage. Using BAC-FISH mapping, we verified the breakpoints positions in a total of eight karyotypically different cervid species from subfamilies Cervinae and Capreolinae and specified the centromere-telomere orientation of their evolutionarily rearranged chromosomes.

2. Materials and Methods

2.1. Samples and Karyotype Analysis

Samples of whole peripheral blood of cattle (*Bos taurus*) and eight deer species including the red deer (*C. elaphus*) were obtained from captive born animals held in the Prague zoological garden and/or in deer enclosures in Bila Lhota and Frycovice (Czech Republic). The analysed species are listed in Table 1. Taxonomic nomenclature published by Groves and Grubb (2011) was used in this study [22].

Table 1. List of analysed species.

Species	Latin Name	Abbrev.	2n	FNa	Bia	X	Fused BTA Orthologues
Red deer	*Cervus elaphus*	CEL	68	68	2	A	17/19
White-lipped deer	*Cervus albirostris*	CAL	66	68	4	A	17/19, 25/6prox
Rusa deer	*Rusa timorensis*	RTI	60	68	10	A	17/19, 5prox/22, 2dist/7, 5dist/8prox, 5prox/22, 18/3
Eld's deer	*Rucervus eldii*	REL	58	68	12	A	17/19, 2dist/7, 5dist/8prox, 5prox/10, 18/1prox, 22/1dist
Roe deer	*Capreolus capreolus*	CCA	70	68	0	B	
Reindeer	*Rangifer tarandus*	RTA	70	70	2	B	
Moose	*Alces alces*	AAL	68	70	4	B	29/17
White-tailed deer	*Odocoileus virginianus*	OVI	70	70	2	B	

2n—diploid number; FNa—fundamental number of autosomal arms; Bia—number of bi-armed autosomes; BTA—Bos Taurus; A—acrocentric; B—bi-armed. The evolutionary chromosome fusions were detected previously using bovine whole chromosome painting probes [13].

Peripheral blood lymphocytes were cultured, harvested and fixed according to the previously described protocols [23]. Metaphase chromosome spreads for the karyotype and FISH analysis were prepared according to the procedures described previously [24]. GTG-banded karyotypes of *B. taurus* and *C. elaphus* were prepared using the standard trypsin/Giemsa method [25]. The karyotype of *C. elaphus* was arranged in accordance with the previously published deer karyotypes [13,26].

2.2. Chromosome Orthology and Breakpoint Site Prediction

Orthology between the red deer and bovine chromosomes was assessed by a comparison of *B. taurus* ARS-UCD1.2 and *C. elaphus hippelaphus* CerEla1.0 annotated genome assemblies available in the NCBI database (Accessed on 15 May 2020). Predicted locations of protein coding genes in the CerEla1.0 genome assembly (https://www.ncbi.nlm.nih.gov/genome/browse/#!/proteins/10790/321837%7CCervus%20elaphus%20hippelaphus/ accessed on 15 May 2020) were compared with positions of the corresponding genes in the bovine genome (https://www.ncbi.nlm.nih.gov/gene/advanced accessed on 15 May 2020). Briefly, we selected predicted protein coding genes separated by a distance of approximately 5 Mb along the length of the *C. elaphus* CerEla1.0 C-scaffolds and searched for their positions in the bovine ARS-UCD1.2 genome assembly. The 5 Mb distance was chosen to enable a reliable distinguishing of the mutual positions of BAC probes mapping to these regions when any incongruences would need to be solved by a dual colour BAC-FISH. To specify the breakpoints of the evolutionary chromosome fissions of ancestral chromosomes corresponding to bovine BTA1, 2, 5, 6, 8 and 9 [13,14], we predicted the putative ancestral breakpoint sites on the basis of the flanking gene positions in the deer and bovine genome assembly. The real positions of the evolutionary breakpoints were narrowed using a set of BAC probes and verified in all deer species available for this study.

2.3. FISH Probes

BAC clones specific to proximal and distal chromosome regions, to regions flanking the predicted evolutionary breakpoint sites in Cervidae, and to regions showing incongruences between the deer and bovine genome assembly were selected from the CHORI-240 bovine BAC library (BACPAC Genomics, Emeryville, CA, USA) on the basis of their location along the bovine chromosomes in the ARS-UCD1.2 genome assembly. The chromosome positions of cervid evolutionary breakpoints were further narrowed using additional BAC clones located in neighbouring positions. The BAC clones used in this study are listed in Supplementary Tables S1–S3. The BAC DNA was isolated using Wizard Plus SV Minipreps DNA Purification System (Promega, Madison, WI, USA), labelled with Green-dUTP (Abbott, Abbott Park, IL, USA), biotin-16-dUTP (Roche, Mannheim, Germany), or digoxigenin-11-dUTP (Roche) using BioPrime Array CGH Genomic Labeling Module (Invitrogen, Carlsbad, CA, USA) and used for FISH.

2.4. FISH

A hybridization mixture containing 50% formamide, 2 × SSC, 10% dextran sulfate, 0.7 µg salmon sperm, 1.3 µg Bovine Hybloc DNA (Applied Genetics Laboratories, Melbourne, FL, USA) and 200 ng of the labeled DNA probe was prepared. Ten µL of the mixture were denatured at 75 °C for 10 min, preannealed at 37 °C for at least 30 min, and applied on slides with metaphase chromosomes denatured by 0.07 M NaOH as previously described [27]. After hybridization in a humid chamber at 37 °C overnight, the slides were washed in 0.7 × SSC at 72 °C for 2 min. The BAC probes labeled with biotin-16-dUTP or digoxigenin-11-dUTP were detected with Avidin-Cy3 (Amersham Pharmacia Biotech, Piscataway, NJ, USA), Streptavidine-Cy5 (Invitrogen/Molecular Probes, Camarillo, CA, USA) and antidigoxigenin-rhodamine (Roche) according to manufacturers' instructions. If we used a combination of two probes labelled/detected by the same fluorochrome for the same chromosome, we performed two rounds of FISH, so that the position of each probe could be reliably determined. The slides were mounted in Vectashield mounting medium containing 1.5 mg DAPI (Vector Laboratories) and analysed using Zeiss Axio Imager.Z2 fluorescence microscope (Carl Zeiss Microimaging GmbH, Jena Germany) equipped with appropriate fluorescent filters and the Metafer Slide Scanning System (MetaSystems, Altlussheim, Germany). Images of well-spread metaphase cells were captured by CoolCube CCD camera (MetaSystems) and analysed using Isis3 software (MetaSystems). The reliability of the BAC probes was confirmed by their hybridization on bovine chromosomes prior to FISH in deer.

3. Results

Comparing chromosomal positions of the predicted genes annotated to the CerEla1.0 C-scaffolds with their locations in the bovine ARS-UCD1.2 genome assembly, we assigned all red deer C-scaffolds to their bovine orthologues (Supplementary Table S4). Then we verified the deer-bovine chromosome orthology by BAC-FISH, which also enabled reliable physical anchoring of CerEla1.0 C-scaffolds to *C. elaphus* karyotype. Using BAC probes, we observed identical physical centromere-telomere orientation of orthologous red deer and bovine chromosomes. However, the orientation of CerEla1.0 C-scaffolds 2, 6, 8, 11, 12, 16, and 22 in the NCBI database was found to be reversed, and the deer chromosome CEL4 was found rearranged, when compared with the corresponding CerEla1.0 C-scaffold 19. The orthology between the G-banded red deer and cattle karyotypes are displayed in Figure 1 and Supplementary Figure S1. The relationships among the CerEla1.0 C-scaffolds and the red deer and cattle chromosomes are summarized in Table 2. The comparative FISH results in cattle and the red deer are documented in Figure 2. Karyotypes of the additional studied cervid species with indicated homologies with *B. taurus* are displayed in Supplementary Figure S2.

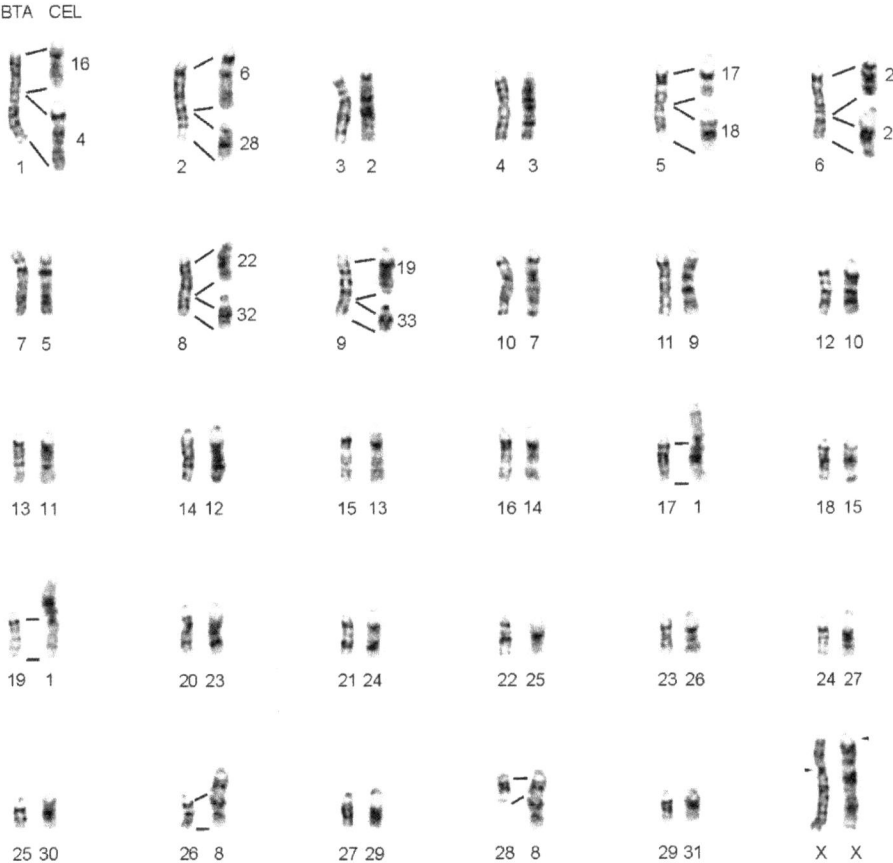

Figure 1. The orthology between the G-banded chromosomes of cattle (*B. taurus*, BTA) and red deer (*C. elaphus*, CEL). Arrowheads indicate the position of centromeres on the X chromosomes.

Table 2. CerEla1.0 and ARS-UCD1.2 genome assembly comparisons.

Red Deer (CerEla1.0)			CEL Chr	Cattle (ARS-UCD1.2)			Comments *
Pseudochr	INSDC	Size (Mb)		BTA Chr	RefSeq	Size (Mb)	
1	CM0080008.1	104.5	13	15	NC_037342.1	85.01	
2	CM0080009.1	63.26	31	29	NC_037356.1	51.1	Reverse
3	CM0080010.1	88.46	17	5prox (1–70 Mb)	NC_037332.1	120.09	1–55 Mb of BTA5
4	CM0080011.1	81.2	15	18	NC_037345.1	65.82	
5	CM0080012.1	178.03	1	17/19	NC_037344.1	73.17	
					NC_037346.1	63.45	
6	CM0080013.1	73.11	21	6dist (64–118 Mb)	NC_037333.1	117.81	Reverse, 70–118 Mb of BTA6
7	CM0080014.1	66.84	26	23	NC_037350.1	52.5	
8	CM0080015.1	55.92	28	2dist (94–136 Mb)	NC_037329.1	136.23	Reverse, 80–136 Mb of BTA2
9	CM0080016.1	141.95	5	7	NC_037334.1	110.68	
10	CM0080017.1	55.94	30	25	NC_037352.1	42.35	
11	CM0080018.1	140.39	9	11	NC_037338.1	106.98	Reverse
12	CM0080019.1	127.78	7	10	NC_037337.1	103.31	Reverse
13	CM0080020.1	89.79	24	21	NC_037348.1	69.86	
14	CM0080021.1	103.59	14	16	NC_037343.1	81.01	

Table 2. Cont.

Red Deer (CerEla1.0)			CEL Chr	Cattle (ARS-UCD1.2)			Comments *
Pseudochr	INSDC	Size (Mb)		BTA Chr	RefSeq	Size (Mb)	
15	CM0080022.1	125.28	8	28/26	NC_037355.1	45.94	
					NC_037353.1	51.99	
16	CM0080023.1	62.95	32	8dist (64–112 Mb)	NC_037335.1	113.32	Reverse, 69–112 Mb of BTA8
17	CM0080024.1	79.72	20	6prox (1–63 Mb)	NC_037333.1	117.81	1–66 Mb of BTA6
18	CM0080025.1	152.66	3	4	NC_037331.1	120	
19	CM0080026.1	127.24	4	1dist (59–158 Mb)	NC_037328.1	158.53	Rearranged, 57–158 Mb of BTA1
20	CM0080027.1	149.34	2	3	NC_037330.1	121.01	
21	CM0080028.1	107.36	12	14	NC_037341.1	82.4	
22	CM0080029.1	63.92	18	5dist (71–121 Mb)	NC_037332.1	120.09	Reverse, 60–121 Mb of BTA5
23	CM0080030.1	109.47	11	13	NC_037340.1	83.47	
24	CM0080031.1	78.16	25	22	NC_037349.1	60.77	
25	CM0080032.1	96.54	23	20	NC_037347.1	71.97	
26	CM0080033.1	55.1	33	9dist (64–106 Mb)	NC_037336.1	105.45	65–106 Mb of BTA9
27	CM0080034.1	84.64	27	24	NC_037351.1	62.32	
28	CM0080035.1	82.07	19	9prox (1–62 Mb)	NC_037336.1	105.45	1–64 Mb of BTA9
29	CM0080036.1	80.17	22	8prox (1–63 Mb)	NC_037335.1	113.32	1–64 Mb of BTA8
30	CM0080037.1	117.8	10	12	NC_037339.1	87.22	
31	CM0080038.1	75.46	16	1prox (1–58 Mb)	NC_037328.1	158.53	1–51 Mb of BTA1
32	CM0080039.1	60.01	29	27	NC_037354.1	45.61	
33	CM0080040.1	121.43	6	2prox (1–92 Mb)	NC_037329.1	136.23	1–71 Mb of BTA2
X	CM008041.1	181.54	X	X	NC_037357.1	139.01	
Y	CM008042.1	4.03	-	-	-	-	

* Reverse—inversed centromere-telomere orientation of the CerEla1.0 sequence; Rearranged—intrachromosomal rearrangement. Factual span on the BTA orthologue verified by BAC-FISH.

Minor differences in the gene order revealed between CerEla1.0 C-scaffolds 4, 5, 6, 11, 12, 18, 19, 23, 27, 33 and X and their bovine orthologues in the ARS-UCD1.2 genome assembly are highlighted in Supplementary Table S4. We clarified the incongruences in eight of these regions on six red deer chromosomes, i.e., where the order of the BAC probes mapping to the incongruent regions was clearly visible. We observed an identical order of the BAC-FISH signals on cattle and red deer in all of the regions (Figure 3). Interesting results were obtained using the BAC probe CH240-134N9 targeted to the incongruent proximal region of the C-scaffold 11 corresponding to the distal part (82.9 Mb) of BTA11 according to CerEla1.0 and ARS-UCD1.2 comparisons, and to orthologous *C. elaphus* chromosome CEL9. Instead of BTA11 and CEL9, this probe hybridized to a distal part of other chromosome in both cattle and red deer. This chromosome was subsequently identified as BTA29, and CEL31, respectively, by FISH with the BAC probe CH240-384F12 specific to the proximal part (5.8 Mb) of BTA29 orthologous to CEL31 (Figure 3B).

Regarding the X chromosome, we found that the sequences spanning 1–86 Mb of the CerEla1.0 X chromosome C-scaffold copy the gene order of the bovine X chromosome. However, a different order of the evolutionary conserved X chromosome segments was previously reported in studies using BAC-FISH in Cervidae [13,28].

Positions of the evolutionary chromosome breakpoints in chromosomes orthologous to BTA1, 2, 5, 6, 8 and 9 in the cervid ancestor were predicted on the basis of the genes located in the most proximal and distal positions of the corresponding CerEla1.0 C-scaffolds and, thus, flanking the assumed breakpoints. However, we revealed that the real breakpoints were located in a slightly different positions by a physical FISH-mapping with a series of BAC probes distributed along the chromosomes in the proximity of the predicted breakpoints (Figures 4 and 5). The subsequent analysis of the breakpoint positions in additional cervid species showed similar results in all deer species analysed in this study (Supplementary Figure S3).

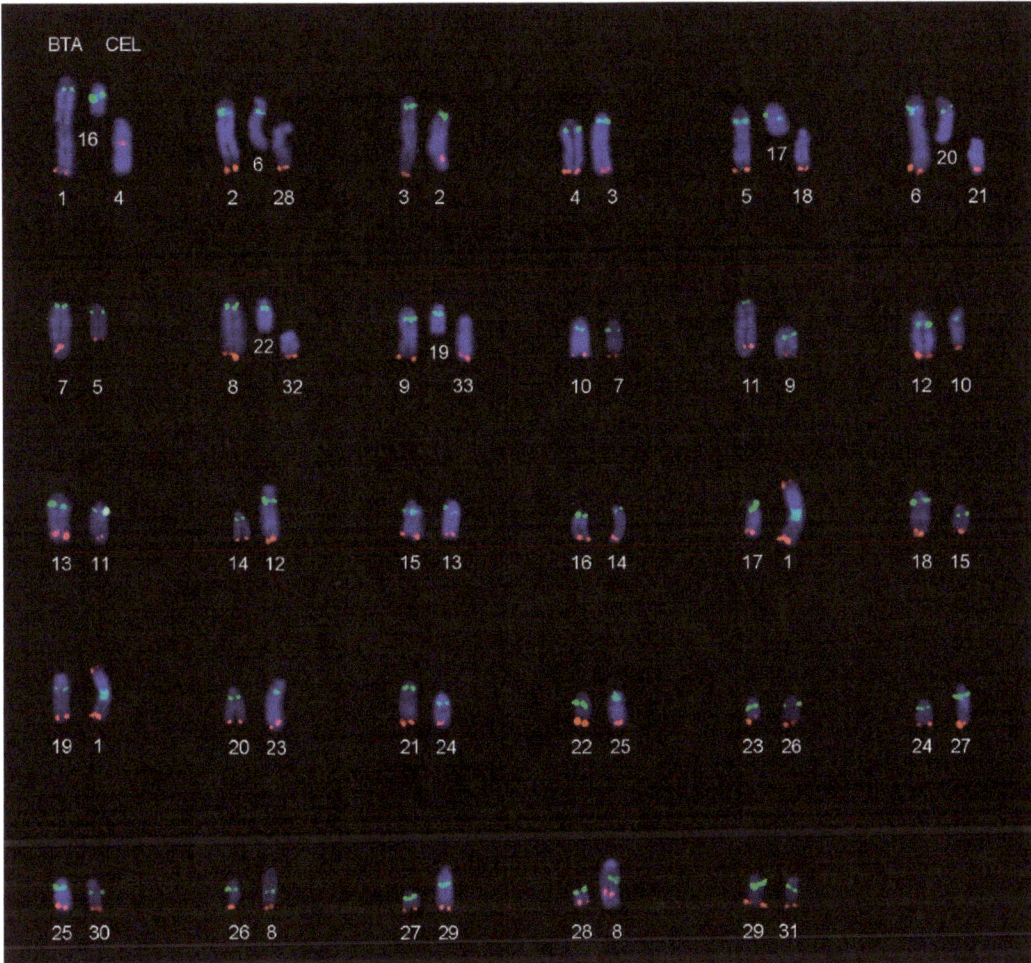

Figure 2. Centromere-telomere orientation of orthologous bovine (*B. taurus*, BTA) and red deer (*C. elaphus*, CEL) chromosomes confirmed by BAC-FISH. Green signal—proximal BAC probe; red signal—distal BAC probe.

The evolution of the BTA1 orthologue in the cervid lineage involved an initial fission followed by intrachromosomal rearrangements of one of the newly formed chromosomes. Two differentially rearranged types of the chromosome orthologous to the distal part of the BTA1 were observed in this study: An acrocentric chromosome common to Cervinae and *C. capreolus* and a submetacentric chromosome observed in the remaining Capreolinae in this study (*R. tarandus*, *A. alces* and *O. virginianus*) that was most probably derived from the previous by a pericentric inversion (Figure 5).

Using CerEla1.0 and ARS-UCD1.2 genome assembly comparisons, the fusion site of the ancestral chromosomes corresponding to BTA17 and BTA19, which roughly represents the position of centromere, was found at 95 Mb of the deer C-scaffold 5 (CEL1) length. The evolutionary fission, giving rise to bovine separated BTA28 and BTA26, was located to 60 Mb of the CerEla1.0 C-scaffold 15 (CEL8).

Finally, we used the bovine BAC probes to determine the centromere-telomere orientation of their evolutionarily fused chromosomes in seven additional deer species with rearranged karyotypes (Table 1). Except for the tandem fusion of BTA28;26 common to all

Cervidae, the rearranged chromosomes were formed by evolutionary centric fusions in all studied species (Figure 6).

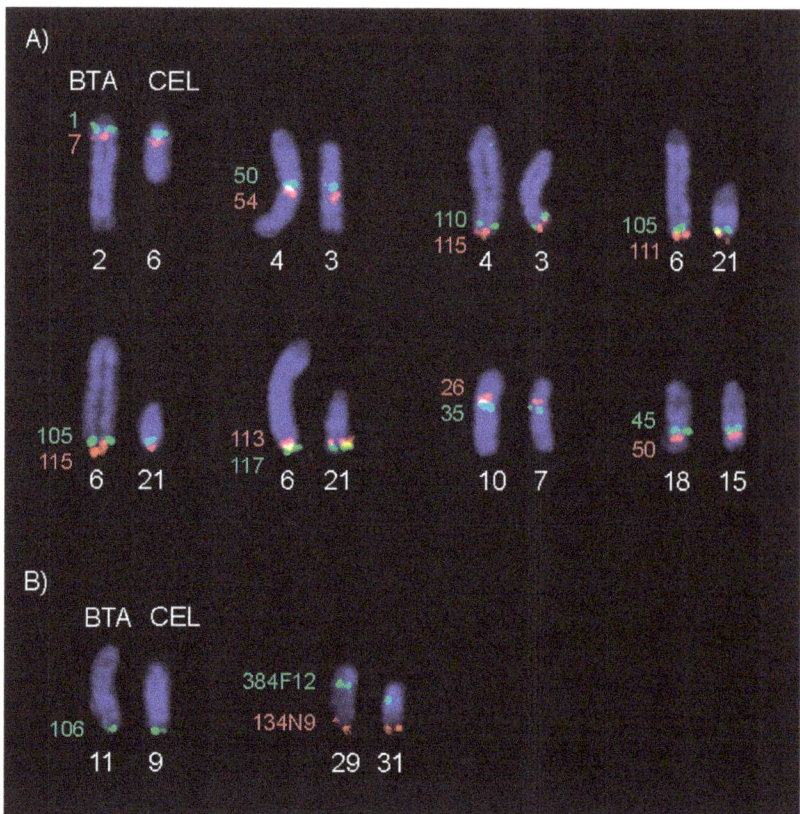

Figure 3. BAC-FISH mapping of the selected regions showing different gene order in CerEla1.0 and ARS-UCD1.2 genome assembly. Numbers indicate the BAC positions (Mb) on individual BTA chromosomes. (**A**,**B**) Identical signal order on orthologous bovine and red deer chromosomes. (**B**) Signal of the BAC probe CH240-134N9 on BTA29 and CEL31 instead of BTA11 and CEL9.

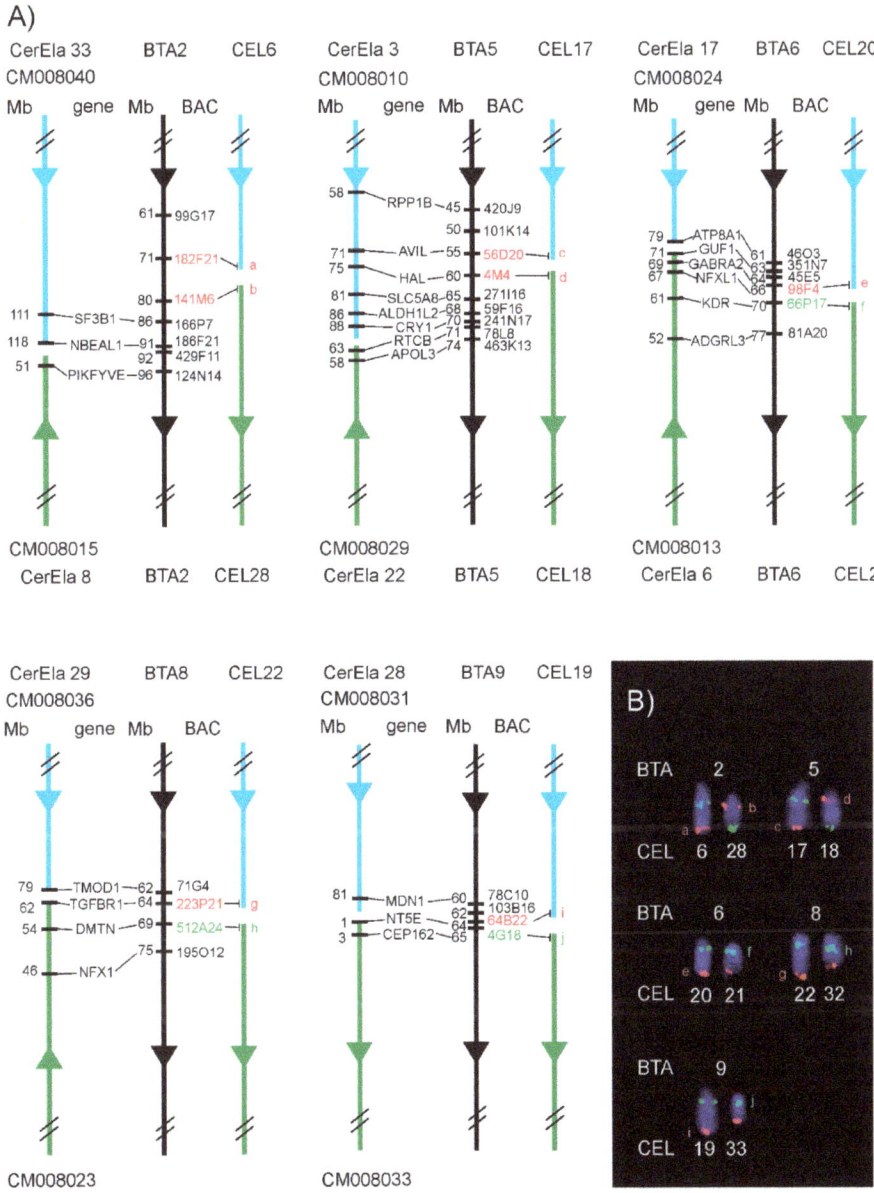

Figure 4. Evolutionary breakpoints on BTA2, 5, 6, 8 and 9 orthologues. (**A**) Schematic presentation of the CerEla1.0 C-scaffolds and BTA and CEL chromosomes with indicated positions of BAC clones used for the verification of the breakpoint positions. The BAC gene content and the position of the genes on the CerEla1.0 C-scaffolds is also shown. Notice differences in the assumed breakpoint positions on the CerEla1.0 C-scaffolds and the positions of breakpoints detected by FISH on the red deer chromosomes. (**B**) BAC-FISH signals at proximal, breakpoint and distal positions on *C. elaphus* chromosomes orthologous to BTA2, 5, 6, 8 and 9. Positions of selected BAC probes indicating the approximate evolutionary breakpoints are marked by letters (a–j): a—182F21, b—141M6, c—56D20, d—4M4, e—98F4, f—66P17, g—223P21, h—512A24, i—64B22, j—4G18. The unmarked FISH signals correspond to the proximal (green); and distal (red) BAC probes.

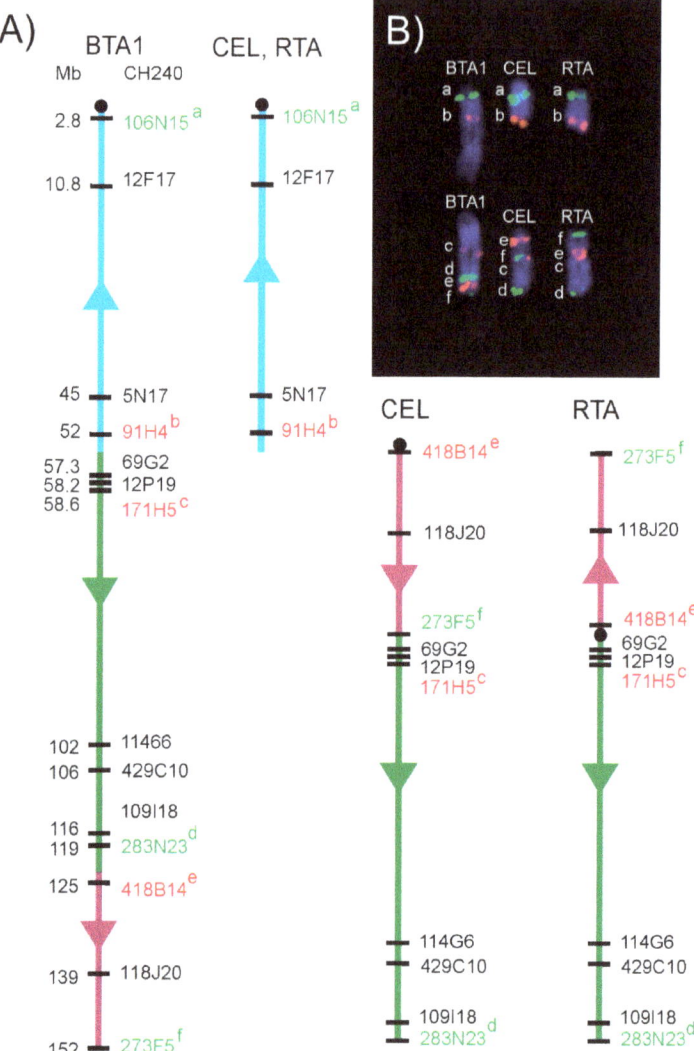

Figure 5. Evolutionary rearrangements of the BTA1 orthologue in Cervidae. (**A**) Schematic presentation. (**B**) BAC-FISH results in *B. taurus* (BTA), *C. elaphus* (CEL) and *R. tarandus* (RTA) using selected individual BAC probes indicated by letters (a–f): a—106N15, b—91H4, c—171H5, d—283N23, e—418B14, f—273F5.

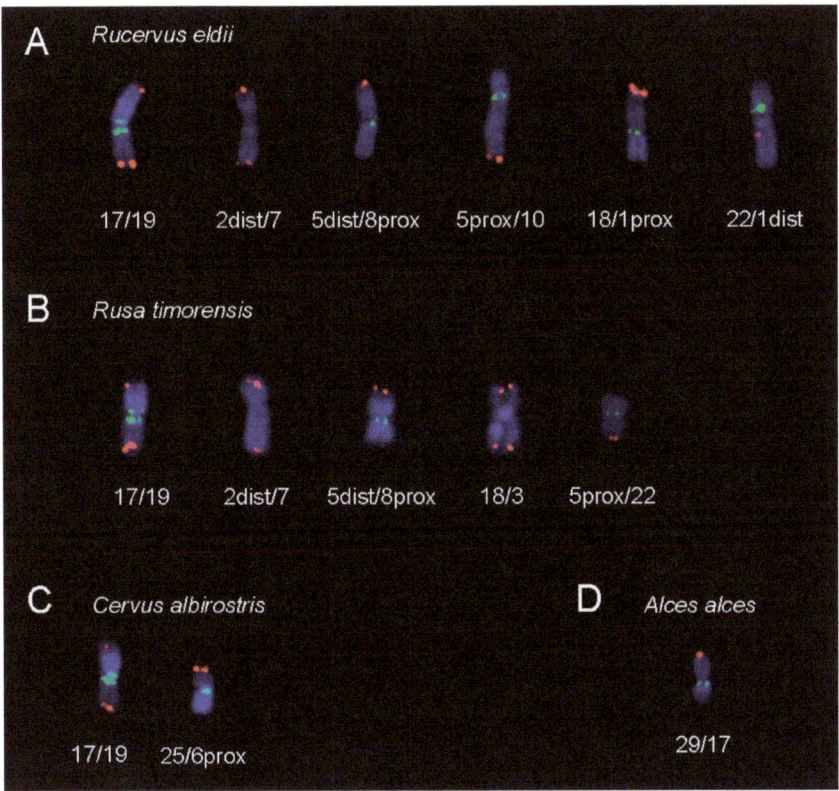

Figure 6. Centromere-telomere orientation of the fused chromosomes in (**A**) *R. eldii*, (**B**) *R. timorensis*, (**C**) *C. albirostris* and (**D**) *A. alces*. Green signal—proximal BAC probe; red signal—distal BAC probe.

4. Discussion

The recent publishing of the *C. elaphus* whole genome assembly (CerEla1.0) [16] brought a great resource for a research in the field of deer evolution, conservation and population genetics. However, the high automation in the genomic assembly construction may lead to errors. A verification and further improvements provided by molecular genetic and cytogenetic approaches are recommended for all newly established genome assemblies [18–21]. Inter- and intraspecies assembly comparisons supported by FISH enabled the detection and correction of misassembled sequences in genome assemblies of economically important bovid species (cattle, *Bos taurus*, sheep, *Ovis aries* and goat, *Capra hircus*) [29,30]. The combination of bioinformatic comparisons and BAC-FISH allowed identification of cryptic divergences between cattle and goat [15]. Using universal BAC sets, multiple scaffolds can be anchored to chromosomes of various species, as it was shown in birds [20].

In this study, we focused on the verification of chromosome relationships among *C. elaphus* CerEla1.0 and *B. taurus* ARS-UCD1.2 genome assemblies and karyotypes of both species. Using bovine BAC probes, we physically anchored the CerEla1.0 C-scaffolds to *C. elaphus* and *B. taurus* karyotype (Figure 1). Similar approach exploiting BAC-FISH mapping technique was previously successfully used for an integration of cytogenetic landmarks or upgrading draft genome sequences to chromosomal level in other species [20,31]. The C-scaffolds of the CerEla1.0 genome assembly had been constructed according to the reference deer linkage map [17] and the well-established bovine (*B. taurus*) Btau_5.0.1 genome assembly [16]. The order, orientation and schematic length of the C-scaffolds in the NCBI database comply with the deer genetic linkage map [17] but do not correspond with their sequence length in Mb, nor the position of the chromosomes in the red deer karyotype [5,26,32].

To document the results of this study, we arranged the G-banded red deer karyotype with regard to the chromosome morphology, physical lengths and G-banding patterns. Our G-banding and BAC-FISH showed concordant centromere-telomere orientation of the orthologous chromosomes in C. elaphus and B. taurus karyotypes. In compliance with the published paper on the CerEla1.0 assembly [16], we observed that the CerEla1.0 C-scaffolds 2, 6, 8, 11, 12, 16, 19 and 22 are presented in reversed centromere-telomere orientation in the NCBI database compared with the physical orientation of the red deer and bovine chromosomes.

Comparing the gene order in the CerEla1.0 and ARS-UCD1.2 genome assembly, we observed differences in several CerEla1.0 C-scaffolds. Bana et al. [16] suggested that these red deer genomic regions represent inverted segments. We analysed eight of these regions by BAC-FISH and observed identical BAC probes order in the orthologous bovine and deer chromosomes in all studied regions (Figure 3). Nevertheless, we revealed that the BAC probe CH240-134N9, selected from the position 82.9 Mb of the BTA11 in ARS-UCD1.2 genome assembly, hybridised to a distal part of BTA29 and, correspondingly, to the BTA29 orthologue in the red deer (CEL31). Either the chromosome position of this BAC in the NCBI database is incorrect, or the region covered by this BAC probe in the bovine ARS-UCD1.2 genome assembly and probably the wider region at the start of the CerEla1.0 C-scaffold 11 showing several incongruences with ARS-UCD1.2 (Supplementary Table S4), actually represent sequences of the chromosome BTA29 and CEL31, respectively. The above-mentioned regions of the CerEla1.0 assembly need further thorough revision.

In the published paper on the CerEla1.0 de novo genome assembly, the C-scaffold 33 was supposed to comprise sequences orthologous to parts of chromosomes BTA2 and BTA22 [16]. However, the bovine counterparts of all genes predicted to the CerEla1.0 C-scaffold 33 and selected for the CerEla1.0 - ARS-UCD1.2 comparisons in this study were found on BTA2.

Our comparisons of the CerEla1.0 C-scaffold X with the bovine chromosome X in ARS-UCD1.2 showed that, despite several smaller discrepancies, the gene order on the CerEla1.0 C-scaffold X corresponds to that on the bovine X chromosome. However, it was previously published that cervid X chromosomes were shaped by complex evolutionary rearrangements, including neocetromere formation, that differentiated them to two distinct types characteristic for Cervinae and Capreolinae [13,28]. With regard to the previously published findings on the X chromosome structure in Cervidae [13,28], the first 86 Mb of the CerEla1.0 X chromosome C-scaffold need to be revised accordingly.

Regarding another evolutionary chromosome changes, it is known that karyotypes of the current deer species derived from the pecoran ancestral karyotype (2n = 58) by fissions of six ancestral chromosomes orthologous to BTA1, 2, 5, 6, 8, 9 [12–14,17]. We used BAC probes selected on the basis of CerEla1.0 and ARS-UCD1.2 comparisons to hybridise to positions flanking the predicted evolutionary breakpoints, with the aim to physically verify the breakpoint sites. We revealed that the factual breakpoints differed from those predicted on the basis of CerEla1.0 C-scaffolds by up to 10 Mb, showing that the sequence span of the CerEla1.0 C-scaffolds needs to be properly adjusted. The newly assessed breakpoint locations were proved in all analysed species (four Cervinae and four Capreolinae) in this study.

We also showed that the evolutionary history of the BTA1 orthologue in Cervidae was more complicated than a simple fission and involved also intrachromosomal rearrangements, as was previously suggested [13,16]. The actual evolutionary breakpoint sites on the ancestral BTA1 orthologue, approximated by the set of BAC probes used in this study, diverged from those predicted on the basis of the CerEla1.0 and ARS-UCD1.2 genome assembly comparisons, neither they corresponded to the schematic presentation of the B. taurus and C. elaphus chromosome differences shown in Bana et al. (2018). Using BAC probes at positions flanking the evolutionary breakpoints, we showed that the primary evolutionary fission of the ancestral chromosome orthologous to BTA1 occurred between 52 and 57 Mb of the BTA1 length. This led to the formation of two neochromosomes with different lengths. The smaller neochromosome orthologous to the proximal part of BTA1 corresponds to CEL16 and CerEla1.0 C-scaffold 31 is present in both Cervinae and Capreolinae. This indicates that this fission of the ancestral BTA1 orthologue together with fissions of BTA2, 5, 6, 8 and 9 orthologues probably represent a defining event of the karyotype evolution of Cervidae. The larger neochromosome orthologous to the distal part of BTA1 then underwent an intrachromosomal rearrangement with a breakpoint between 119 and 125 Mb of the BTA1 length in the common ancestor of *C. capreolus* and the current Cervinae. This rearrangement was followed by a pericentric inversion of the proximal part of the rearranged chromosome during a separate evolution of the lineage leading to *R. tarandus*, *A. alces* and *O. virginianus* (Figure 5).

Because the BAC-FISH was proved to be an advantageous and sensitive tool for karyotype evolution studies [7,11,13–15,33–36], we used this method for verification of the evolutionary chromosomal rearrangements in Cervidae. The four species of Cervini analysed in this study share the fusion of BTA17;19 previously described on the basis of banding patterns and chromosome painting [7,13,14,37]. Using BAC probes, we proved that the ancestral chromosomes fused by their centromeres. Apart from the BTA17;19, five other centric fusions were proved in *R. eldii*, four in *R. timorensis* and one in *C. albirostris* by BAC-FISH in this study (Figure 6). As for Capreolini, the centric fusion BTA29;17 was confirmed in *A. alces* in this study. The chromosomes involved in the above mentioned fusions were previously identified by FISH with painting probes but their orientation in fused chromosomes could not be further specified by whole chromosome probes [7,13,14].

In general, our analysis of chromosome evolution in the studied cervid species showed that centric fusions probably represented the main evolutionary mechanism shaping their karyotypes. In species analysed in this study, only the chromosome comprising BTA28;26 orthologues (CEL8) was shown to be formed by a tandem (centromere to telomere) fusion. The fact that the BTA28;26 fusion is common to all Cervidae and characteristic for all pecoran species excluding Bovidae [12–14,17] suggests that this chromosome probably represents an ancestral chromosome which underwent a fission at the origin of the Bovidae lineage [38]. Centric fusions are generally characteristic for the karyotype evolution in the family Bovidae [38,39]. However, in Cervidae, centric and tandem fusions dominate differentially in individual clades. In the subfamily Cervinae, centric fusions are relatively common in the tribe Cervini but the karyotypes of Muntjacini were diversified by extensive tandem fusions [4,5,11,33,38]. Among Capreolinae, presumed centric fusions occurred in the karyotype evolution of *Ozotoceros bezoarticus*, *Blastocerus dichotomus* and *A. alces* [3,4] (the latter one was proved in this study). On the other hand, both centric and tandem fusions were involved in the karyotype diversification of South-American Capreolinae species of the genus *Mazama* [40–42]. This suggests that the karyotype evolution has been driven by different mechanisms in the individual cervid lineages and cytogenetic studies employing BAC-FISH for the detailed differentiation of the evolutionary rearrangements can help in future studies focused on the reconstruction of the cervid phylogeny.

5. Conclusions

In this study, we verified the red deer-cattle chromosome relationships, anchored the CerEla1.0 C-scaffolds to the red deer and cattle karyotype and proved the centromere-telomere orientation of the CerEla1.0 C-scaffolds. We indicated necessary adjustments to the CerEla1.0 genome assembly, including better specification of the sequence span of the chromosomes that underwent evolutionary chromosome fissions. Finally, we proved the location of the cervid evolutionary fissions and orientation of the fused chromosomes in a total of eight cervid species. Our results can serve as a basis for the CerEla1.0 genome assembly improvement, supporting, thus, future research in Cervidae.

Supplementary Materials: The following are available online at https://www.mdpi.com/article/10.3390/ani11092614/s1, Figure S1. The red deer (C. elaphus, CEL) karyotype with indicated orthology with chromosomes of cattle (*B. taurus*, BTA). Figure S2. G-banded karyotypes of (A) *C. albirostris* (CAL), (B) *R. timorensis* (RTI), (C) *R. eldii* (REL), (D) *R. tarandus* (RTA), (E) *A. alces* (AAL) and (F) *O. virginianus* (OVI) with indicated orthology with chromosomes of *Bos taurus* (BTA). The karyotype of RTI was previously published in Frohlich et al. (2017) [13]; the karyotypes of CAL and REL were published in O'Brien et al. (2020) [26]. Figure S3. Evolutionary chromosome breakpoints in BTA2, 5, 6, 8 and 9 orthologues in the analysed cervid species. *C. albirostris* (CAL), *R. timorensis* (RTI), *R. eldii* (REL), *C. capreolus* (CCA), *R. tarandus* (RTA), *A. alces* (AAL) and *O. virginianus* (OVI). Positions of BAC probes indicating the approximate evolutionary breakpoints are marked by letters (a–j): a—182F21, b—141M6, c—56D20, d—4M4, e—98F4, f—66P17, g—223P21, h—512A24, i—64B22, j—4G18. Table S1: BACs for the physical analysis of the cattle-red deer chromosome orthology and centromere-telomere orientation. Table S2: BACs for the analysis of incongruences between CerEla1.0 and ARS-UCD1.2. Table S3: BACs for the specification of evolutionary chromosome breakpoints in Cervidae. Table S4: Comparison of the CerEla1.0 and ARS-UCD1.2 genome assembly with indicated incongruences.

Author Contributions: Conceptualization, M.V., S.K. and J.F.; methodology, S.K., M.V., H.C. and J.F.; investigation, M.V., S.K., H.C. and J.F.; writing—original draft preparation, M.V., J.F., S.K. and H.C.; writing—review and editing, M.V. and J.R.; visualization, M.V., H.C.; supervision, J.R.; funding acquisition, M.V. and J.R. All authors have read and agreed to the published version of the manuscript.

Funding: This research was funded by Czech Science Foundation grant number 20-22517J, Ministry of Agriculture of the Czech Republic, grant No. RO 0520 and Ministry of Education, Youth and Sports of the Czech Republic under the project CEITEC 2020, grant No. LQ1601.

Institutional Review Board Statement: The study was conducted according to the guidelines of the Declaration of Helsinki. All procedures performed in this study were in accordance with the ethical standards of the Veterinary Research Institute (Brno, Czech Republic), which complies with the Czech and European Union Legislation for the protection of animals used for scientific purposes. According to these regulations ethics approval was not required, as the biological material (blood/tissue) was obtained post-mortem from animals upon animal slaughter in abattoir or which died during the hunting. A ZOO veterinarian collected the blood from living animals during other medical procedures. All collaborating ZOOs have license issued by the Ministry of the Environment of the Czech Republic (Act No 162/2003 Coll.).

Informed Consent Statement: Not applicable.

Data Availability Statement: All data is contained within the manuscript and Supplementary Materials. The FISH images are available from the authors upon request.

Acknowledgments: The authors wish to thank the Prague Zoo veterinarian R. Vodicka and the deer enclosure keepers S. Zbanek (Bila Lhota) and L. Volný (Sovinec, Frycovice) for providing blood samples from the studied species.

Conflicts of Interest: The authors declare no conflict of interest. The funders had no role in the design of the study; in the collection, analyses, or interpretation of data; in the writing of the manuscript, or in the decision to publish the results.

References

1. Wilson, D.E.; Reeder, D.M. *Mammal Species of the World: A Taxonomic and Geographic Reference*; Johns Hopkins University Press: Baltimore, MD, USA, 2005; ISBN 978-0-8018-8221-0.
2. Wurster, D.H.; Benirschke, K. Indian Muntjac, Muntiacus Muntjak: A Deer with a Low Diploid Chromosome Number. *Science* **1970**, *168*, 1364–1366. [CrossRef]
3. Nietzel, H. Chromosome Evolution of Cervidae: Karyotypic and Molecular Aspects. In *Cytogenetics: Basic and Applied Aspects*; Obe, G., Basler, A., Eds.; Springer: Berlin/Heidelberg, Germany, 1987; ISBN 978-3-642-72804-4.
4. Fontana, F.; Rubini, M. Chromosomal Evolution in Cervidae. *BioSystems* **1990**, *24*, 157–174. [CrossRef]
5. Huang, L.; Chi, J.; Nie, W.; Wang, J.; Yang, F. Phylogenomics of Several Deer Species Revealed by Comparative Chromosome Painting with Chinese Muntjac Paints. *Genetica* **2006**, *127*, 25–33. [CrossRef] [PubMed]
6. Rubini, M.; Negri, E.; Fontana, F. Standard Karyotype and Chromosomal Evolution of the Fallow Deer (*Dama dama* L.). *Cytobios* **1990**, *64*, 155–161. [PubMed]

7. Bonnet-Garnier, A.; Claro, F.; Thévenon, S.; Gautier, M.; Hayes, H. Identification by R-Banding and FISH of Chromosome Arms Involved in Robertsonian Translocations in Several Deer Species. *Chromosome Res.* **2003**, *11*, 649–663. [CrossRef] [PubMed]
8. Duarte, J.M.B.; Jorge, W. Morphologic and Cytogenetic Description of the Small Red Brocket (Mazama Bororo Duarte, 1996) in Brazil. *Mammalia* **2009**, *67*, 403–410. [CrossRef]
9. Yang, F.; O'Brien, P.C.; Wienberg, J.; Ferguson-Smith, M.A. A Reappraisal of the Tandem Fusion Theory of Karyotype Evolution in Indian Muntjac Using Chromosome Painting. *Chromosome Res.* **1997**, *5*, 109–117. [CrossRef]
10. Chi, J.; Fu, B.; Nie, W.; Wang, J.; Graphodatsky, A.S.; Yang, F. New Insights into the Karyotypic Relationships of Chinese Muntjac (Muntiacus Reevesi), Forest Musk Deer (*Moschus berezovskii*) and Gayal (*Bos frontalis*). *Cytogenet. Genome Res.* **2005**, *108*, 310–316. [CrossRef] [PubMed]
11. Chi, J.X.; Huang, L.; Nie, W.; Wang, J.; Su, B.; Yang, F. Defining the Orientation of the Tandem Fusions That Occurred during the Evolution of Indian Muntjac Chromosomes by BAC Mapping. *Chromosoma* **2005**, *114*, 167–172. [CrossRef] [PubMed]
12. Dementyeva, P.V.; Trifonov, V.A.; Kulemzina, A.I.; Graphodatsky, A.S. Reconstruction of the Putative Cervidae Ancestral Karyotype by Chromosome Painting of Siberian Roe Deer (*Capreolus pygargus*) with Dromedary Probes. *Cytogenet. Genome Res.* **2010**, *128*, 228–235. [CrossRef] [PubMed]
13. Frohlich, J.; Kubickova, S.; Musilova, P.; Cernohorska, H.; Muskova, H.; Vodicka, R.; Rubes, J. Karyotype Relationships among Selected Deer Species and Cattle Revealed by Bovine FISH Probes. *PLoS ONE* **2017**, *12*, e0187559. [CrossRef]
14. Bonnet, A.; Thévenon, S.; Claro, F.; Gautier, M.; Hayes, H. Cytogenetic Comparison between Vietnamese Sika Deer and Cattle: R-Banded Karyotypes and FISH Mapping. *Chromosome Res.* **2001**, *9*, 673–687. [CrossRef]
15. De Lorenzi, L.; Planas, J.; Rossi, E.; Malagutti, L.; Parma, P. New Cryptic Karyotypic Differences between Cattle (Bos Taurus) and Goat (*Capra hircus*). *Chromosome Res.* **2015**, *23*, 225–235. [CrossRef]
16. Bana, N.Á.; Nyiri, A.; Nagy, J.; Frank, K.; Nagy, T.; Stéger, V.; Schiller, M.; Lakatos, P.; Sugár, L.; Horn, P.; et al. The Red Deer Cervus Elaphus Genome CerEla1.0: Sequencing, Annotating, Genes, and Chromosomes. *Mol. Genet. Genom.* **2018**, *293*, 665–684. [CrossRef]
17. Slate, J.; Van Stijn, T.C.; Anderson, R.M.; McEwan, K.M.; Maqbool, N.J.; Mathias, H.C.; Bixley, M.J.; Stevens, D.R.; Molenaar, A.J.; Beever, J.E.; et al. A Deer (Subfamily Cervinae) Genetic Linkage Map and the Evolution of Ruminant Genomes. *Genetics* **2002**, *160*, 1587–1597. [CrossRef]
18. Alkan, C.; Sajjadian, S.; Eichler, E.E. Limitations of Next-Generation Genome Sequence Assembly. *Nat. Methods* **2011**, *8*, 61–65. [CrossRef]
19. Ariyadasa, R.; Stein, N. Advances in BAC-Based Physical Mapping and Map Integration Strategies in Plants. *J. Biomed. Biotechnol.* **2012**, *2012*, 184854. [CrossRef]
20. Damas, J.; O'Connor, R.; Farré, M.; Lenis, V.P.E.; Martell, H.J.; Mandawala, A.; Fowler, K.; Joseph, S.; Swain, M.T.; Griffin, D.K.; et al. Upgrading Short-Read Animal Genome Assemblies to Chromosome Level Using Comparative Genomics and a Universal Probe Set. *Genome Res.* **2017**, *27*, 875–884. [CrossRef] [PubMed]
21. Lewin, H.A.; Graves, J.A.M.; Ryder, O.A.; Graphodatsky, A.S.; O'Brien, S.J. Precision Nomenclature for the New Genomics. *Gigascience* **2019**, *8*, giz086. [CrossRef]
22. Groves, C.; Grubb, P. *Ungulate Taxonomy*, 1st ed.; Johns Hopkins University Press: Baltimore, MD, USA, 2011; ISBN 978-1-4214-0093-8.
23. Cernohorska, H.; Kubickova, S.; Vahala, J.; Robinson, T.J.; Rubes, J. Cytotypes of Kirk's Dik-Dik (Madoqua Kirkii, Bovidae) Show Multiple Tandem Fusions. *Cytogenet. Genome Res.* **2011**, *132*, 255–263. [CrossRef] [PubMed]
24. Cernohorska, H.; Kubickova, S.; Vahala, J.; Rubes, J. Molecular Insights into X;BTA5 Chromosome Rearrangements in the Tribe Antilopini (Bovidae). *Cytogenet. Genome Res.* **2012**, *136*, 188–198. [CrossRef]
25. Seabright, M. A Rapid Banding Technique for Human Chromosomes. *Lancet* **1971**, *2*, 971–972. [CrossRef]
26. O'Brien, S.J.; Graphodatsky, A.S.; Perelman, P.L. (Eds.) *Atlas of Mammalian Chromosomes*, 2nd ed.; Wiley-Blackwell: Hoboken, NJ, USA, 2020; ISBN 978-1-119-41803-0.
27. Vozdova, M.; Kubickova, S.; Cernohorska, H.; Fröhlich, J.; Vodicka, R.; Rubes, J. Comparative Study of the Bush Dog (Speothos Venaticus) Karyotype and Analysis of Satellite DNA Sequences and Their Chromosome Distribution in Six Species of Canidae. *Cytogenet. Genome Res.* **2019**, *159*, 88–96. [CrossRef] [PubMed]
28. Proskuryakova, A.A.; Kulemzina, A.I.; Perelman, P.L.; Makunin, A.I.; Larkin, D.M.; Farré, M.; Kukekova, A.V.; Lynn Johnson, J.; Lemskaya, N.A.; Beklemisheva, V.R.; et al. X Chromosome Evolution in Cetartiodactyla. *Genes* **2017**, *8*, 216. [CrossRef]
29. Partipilo, G.; D'Addabbo, P.; Lacalandra, G.M.; Liu, G.E.; Rocchi, M. Refinement of Bos Taurus Sequence Assembly Based on BAC-FISH Experiments. *BMC Genom.* **2011**, *12*, 639. [CrossRef] [PubMed]
30. Lorenzi, L.D.; Parma, P. Identification of Some Errors in the Genome Assembly of Bovidae by FISH. *CGR* **2020**, *160*, 85–93. [CrossRef] [PubMed]
31. BAC Resource Consortium, T.; Cheung, V.G.; Nowak, N.; Jang, W.; Kirsch, I.R.; Zhao, S.; Chen, X.-N.; Furey, T.S.; Kim, U.-J.; Kuo, W.-L.; et al. Integration of Cytogenetic Landmarks into the Draft Sequence of the Human Genome. *Nature* **2001**, *409*, 953–958. [CrossRef]
32. Herzog, S. The Karyotype of the Red Deer (*Cervus elaphus* L.). *Caryologia* **1987**, *40*, 299–305. [CrossRef]
33. Huang, L.; Chi, J.; Wang, J.; Nie, W.; Su, W.; Yang, F. High-Density Comparative BAC Mapping in the Black Muntjac (Muntiacus Crinifrons): Molecular Cytogenetic Dissection of the Origin of MCR 1p+4 in the X1X2Y1Y2Y3 Sex Chromosome System. *Genomics* **2006**, *87*, 608–615. [CrossRef]

34. Cernohorska, H.; Kubickova, S.; Kopecna, O.; Kulemzina, A.I.; Perelman, P.L.; Elder, F.F.B.; Robinson, T.J.; Graphodatsky, A.S.; Rubes, J. Molecular Cytogenetic Insights to the Phylogenetic Affinities of the Giraffe (Giraffa Camelopardalis) and Pronghorn (Antilocapra Americana). *Chromosome Res.* **2013**, *21*, 447–460. [CrossRef]
35. Cernohorska, H.; Kubickova, S.; Kopecna, O.; Vozdova, M.; Matthee, C.A.; Robinson, T.J.; Rubes, J. Nanger, Eudorcas, Gazella, and Antilope Form a Well-Supported Chromosomal Clade within Antilopini (Bovidae, Cetartiodactyla). *Chromosoma* **2014**, *124*, 235–247. [CrossRef] [PubMed]
36. Kiazim, L.G.; O'Connor, R.E.; Larkin, D.M.; Romanov, M.N.; Narushin, V.G.; Brazhnik, E.A.; Griffin, D.K. Comparative Mapping of the Macrochromosomes of Eight Avian Species Provides Further Insight into Their Phylogenetic Relationships and Avian Karyotype Evolution. *Cells* **2021**, *10*, 362. [CrossRef] [PubMed]
37. Gallagher, D.S.; Davis, S.K.; De Donato, M.; Burzlaff, J.D.; Womack, J.E.; Taylor, J.F.; Kumamoto, A.T. A Molecular Cytogenetic Analysis of the Tribe Bovini (Artiodactyla: Bovidae: Bovinae) with an Emphasis on Sex Shromosome Morphology and NOR Distribution. *Chromosome Res.* **1999**, *7*, 481–492. [CrossRef]
38. Rubes, J.; Musilova, P.; Kopecna, O.; Kubickova, S.; Cernohorska, H.; Kulemsina, A.I. Comparative Molecular Cytogenetics in Cetartiodactyla. *Cytogenet. Genome Res.* **2012**, *137*, 194–207. [CrossRef] [PubMed]
39. Gallagher, D.S., Jr.; Womack, J.E. Chromosome Conservation in the Bovidae. *J. Hered.* **1992**, *83*, 287–298. [CrossRef] [PubMed]
40. Abril, V.V.; Duarte, J.M.B. Chromosome Polymorphism in the Brazilian Dwarf Brocket Deer, Mazama Nana (Mammalia, Cervidae). *Genet. Mol. Biol.* **2008**, *31*, 53–57. [CrossRef]
41. Abril, V.V.; Carnelossi, E.A.G.; González, S.; Duarte, J.M.B. Elucidating the Evolution of the Red Brocket Deer Mazama Americana Complex (Artiodactyla; Cervidae). *Cytogenet. Genome Res.* **2010**, *128*, 177–187. [CrossRef]
42. Duarte, J.M.B.; González, S. *Neotropical Cervidology: Biology and Medicine of Latin American Deer*; Funep: Jaboticabal, Brazil; IUCN: Gland, Switzerland, 2010; ISBN 978-85-7805-046-7.

Article

Karyotype Evolution and Genomic Organization of Repetitive DNAs in the Saffron Finch, *Sicalis flaveola* (Passeriformes, Aves)

Rafael Kretschmer [1], Benilson Silva Rodrigues [2], Suziane Alves Barcellos [3], Alice Lemos Costa [3], Marcelo de Bello Cioffi [4], Analía del Valle Garnero [3], Ricardo José Gunski [3], Edivaldo Herculano Corrêa de Oliveira [5,6] and Darren K. Griffin [1,*]

1 School of Biosciences, University of Kent, Canterbury CT2 7NJ, UK; rafa.kretschmer@hotmail.com
2 Instituto Federal do Pará, Abaetetuba 8440-000, Brazil; benilson.rodrigues@gmail.com
3 Laboratório de Diversidade Genética Animal, Universidade Federal do Pampa, São Gabriel 97300-162, Brazil; suzianebarcellos@gmail.com (S.A.B.); alicelemoscosta14bio@gmail.com (A.L.C.); analiagarnero@unipampa.edu.br (A.d.V.G.); ricardogunski@unipampa.edu.br (R.J.G.)
4 Centro de Ciências Biológicas e da Saúde, Laboratório de Citogenética de Peixes, Departamento de Genética e Evolução, Universidade Federal de São Carlos, São Carlos 13565-905, Brazil; mbcioffi@ufscar.br
5 Instituto de Ciências Exatas e Naturais, Universidade Federal do Pará, Belém 66075-110, Brazil; ehco@ufpa.br
6 Laboratório de Cultura de Tecidos e Citogenética, SAMAM, Instituto Evandro Chagas, Ananindeua 67030-000, Brazil
* Correspondence: d.k.griffin@kent.ac.uk; Tel.: +44-1227-823022

Citation: Kretschmer, R.; Rodrigues, B.S.; Barcellos, S.A.; Costa, A.L.; Cioffi, M.d.B.; Garnero, A.d.V.; Gunski, R.J.; de Oliveira, E.H.C.; Griffin, D.K. Karyotype Evolution and Genomic Organization of Repetitive DNAs in the Saffron Finch, *Sicalis flaveola* (Passeriformes, Aves). *Animals* **2021**, *11*, 1456. https://doi.org/10.3390/ani11051456

Academic Editors: Leopoldo Iannuzzi and Pietro Parma

Received: 19 April 2021
Accepted: 18 May 2021
Published: 19 May 2021

Publisher's Note: MDPI stays neutral with regard to jurisdictional claims in published maps and institutional affiliations.

Copyright: © 2021 by the authors. Licensee MDPI, Basel, Switzerland. This article is an open access article distributed under the terms and conditions of the Creative Commons Attribution (CC BY) license (https:// creativecommons.org/licenses/by/ 4.0/).

Simple Summary: Detailed chromosome studies of birds, addressing both macrochromosomes and microchromosomes, have been reported only for few species. Hence, in this study, we performed investigations of chromosome evolution in the Saffron finch (*Sicalis flaveola*), a semi-domestic species, tolerant of human proximity and nesting in roof spaces. We also explored the organization of simple short repeats (SSR) in the genome of this species. Our results revealed that most of the Saffron finch chromosomes remained highly conserved when compared to the avian ancestral karyotype and that the SSR accumulated mainly in the microchromosomes and the short arms of Z (sex) chromosome. Finally, we compared our results with other avian species, contributing to a better understanding of the chromosome organization and evolution of the Saffron finch genome.

Abstract: The Saffron finch (*Sicalis flaveola*), a semi-domestic species, is tolerant of human proximity and nesting in roof spaces. Considering the importance of cytogenomic approaches in revealing different aspects of genomic organization and evolution, we provide detailed cytogenetic data for *S. flaveola*, including the standard Giemsa karyotype, C- and G-banding, repetitive DNA mapping, and bacterial artificial chromosome (BAC) FISH. We also compared our results with the sister groups, Passeriformes and Psittaciformes, bringing new insights into the chromosome and genome evolution of birds. The results revealed contrasting rates of intrachromosomal changes, highlighting the role of SSR (simple short repetition probes) accumulation in the karyotype reorganization. The SSRs showed scattered hybridization, but brighter signals were observed in the microchromosomes and the short arms of Z chromosome in *S. flaveola*. BACs probes showed conservation of ancestral syntenies of macrochromosomes (except GGA1), as well as the tested microchromosomes. The comparison of our results with previous studies indicates that the great biological diversity observed in Passeriformes was not likely accompanied by interchromosomal changes. In addition, although repetitive sequences often act as hotspots of genome rearrangements, Passeriformes species showed a higher number of signals when compared with the sister group Psittaciformes, indicating that these sequences were not involved in the extensive karyotype reorganization seen in the latter.

Keywords: Thraupidae; micro and macrochromosomes; inter and intrachromosomal rearrangements; genetic organization; SSRs

1. Introduction

The tanagers (Passeriformes: Thraupidae) exhibit a range of plumage colors and patterns, behaviors, morphologies, and habitats [1]. According to Gill et al. [2], the tanagers are composed of approximately 380 species, representing 4% of the members of the order Passeriformes. Given the extensive diversity found among tanagers, their taxonomic classification has been problematic [1,3]. For instance, the genus *Sicalis* has already been the subject of several taxonomic studies due to controversies on its permanence in Emberizidae [4] or Thraupidae [5]. *Sicalis flaveola*, the subject of this study, is popularly known as the Saffron finch and has an extremely large range in South America [6]. It is a semi-domestic species, tolerant of humans, and frequently nesting in the roof eaves of suburban houses in Eastern Ecuador, Western Peru, Eastern and Southern Brazil (where it is commonly referred to as the "canário-da-terra" or "native canary"—despite not, taxonomically, being a canary).

Cytogenetic studies in tanager species are still scarce and based mostly on conventional staining (Giemsa) [7]. Although only 11% of Thraupidae species have been karyotyped, high chromosomal similarities were observed among them, which approximately 63% of karyotyped species showing 2n = 78 chromosomes [7]. However, some deviations have been described, such as 2n = 72 in *Oryzoborus maximiliani* [8], and 2n = 88 in *Saltator coerulescens* [9]. Molecular cytogenetic studies are even more scarce, with only two species—*Saltator aurantiirostris* and *Saltator similis*, both with 2n = 80—analyzed by comparative chromosome painting using *Gallus gallus* (GGA) and *Leucopternis albicollis* (LAL) probes [10]. Both species presented macrochromosome conservation, except for centric fission of chromosome GGA1, which has been found in all passerines thus far analyzed [11].

Despite the low rate of interchromosomal rearrangements in Passeriformes species, a high rate of intrachromosomal rearrangements, such as inversions, have been described, both *in silico* [12,13] and following *in situ* experiments [10,14–18]. The most phylogenetically informative finding is a series of intrachromosomal rearrangements involving paracentric and pericentric inversions in the syntenic group corresponding to GGA1q, including oscines and suboscines [10,14–18]. Therefore, these studies suggested that this complex pattern of intrachromosomal rearrangements was already present in the common ancestor of Passeriformes.

Microchromosomes correspond to approximately 25% of the avian genome [19], and around 50% of avian genes are on these chromosomes [20]. Because of technical limitations, however, most of the molecular cytogenetics studies in Passeriformes have focused only on the comparison of homology with chicken macrochromosomes [11]. For instance, up to now, only four Passeriformes species had their karyotype analyzed in detail, i.e., macro and microchromosomes: *Taeniopygia guttata*, *Turdus merula*, and *Serinus canaria* [17,21,22] from oscines suborder, and *Willisornis vidua* from suboscines suborder [23]. The results revealed that the microchromosomes were not involved in interchromosomal events in the oscines species. However, the chicken microchromosome 17 was found fused to a macrochromosome of *W. vidua*. Interchromosomal rearrangements involving these small elements are rare in birds but have been found only extensively in Falconiformes and Psittaciformes [22,24–27]. In addition, microchromosome fusions have been found in Cuculiformes, Suliformes, and Caprimulgiformes species [28,29], and future studies are necessary to investigate if it is a species-specific feature or if it is shared with other members of these orders. Future studies are also necessary for other Passeriformes members, considering the great diversity in the number of species.

Cytogenomic studies using other types of chromosomal markers, such as repetitive sequences, are also scarce in birds. Repetitive DNA plays an important role in the chromosome structure and genome organization [30,31]. Furthermore, they often serve as hotspots of genome rearrangements and evolutionary innovation [32]. These sequences are classified into distinct categories. Among them are the microsatellites, which represent the most variable types of DNA sequences [33]. To this end, it is essential to know how these elements are organized in the genome. Despite the significance of simple short repetition probes (SSR), data concerning the mapping of these sequences by fluorescent *in situ* hybridization (FISH) are available for a few species of birds and results so far have shown the involvement of amplification of these elements in atypical sex chromosomes, in which the repetitive DNA amount was related to the enlargement of these elements in some cases [34–40].

The karyotype of *S. flaveola* has been investigated only by giemsa staining, revealing a diploid number of 80 chromosomes [41–43]. In the present study, we provide the detailed cytogenetic data for the Saffron finch, *S. flaveola*, including the standard Giemsa karyotype, C- and G-banding, repetitive DNA mapping, and bacterial artificial chromosome (BAC) FISH, bringing new insights into the chromosome and genome evolution of birds, especially tanagers and Passeriformes.

2. Materials and Methods

2.1. Animals and Chromosome Preparations

Fibroblast cell lines were established from 1 male and 3 female embryos of *S. flaveola*, selected after sexing by Giemsa staining, chromosome banding, and FISH results using BAC for chicken Z and W chromosomes. The cells were cultivated in Dulbecco's Modified Eagle's Medium (DMEM) supplemented with 15% fetal bovine serum, 2% penicillin-streptomycin, and 1% L-glutamine at 37 °C, according to Sasaki et al. [44]. Metaphase chromosomes were obtained by standard protocols: treatment with colcemid (1 h), hypotonic solution (0.075 M KCl, 15 min), and fixation with 3:1 methanol/acetic acid. The embryos were collected in their natural environment in São Gabriel city, Rio Grande do Sul State, Brazil, following the procedures approved by the "Biodiversity Authorization and Information System", permission numbers 44173-1 and 33860-4. The experiments using animals were approved by the Ethics Committee on Animal Experimentation (CEUA) of the Universidade Federal do Pampa under no. 026/2012 and 018/2014.

2.2. Diploid Number, C and G-Banding

For the karyotype description and diploid number, an average of 30 metaphases in conventional staining (5% Giemsa in 0.07 M phosphate buffer, pH 6.8) were analyzed per specimen. Chromosomes were arranged and classified according to the nomenclature of Guerra [45]. Blocks of constitutive heterochromatic were detected by C-banding [46]. G-banding patterns were performed according to Schnedl [47], with modifications proposed by Costa et al. [48].

2.3. Fluorescence In Situ Hybridization (FISH) with Simple Short Repeat Probes (SSR) and Bacterial Artificial Chromosomes (BAC) Probes

Six simple short repeat probes (SSR) were used: $(CA)_{15}$, $(CAA)_{10}$, $(CAC)_{10}$, $(CAG)_{10}$, $(GAA)_{10}$, and $(GAG)_{10}$. Probes were directly labeled with Streptavidin-Cy3 during their synthesis and the hybridization procedures followed Kubat et al. [49].

A total of 64 bacterial artificial chromosomes (BAC) probes from *G. gallus* (GGA, CH261) or *Taeniopygia guttata* (TGMCBA), corresponding to GGA1-28 (except GGA16) and Z and W sex chromosomes were selected and applied to the metaphases of *S. flaveola*. Two BAC clones corresponded to pairs GGA4-28 were used (Table S1). However, a higher number of BAC clones were used to pairs GGA1, 2, and 3 in order to detect intrachromosomal rearrangements (Table S1). Isolation, amplification, labeling, and hybridization of BAC clones were performed according to O'Connor et al. [22]. Probes were labeled with Texas red (red) or FITC (green).

2.4. Microscopic Analysis and Image Capturing

For conventional experiments, the slides were analyzed using an Olympus DP53 optical microscope. Images of repetitive DNAs FISH experiments were analyzed and captured using a Zeiss Imager 2 microscope with Axiovision 4.8 software (Zeiss, Germany). Images of BAC FISH experiments were captured using a CCD camera and SmartCapture (Digital Scientific UK) system coupled on an Olympus BX61 epifluorescence microscope. Final image processing was performed using Adobe Photoshop 7.0. At least 15 metaphase spreads were analyzed to confirm the chromosomal morphologies and FISH results.

3. Results

3.1. Karyotype Description, C and G-Banding

The results showed a diploid number of 2n = 80 in *S. flaveola*, with 11 pairs of macrochromosomes, including the sex chromosomes, and 28 pairs of microchromosomes, as previously proposed [41–43]. Pairs 1, 4, and Z were submetacentric, while the remaining ones were acrocentric (Figure 1). C-banding revealed huge blocks of constitutive heterochromatin in three pairs of microchromosomes, in the short arms of chromosome Z, in the centromere of most macro and microchromosomes, and in the W chromosome, which is heterochromatic in most of its length (Figure 2A).

Figure 1. Complete karyotype of a female specimen of *Sicalis flaveola* 2n = 80.

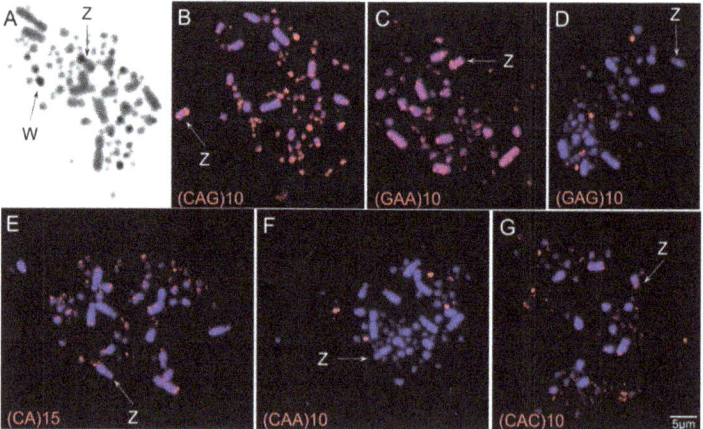

Figure 2. C-banding patterns (**A**) and hybridization of simple short repeats (**B**–**G**) onto metaphases of a female individual of *Sicalis flaveola*. The chromosome probes used are indicated on the left bottom, and the sex chromosomes (Z and W) are indicated by arrows.

3.2. Chromosomal Distribution of Simple Short Repeats (SSRs)

In general, the SSRs tested here showed scattered hybridization, but a general higher accumulation was observed in the microchromosomes and the short arm of the Z chromosome (Figure 2B–G). The W chromosome, on the other hand, showed dispersed signals, like the autosomes. Specifically, sequences $(CA)_{15}$, $(GAA)_{10}$, $(CAG)_{10}$, and $(CAC)_{10}$ showed scattered signals in all chromosomes but with strong signals on the telomere regions of macrochromosomes and in the microchromosomes (Figure 2B,C,E,G). $(GAA)_{10}$, $(CAC)_{10}$, and $(CAG)_{10}$ also produced signals in the short arms of chromosome Z (Figure 2B,C,G). $(GAG)_{10}$ and $(CAA)_{10}$ produced bright signals in two microchromosome pairs and slight signals in an additional pair of microchromosomes (Figure 2B,F). $(GAG)_{10}$ also showed signals on the Z chromosome (Figure 2B).

3.3. Chromosomal Homology Between Chicken and Sicalis flaveola

The chromosomal mapping of BAC clones corresponding to chicken chromosomes GGA1-28, except 16 and 25, and sex chromosomes Z and W evidenced the syntenic conservation of these chromosomes in *S. flaveola* (SFL), with exception of GGA1, which was split into two pairs (SFL 2 and 4) due to centric fission (Figures 3–6). The *S. flaveola* homologous chromosomes to GGA16 and 25 could not be identified because there were no BAC probes to GGA16, and the probes from GGA25 did not produce signals. Chicken chromosome 4 revealed the GGA4q and 4p as separated chromosomes in *S. flaveola* (SFL5 and 12), as in the putative Neognathae karyotype [29]. The analysis of different BAC clones corresponding to GGA1 revealed that intrachromosomal rearrangements occurred in SFL2, homologous to GGA1q. On the other hand, no evidence of this type of rearrangement was observed in the pairs homologous to GGA1p, 2, and 3 (Figure 5). The homology map between *G. gallus* and *S. flaveola* is shown in Figure 6.

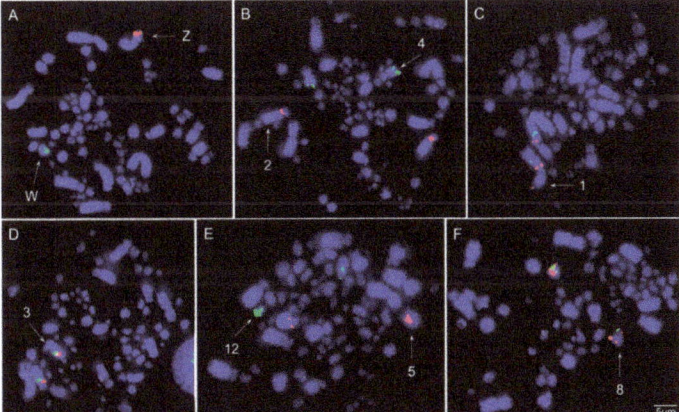

Figure 3. Representative FISH experiments using chicken (CH261) and zebra finch (TGMCBA) macrochromosomes BAC probes in *Sicalis flaveola*: (**A**) chicken macrochromosome Z TGMCBA-270I9 (red) and CH261-94E12 (green); (**B**) chicken macrochromosome 1 TGMCBA-146O14 (red) and TGMCBA-206D5 (green); (**C**) chicken macrochromosome 2 TGMCBA-340P4 (red) and TGMCBA-78C11 (green); (**D**) chicken macrochromosome 3 CH261-130M12 (red) and CH261-97P20 (green); (**E**) chicken macrochromosome 4 CH261-89P6 (red) and CH261-71L6 (green); (**F**) chicken macrochromosome 7 CH261-180H18 (red) and CH261-56K7 (green).

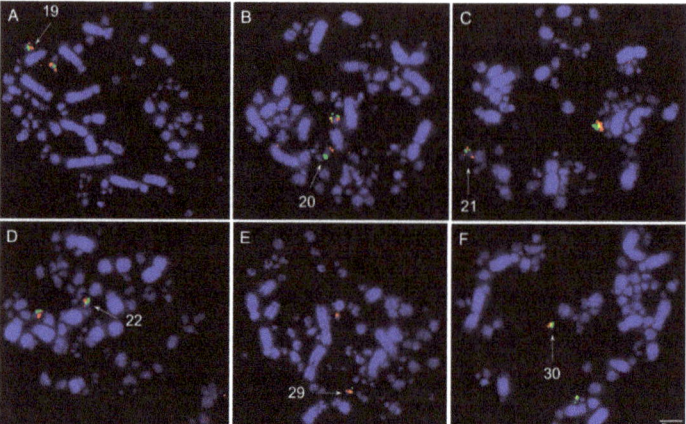

Figure 4. Representative FISH experiments using chicken (CH261) and zebra finch (TGMCBA) microchromosomes BAC probes in *Sicalis flaveola*: (**A**) chicken microchromosome 17 CH261-42P16 (red) and TGMCBA-375I5 (green); (**B**) chicken microchromosome 18 CH261-72B18 (red) and CH26-60N6 (green); (**C**) chicken microchromosome 19 CH261-10F1 (red) and CH261-50H12 (green); (**D**) chicken microchromosome 20 TGMCBA-250E3 (red) and TGMCBA-341F20 (green); (**E**) chicken microchromosome 27 CH261-28L10 (red) and CH261-66M16 (green); (**F**) chicken microchromosome 28 CH261-72A10 (red) and CH261-64A15 (green).

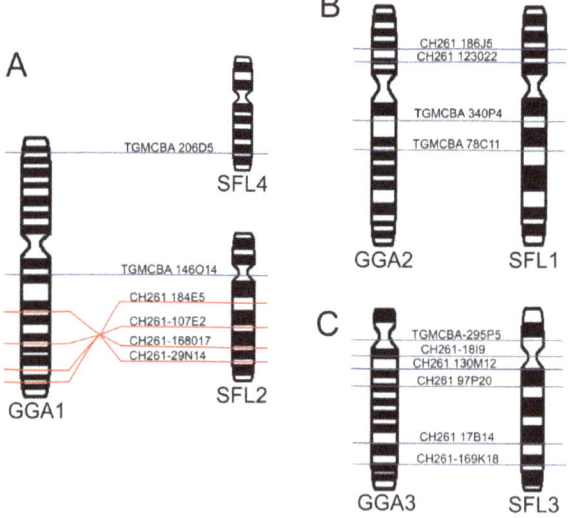

Figure 5. Schematic representation of BAC clones from *Gallus* (CH261) or *Taeniopygia guttata* (TGMCBA) homologous to *G. gallus* chromosome 1 (GGA 1) (**A**), chromosome 2 (GGA 2) (**B**), and chromosome 3 (GGA 3) (**C**) in *Sicalis flaveola* (SFL). Ideograms are represented with G-banding patterns. G-banding data from *G. gallus* followed Ladjali-Mohammedi et al. [50].

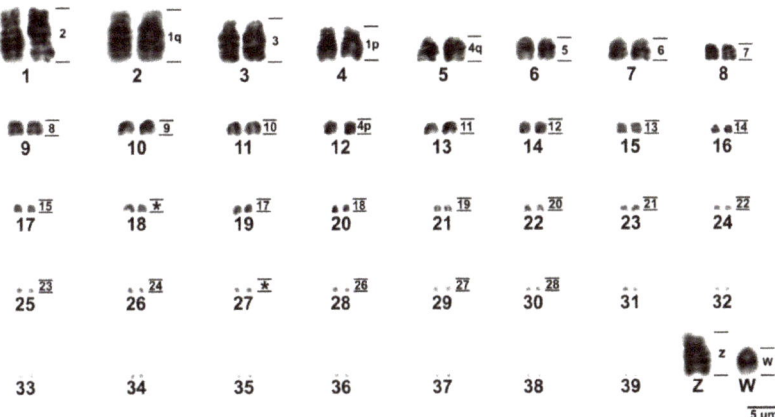

Figure 6. G-banded karyotype of *Sicalis flaveola* and homologous chromosomal segments with *Gallus* chromosomes (right). * Asterisks indicate the probable chromosomes corresponding to GGA16 and GGA25.

4. Discussion

We described here a detailed karyotype description for *S. flaveola*, a representative member of the Thraupidae family, and compared our results with previous studies in Passeriformes, especially Thraupidae. Our results confirmed a typical avian karyotype, with 80 chromosomes, divided into 11 pairs of macrochromosomes, including the Z and W sex chromosomes, and 28 pairs of microchromosomes, corroborating the previous karyotype description [41–43]. This pattern of karyotype is also typical for Passeriformes and Thraupidae species [7].

Despite the constancy of the 2n among Thraupidae species, their C-positive heterochromatin distribution shows distinct patterns among them. In *S. flaveola*, we found C-banding positive in three pairs of microchromosomes, in the centromere of the seventh pair, in the W, and in the entire short arms of the Z. Interestingly, in four other species of Thraupidae, *S. similis*, *S. aurantiirostris*, *Ramphocelus carbo*, and *Tangara cayana*, a block of constitutive heterochromatin was also found in the short arms of the Z chromosome [10,51]. The only exception so far is *Tachyphonus rufus*, in which this block was not found [51]. *R. carbo* and *T. rufus* are a member from the same subfamily (Tachyphoninae). Hence, it is likely that the block of constitutive heterochromatin is a common trait of Thraupidae family, and it was eliminated in *T. rufus*. However, the block of constitutive heterochromatin on the short arms of Z chromosome is not restricted to Thraupidae since a similar pattern has been observed in Passeridae [52,53] and Estrildidae [54] species. Future studies are necessary to investigate if this block of constitutive heterochromatin has a common or independent origin in Thraupidae, Passeridae, and Estrildidae. Nevertheless, these findings highlight that the accumulation/elimination of constitutive heterochromatin in the sex chromosomes is an active process during the chromosomal evolution of Passeriformes. Such role of heterochromatin in the differentiation of sex chromosomes is widely reported in many other groups, including mammals [55], fishes [56,57], plants [58], reptiles [59], among others.

Overall, the SSR probes tested showed scattered hybridization, but brighter signals were observed in the microchromosomes and the short arms of the Z chromosome. They were preferentially associated with heterochromatic regions, corroborating the hypothesis that repetitive DNAs are found in condensed and inactive regions of the genome [11,33]. In addition, previous studies also mentioned that most SSRs are incorporated into non-coding DNA, although they can be found in coding regions, suggesting that these sequences may affect the structure and function of proteins [60].

A distinct pattern of SSRs hybridization has been described in the sex chromosomes among birds. In general, SSRs accumulate in the W chromosomes [28,36–39], except for Piciformes, in which no specific hybridization signal was observed in this chromosome [35,61]. The Piciformes, in contrast, showed extensive SSRs hybridization signals in the Z chromosome of all species tested so far, which were proposed as the main cause of its enlargement [35,61]. SSRs hybridization signals have also been found in the Z chromosome of Passeriformes and Psittaciformes, but it is not a general rule in species from these orders [36,38]. For instance, in Psittaciformes, it has been found in *Myiopsitta monachus*, but no evidence of SSRs accumulation in *Amazona aestiva* has been observed [36]. Similarly, in Passeriformes, it has been described in *Progne tapera*, but not in *Progne chalybea* and *Pygochelidon cyanoleuca* [38]. Therefore, these studies highlight the role of species-specific repetitive DNAs accumulation in the avian sex chromosomes.

The chromosomal mapping of BAC clones indicated a high degree of inter-chromosomal karyotype conservation between *G. gallus* and *S. flaveola*, due to the unique interchromosomal rearrangement that was detected, involving centric fission of the ancestral chromosome 1 (GGA1). This fission is widely reported in Passeriformes species and is therefore considered a synapomorphy for the group [11]. Chicken chromosome 4 hybridized two chromosome pairs in *S. flaveola* (SFL5 and 12). However, this is the ancestral state, as proposed to the putative Neognathae karyotype [29]. Furthermore, intrachromosomal rearrangements already detected in previous studies in other species of Passeriformes were also observed in the chromosome homologous to GGA1q (SFL2). These paracentric and pericentric inversions occurred in the GGA1q chromosome in different Passeriformes species, both oscines and suboscines [10,15–18]. Hence, our study reinforces the hypothesis that these intrachromosomal rearrangements were already present in the common ancestral of Passeriformes [15,16].

Most of the cytogenetic studies on birds address only the macrochromosomes, limiting our understanding of the GGA1-9 [11]. Here we provided, for the first time, a detailed analysis of the microchromosomes in Thraupidae species. Our results revealed that the microchromosomes GGA10-28 (except GGA16 that does not have BAC clones, and GGA25, which in turn do not hybridize in Passeriformes species), are conserved as individual chromosomes in *S. flaveola*. Similar results were found recently in other oscines species, *Taeniopygia guttata*, *Turdus merula*, and *Serinus canaria* [22]. These data indicate that not only the macrochromosomes but also the microchromosomes are highly conserved among Passeriformes. Thus, we suggest that the ancestral pattern of microchromosome organization was already present in the last common ancestral to Passeriformes.

The order Passeriformes represents approximately 60% of the avian species [2], and no other avian clade has evolved such great diversity in terms of number of species, morphological and ecological diversification [62]. Interestingly, this diversity was not accompanied by interchromosomal reorganization (Figure 7 and Table S2). On the other hand, parrots (Psittaciformes), the sister group of the Passeriformes [63–65], which represent approximately 3.6% of the avian species [2], underwent a high rate of interchromosomal rearrangements, involving fusions of macrochromosomes (Figure 7 and Table S2) and microchromosomes [22,25,27]. This may indicate that the maintenance of the ancestral pattern of karyotype in Passeriformes was crucial to the successful diversification seen in this clade. Intrachromosomal rearrangements, such as inversions, have been extensively described in both Passeriformes and Psittaciformes species. For instance, 125 and 134 intrachromosomal changes have been described in *T. guttata* and *Melopsittacus undulatus*, respectively [66]. Although intrachromosomal rearrangements are considered as one of the most prominent adaptation mechanisms [67–69], this type of rearrangement does not explain the great difference in terms of the number of species between Passeriformes and Psittaciformes, since both orders underwent a similar amount of intrachromosomal changes.

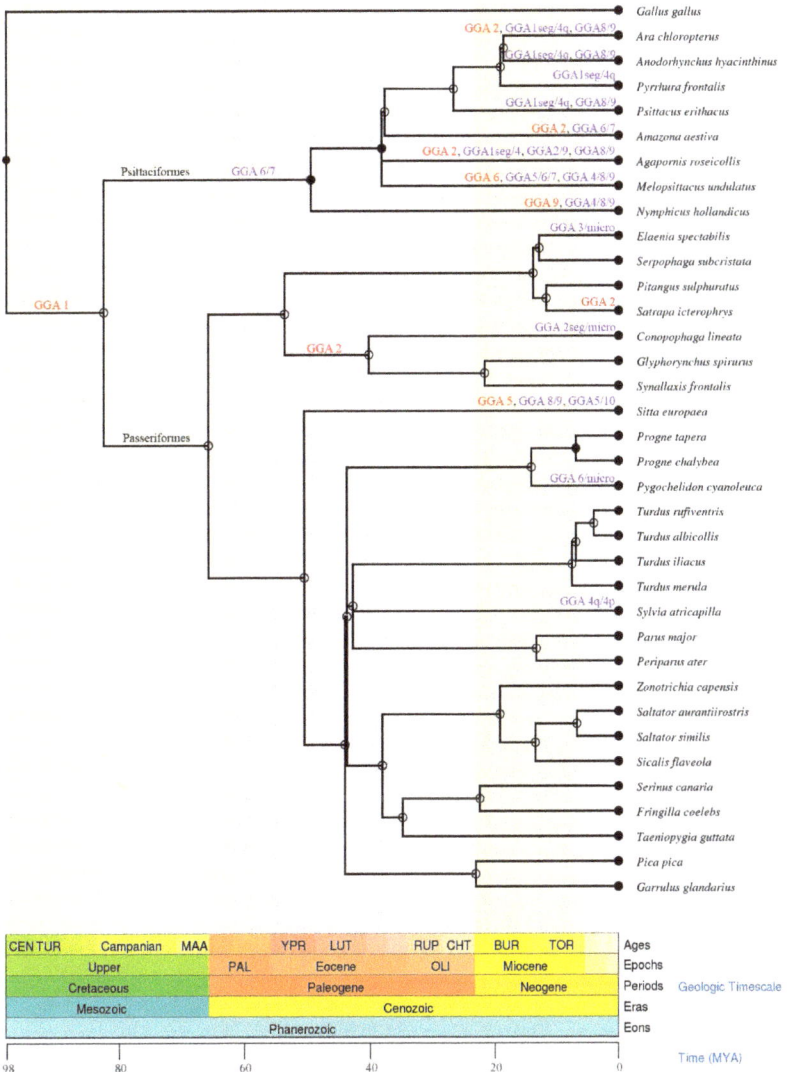

Figure 7. Chromosomal rearrangements in Passeriformes and Psittaciformes species analyzed with chromosome painting with *Gallus gallus* (GGA) probes (GGA1-10) or BACs clones corresponding to these GGA chromosomes. The phylogenetic tree was sourced from TimeTree databases (http://www.timetree.org, accessed on 12 May 2021) [70]. Rearrangements are represented by fissions (red) and fusions (blue). Seg = segment, q = long arm, micro = microchromosome.

Repetitive DNA plays an important role in genome organization and function, and they often serve as hotspots of genome rearrangements [31,32]. Hence, by comparing the chromosomal mapping of microsatellite sequences between the sister clades Passeriformes and Psittaciformes, with low and high rates of chromosomal rearrangements, respectively, we can speculate about the importance of these sequence in the karyotype reorganization and the diversification of these clades. Up to now, only four Passeriformes species (*P. tapera, P. chalybea, P. cyanoleuca,* and *S. flaveola*) and two Psittaciformes ones (*Myiopsitta monachus* and *Amazona aestiva*) have been analyzed with chromosomal mapping of SSRs sequences [36,38] (Table 1). Comparing the three SSRs used in common in these species is clear that these Passeriformes have a higher number of signals than the Psittaciformes species. These findings suggest that the SSRs were not involved in the high difference observed between the karyotype organization of Passeriformes and Psittaciformes. However, we cannot discard the involvement of repetitive sequences in the karyotype reorganization of Psittaciformes since other types of these sequences were not explored, such as transposable elements and satellites DNA.

Table 1. Chromosome mapping comparison of microsatellites among Passeriformes and Psittaciformes species.

Species	SSRs		
	$(CAA)_{10}$	$(CAG)_{10}$	$(CA)_{15}$
S. flaveola, 2n = 80 [1]	Three pairs of micros	Scattered signals in all chromosomes but strong signals on the telomere regions of macros and micros, and in the Zp	Scattered signals in all chromosomes but strong signals on the telomere regions of macros and in the micros
P. cyanoleuca, 2n = 76 [2]	Telomere of Wq	Telomere of Wpq	Telomere of Wq
P. tapera, 2n = 76 [2]	Telomere of 1q	Wq	Telomere of 1q, 2q, Wq; Pericentromeric region of 6, 7, Wq
P. chalybea, 2n = 76 [2]	Telomere of 1pq, 2q, Wq; pericentromeric region of 1pq	-	Telomere of 1pq, 2q, 4q, Wq; pericentromeric region of 1pq, Wpq
M. monachus, 2n = 48 [3]	Wq	Telomere region of 1p, 2q, 3q, 4pq, 5q, 6p, Zp; pericentromeric region of 1q, 7q, Wq; centromeric region 1-9; all micros	-
A. aestiva, 2n = 70 [3]	-	-	-

[1] Present study, [2] Barcellos et al. [38], [3] Furo et al. [36], 2n = diploid number, macro = macrochromosomes, micro = microchromosomes, p = short arms, and q = long arms.

5. Conclusions

In the present study, we demonstrated the most complete cytogenetic analysis to date of a Thraupidae family member, contributing to a better understanding of its chromosome organization and evolution. The BAC probes of G. gallus were applied for the first time in S. flaveola, showing conservation in the ancestral microchromosomes and most macrochromosomes. Taken together, our findings displayed a typical avian karyotype with a high rate of homology with G. gallus, some intrachromosomal rearrangements, scattered SSRs distribution, and an uncommon accumulation of these sequences in the Z chromosome. Our comparison of chromosomal mapping of SSRs between the sister clades Passeriformes and Psittaciformes indicated that these sequences were not involved in the karyotype reorganization of Psittaciformes since Passeriformes species showed a higher number of signals.

Supplementary Materials: The following are available online at https://www.mdpi.com/article/10.3390/ani11051456/s1, Table S1: List of BACs applied to *Sicalis flaveola* (SFL); and Table S2: List of Passeriformes and Psittaciformes species analyzed with *Gallus gallus* (GGA) chromosome painting or BACs clones corresponding to GGA1-10.

Author Contributions: Conceptualization, R.K.; methodology, R.K., A.d.V.G., R.J.G., B.S.R., S.A.B. and A.L.C.; validation, R.K., B.S.R. and A.L.C.; formal analysis, R.K. and B.S.R.; investigation, R.K.; resources, R.K., A.d.V.G., R.J.G., M.d.B.C., E.H.C.d.O. and D.K.G.; data curation, R.K.; writing—original draft preparation, R.K.; writing—review and editing, R.K., A.d.V.G., R.J.G., M.d.B.C., E.H.C.d.O. and D.K.G.; visualization, R.K. and B.S.R.; supervision, D.K.G.; project administration, D.K.G.; funding acquisition, R.K., A.d.V.G., R.J.G., M.d.B.C., E.H.C.d.O. and D.K.G.; All authors have read and agreed to the published version of the manuscript.

Funding: This research was funded by the Conselho Nacional de Desenvolvimento Científico e Tecnológico (CNPq, Proc. PDE 204792/2018-5, PDJ 151056/2020-0, and PQ 307382/2019-2), and the Biotechnology and Biological Sciences Research Council UK (BB/K008226/1).

Institutional Review Board Statement: The study was conducted following the procedures approved by the "Biodiversity Authorization and Information System", permission numbers 44173-1 and 33860-4, and by the Ethics Committee on Animal Experimentation (CEUA) of the Universidade Federal do Pampa under no. 026/2012 and 018/2014.

Data Availability Statement: The data presented in this study are available in the article or supplementary material.

Acknowledgments: The authors are grateful to all our colleagues at the Laboratório de Diversidade Genética Animal from the Universidade Federal do Pampa (RS, Brazil) for their support to collect and perform the cell culture of the specimens analyzed in this study. We are also grateful to Alex Pinheiro de Araújo for the illustration of the *Sicalis flaveola* used in Figure 1.

Conflicts of Interest: The authors declare no conflict of interest.

References

1. Burns, K.J.; Shultz, A.J.; Title, P.O.; Mason, N.A.; Barker, F.K.; Klicka, J.; Lanyon, S.M.; Lovette, I.J. Phylogenetics and diversification of tanagers (Passeriformes: Thraupidae), the largest radiation of Neotropical songbirds. *Mol. Phylogenetics Evol.* **2014**, *75*, 41–77. [CrossRef] [PubMed]
2. Gill, F. *IOC World Bird List*; (v11.1); Gill, F., Donsker, D., Rasmussen, P., Eds.; Gill & Wright: London, UK, 2021. Available online: https://www.worldbirdnames.org/ioc-lists/crossref/ (accessed on 21 December 2020).
3. Ericson, P.G.P.; Irestedt, M.; Johansson, U.S. Evolution, biogeography, and patterns of diversification in passerine birds. *J. Avian Biol.* **2003**, *34*, 3–15. [CrossRef]
4. Ridgely, R.S.; Tudor, G. *Field Guide to the Songbirds of South America: The Passerines*; University of Texas Press: Austin, TX, USA, 2009.
5. Christidis, L.; Boles, W. *Systematics and Taxonomy of Australian Birds*; Csiro Publishing: Clayton, Australia, 2008.
6. BirdLife International. Species Factsheet: Sicalis Flaveola. Available online: http://www.birdlife.org (accessed on 21 December 2020).
7. Degrandi, T.M.; Barcellos, S.A.; Costa, A.L.; Garnero, A.D.; Hass, I.; Gunski, R.J. Introducing the Bird Chromosome Database: An Overview of Cytogenetic Studies in Birds. *Cytogenet. Genome Res.* **2020**, *160*, 199–205. [CrossRef] [PubMed]
8. Goldschmidt, B.; Nogueira, D.M.; Monsores, D.W.; Souza, L.M. Chromosome study in two *Aratinga* species (*A. guarouba* and *A. acuticaudata*) (Psittaciformes). *Braz. J. Genet.* **1997**, *20*. [CrossRef]
9. Cabanne, G.S.; Gunski, R.J.; Contreras, J.R. *Primeros Resultados de Estudios Citogenéticos en Saltator coerulescens y Saltator similis (Aves: Emberizidae)*; VI Jornada de Ciências Naturales del Litoral: Corrientes, Argentina, 1997.
10. Dos Santos, M.S.; Kretschmer, R.; Silva, F.A.O.; Ledesma, M.A.; O'Brien, P.C.M.; Ferguson-Smith, M.A.; Garnero, A.D.V.; de Oliveira, E.H.C.; Gunski, R.J. Intrachromosomal rearrangements in two representatives of the genus *Saltator* (Thraupidae, Passeriformes) and the occurrence of heteromorphic Z chromosomes. *Genetica* **2014**, *143*, 535–543. [CrossRef]
11. Kretschmer, R.; Ferguson-Smith, M.A.; de Oliveira, E.H.C. Karyotype evolution in birds: From conventional staining to chromosome painting. *Genes* **2018**, *9*, 181. [CrossRef] [PubMed]
12. Volker, M.; Backstrom, N.; Skinner, B.M.; Langley, E.J.; Bunzey, S.K.; Ellegren, H.; Griffin, D.K. Copy number variation, chromosome rearrangement, and their association with recombination during avian evolution. *Genome Res.* **2010**, *20*, 503–511. [CrossRef]
13. Warren, W.C.; Clayton, D.F.; Ellegren, H.; Arnold, A.P.; Hillier, L.W.; Künstner, A.; Searle, S.; White, S.; Vilella, A.J.; Fairley, S.; et al. The genome of a songbird. *Nature* **2010**, *464*, 757–762. [CrossRef]
14. Kretschmer, R.; Gunski, R.J.; Garnero, A.D.V.; Furo, I.O.; O'Brien, P.C.; Ferguson-Smith, M.A.; de Oliveira, E.H.C. Molecular cytogenetic characterization of multiple intrachromosomal rearrangements in two representatives of the genus *Turdus* (Turdidae, Passeriformes). *PLoS ONE* **2014**, *9*, e103338. [CrossRef]
15. Kretschmer, R.; de Oliveira, E.H.C.; dos Santos, M.S.; Furo, I.O.; O'Brien, P.C.; Ferguson-Smith, M.A.; Garnero, A.D.V.; Gunski, R.J. Chromosome mapping of the large elaenia (*Elaenia spectabilis*): Evidence for a cytogenetic signature for passeriform birds? *Biol. J. Linn. Soc.* **2015**, *115*, 391–398. [CrossRef]
16. Kretschmer, R.; de Lima, V.L.C.; de Souza, M.S.; Costa, A.L.; O'Brien, P.C.; Ferguson-Smith, M.A.; de Oliveira, E.H.C.; Gunski, R.J.; Garnero, A.D.V. Multidirectional chromosome painting in *Synallaxis frontalis* (Passeriformes, Furnariidae) reveals high chromosomal reorganization, involving fissions and inversions. *Comp. Cytogenet.* **2018**, *12*, 97–110. [CrossRef] [PubMed]
17. Dos Santos, M.S.; Kretschmer, R.; Frankl-Vilches, C.; Bakker, A.; Gahr, M.; O'Brien, P.C.M.; Ferguson-Smith, M.A.; de Oliveira, E.H.C. Comparative cytogenetics between two important songbird models: The zebra finch and the canary. *PLoS ONE* **2017**, *12*, e0170997. [CrossRef] [PubMed]

18. Rodrigues, B.S.; Kretschmer, R.; Gunski, R.J.; Garnero, A.D.V.; O'Brien, P.C.M.; Ferguson-Smith, M.; de Oliveira, E.H.C. Chromosome painting in tyrant flycatchers confirms a set of inversions shared by Oscines and Suboscines (Aves, Passeriformes). *Cytogenet. Genome Res.* **2017**, *153*, 205–212. [CrossRef] [PubMed]
19. Smith, J.; Burt, D.W. Parameters of the chicken genome (*Gallus gallus*). *Anim. Genet.* **1998**, *29*, 290–294. [CrossRef]
20. Smith, J.; Bruley, C.K.; Paton, I.R.; Dunn, I.; Jones, C.T.; Windsor, D.; Morrice, D.R.; Law, A.S.; Masabanda, J.; Sazanov, A.; et al. Differences in gene density on chicken macrochromosomes and microchromosomes. *Anim. Genet.* **2000**, *31*, 96–103. [CrossRef]
21. Guttenbach, M.; Nanda, I.; Feichtinger, W.; Masabanda, J.S.; Griffin, D.K.; Schmid, M. Comparative chromosome painting of chicken autosomal paints 1–9 in nine different bird species. *Cytogenet. Genome Res.* **2003**, *103*, 173–184. [CrossRef]
22. O'Connor, R.E.; Kiazim, L.; Skinner, B.; Fonseka, G.; Joseph, S.; Jennings, R.; Larkin, D.M.; Griffin, D.K. Patterns of microchromosome organization remain highly conserved throughout avian evolution. *Chromosoma* **2019**, *128*, 21–29. [CrossRef]
23. Ribas, T.F.A.; Pieczarka, J.C.; Griffin, D.K.; Kiazim, L.G.; Nagamachi, C.Y.; O'Brien, P.C.M.; Ferguson-Smith, M.A.; Yang, F.; Aleixo, A.; O'Connor, R.E. Analysis of multiple chromosomal rearrangements in the genome of *Willisornis vidua* using BAC-FISH and chromosome painting on a supposed conserved karyotype. *BMC Ecol. Evol.* **2021**, *21*, 34. [CrossRef]
24. Nishida, C.; Ishijima, J.; Kosaka, A.; Tanabe, H.; Habermann, F.A.; Griffin, D.K.; Matsuda, Y. Characterization of chromosome structures of Falconinae (Falconidae, Falconiformes, Aves) by chromosome painting and delineation of chromosome rearrangements during their differentiation. *Chromosome Res.* **2008**, *16*, 171–181. [CrossRef]
25. Nanda, I.; Karl, E.; Griffin, D.K.; Schartl, M.; Schmid, M. Chromosome repatterning in three representative parrots (Psittaciformes) inferred from comparative chromosome painting. *Cytogenet. Genome Res.* **2007**, *117*, 43–53. [CrossRef]
26. Joseph, S.; O'Connor, R.E.; Al Mutery, A.F.; Watson, M.; Larkin, D.M.; Griffin, D.K. Chromosome Level Genome Assembly and Comparative Genomics between Three Falcon Species Reveals an Unusual Pattern of Genome Organisation. *Diversity* **2018**, *10*, 113. [CrossRef]
27. Furo, I.O.; Kretschmer, R.; O'Brien, P.C.M.; Pereira, J.; Garnero, A.D.V.; Gunski, R.J.; O'Connor, R.E.; Griffin, D.K.; Gomes, A.J.B.; Ferguson-Smith, M.A.; et al. Chromosomal evolution in the phylogenetic context of Neotropical Psittacidae with emphasis on a species with high karyotypic reorganization (*Myiopsitta monachus*). *Front. Genet.* **2020**, *11*, 721. [CrossRef] [PubMed]
28. Kretschmer, R.; Gunski, R.J.; Garnero, A.D.V.; de Freitas, T.R.O.; Toma, G.A.; Cioffi, M.B.; de Oliveira, E.H.C.; O'Connor, R.E.; Griffin, D.K. Chromosomal analysis in *Crotophaga ani* (Aves, Cuculiformes) reveals extensive genomic reorganization and an unusual Z-autosome Robertsonian translocation. *Cells* **2021**, *10*, 4. [CrossRef] [PubMed]
29. Kretschmer, R.; de Souza, M.S.; Furo, I.d.O.; Romanov, M.N.; Gunski, R.J.; Garnero, A.d.V.; de Freitas, T.R.O.; de Oliveira, E.H.C.; O'Connor, R.E.; Griffin, D.K. Interspecies Chromosome Mapping in Caprimulgiformes, Piciformes, Suliformes, and Trogoniformes (Aves): Cytogenomic Insight into Microchromosome Organization and Karyotype Evolution in Birds. *Cells* **2021**, *10*, 826. [CrossRef] [PubMed]
30. Yano, C.F.; Poltronieri, J.; Bertollo, L.A.C.; Artoni, R.F.; Liehr, T.; Cioffi, M.B. Chromosomal Mapping of Repetitive DNAs in *Triportheus trifurcatus* (Characidae, Characiformes): Insights into the Differentiation of the Z and W Chromosomes. *PLoS ONE* **2014**, *9*, e90946. [CrossRef]
31. George, C.M.; Alani, E. Multiple cellular mechanisms prevent chromosomal rearrangements involving repetitive DNA. *Crit. Rev. Biochem. Mol. Biol.* **2012**, *47*, 297–313. [CrossRef]
32. Dunn, M.J.; Anderson, M.Z. To Repeat or Not to Repeat: Repetitive Sequences Regulate Genome Stability in *Candida albicans*. *Genes* **2019**, *10*, 866. [CrossRef]
33. Ellegren, H. Microsatellites: Simple sequences with complex evolution. *Nat. Rev. Genet.* **2004**, *5*, 435–445. [CrossRef]
34. Matsubara, K.; O'Meally, D.; Azad, B.; Georges, A.; Sarre, S.D.; Graves, J.A.M.; Matsuda, Y.; Ezaz, T. Amplification of microsatellite repeat motifs is associated with the evolutionary differentiation and heterochromatinization of sex chromosomes in Sauropsida. *Chromosoma* **2016**, *125*, 111–123. [CrossRef]
35. de Oliveira, T.D.; Kretschmer, R.; Bertocchi, N.A.; Degrandi, T.M.; de Oliveira, E.H.C.; Cioffi, M.B.; Garnero, A.D.V.; Gunski, R.J. Genomic organization of repetitive DNA in woodpeckers (Aves, Piciformes): Implications for karyotype and ZW sex chromosome differentiation. *PLoS ONE* **2017**, *12*, e0169987. [CrossRef]
36. Furo, I.O.; Kretschmer, R.; dos Santos, M.S.; Carvalho, C.A.; Gunski, R.J.; O'Brien, P.C.M.; Ferguson-Smith, M.A.; Cioffi, M.B.; de Oliveira, E.H.C. Chromosomal Mapping of Repetitive DNAs in *Myiopsitta monachus* and *Amazona aestiva* (Psittaciformes, Psittacidae), with emphasis on the sex chromosomes. *Cytogenet. Genome Res.* **2017**, *151*, 151–160. [CrossRef] [PubMed]
37. Kretschmer, R.; de Oliveira, T.D.; Furo, I.O.; Silva, F.A.O.; Gunski, R.J.; Garnero, A.V.; Cioffi, M.B.; de Oliveira, E.H.C.; de Freitas, T.R.O. Repetitive DNAs and shrink genomes: A chromosomal analysis in nine Columbidae species (Aves, Columbiformes). *Genet. Mol. Biol.* **2018**, *41*, 98–106. [CrossRef]
38. Barcellos, S.; Kretschmer, R.; de Souza, M.S.; Costa, A.L.; Degrandi, T.M.; dos Santos, M.S.; de Oliveira, E.H.C.; Cioffi, M.B.; Gunski, R.J.; Garnero, A.V. Karyotype Evolution and Distinct Evolutionary History of the W Chromosomes in Swallows (Aves, Passeriformes). *Cytogenet. Genome Res.* **2019**, *158*, 98–105. [CrossRef] [PubMed]
39. Gunski, R.J.; Kretschmer, R.; de Souza, M.S.; Furo, I.O.; Barcellos, S.; Costa, A.L.; Cioffi, M.B.; de Oliveira, E.H.C.; Garnero, A.V. Evolution of bird sex chromosomes narrated by repetitive sequences: Unusual W chromosome enlargement in *Gallinula melanops* (Aves: Gruiformes: Rallidae). *Cytogenet. Genome Res.* **2019**, *158*, 152–159. [CrossRef] [PubMed]

40. de Souza, M.S.; Kretschmer, R.; Barcellos, A.S.; Costa, A.L.; Cioffi, M.B.; de Oliveira, E.H.C.; Garnero, A.D.V.; Gunski, R.J. Repeat Sequence Mapping Shows Different W Chromosome Evolutionary Pathways in Two Caprimulgiformes Families. *Birds* **2020**, *1*, 19–34. [CrossRef]
41. De Lucca, E.J. Cariótipo de oito espécies de Aves. *Rev. Bras. Biol.* **1974**, *34*, 387–392.
42. De Lucca, E.J.; Chamma, L. Estudo do complemento cromossômico de 11 espécies de aves das ordens Columbiformes, Passeriformes e Tinamiformes. *Braz. J. Med. Biol. Res.* **1977**, *10*, 97–105.
43. Carvalho, M.V.P. Estudos Citogenéticos na Família Fringillidae (Passeriformess-Aves). MSc Thesis, Programa de pós-graduação em Genética e Biologia Molecular, Universidade Federal do Rio Grande do Sul, Porto Alegre, Brazil, 1989.
44. Sasaki, M.; Ikeuchi, T.; Makino, S. A feather pulp culture technique for avian chromosomes, with notes on the chromosomes of the peafowl and the ostrich. *Experientia* **1968**, *24*, 1292–1293. [CrossRef] [PubMed]
45. Guerra, M.S. Reviewing the chromosome nomenclature of Levan et al. *Brazil. J. Genet.* **1986**, *9*, 741–743.
46. Sumner, A.T. A simple technique for demonstrating centromeric heterochromatin. *Exp. Cell Res.* **1972**, *75*, 304–306. [CrossRef]
47. Schnedl, W. Analysis of the human karyotype using a reassociation technique. *Chromosoma* **1971**, *34*, 448–454. [CrossRef] [PubMed]
48. Costa, A.L.; Lopes, C.F.; de Souza, M.S.; Barcellos, S.A.; Vielmo, P.G.; Gunski, R.J.; Garnero, A.D.V. Comparative cytogenetics in three species of Wood-Warblers (Aves: Passeriformes: Parulidae) reveal divergent banding patterns and chromatic heterogeneity for the W chromosome. *Caryologia*, in press.
49. Kubat, Z.; Hobza, R.; Vyskot, B.; Kejnovsky, E. Microsatellite accumulation in the Y chromosome in *Silene latifolia*. *Genome* **2008**, *51*, 350–356. [CrossRef]
50. Ladjali-Mohammedi, K.; Bitgood, J.J.; Tixier-Boichard, M.; de Leon, F.P. International system for standardized avian karyotypes (ISSAK): Standardized banded karyotypes of the domestic fowl (*Gallus domesticus*). *Cytogenet. Genome Res.* **1999**, *86*, 271–276. [CrossRef]
51. Correia, V.C.; Garnero, A.D.V.; dos Santos, L.P.; da Silva, R.R.; Barbosa, M.O.; Bonifacio, H.L.; Gunski, R.J. Alta similaridade cariotípica na família Emberizidae (Aves: Passeriformes). *Biosci. J.* **2009**, *25*, 99–111.
52. Fulgione, D.; Aprea, G.; Milone, M.; Odierna, G. Chromosomes and heterochromatin of the Italian sparrow (*Passer italiae*, Vieillot 1817), a taxon of presumed hybrid origins. *Folia Zool.* **2000**, *49*, 199–204.
53. Christid, L. Chromosomal evolution in finches and their allies (families: Ploceidae, Fringillidae, and Emberizidae). *Can. J. Genet. Cytol.* **1986**, *28*, 762–769. [CrossRef]
54. Christid, L. Chromosomal evolution within the family Estrildidae (Aves) II. The Lonchurae. *Genetica* **1986**, *71*, 99–113. [CrossRef]
55. Marchal, J.A.; Acosta, M.J.; Neitzel, H.; Sperling, K.; Bullejos, M.; Díaz de la Guardia, R.; Sanchez, A. X chromosome painting in *Microtus*: Origin and evolution of the giant sex chromosomes. *Chromosome Res.* **2004**, *12*, 767–776. [CrossRef]
56. Cioffi, M.B.; Moreira-Filho, O.; Almeida-Toledo, L.F.; Bertollo, L.A.C. The contrasting role of heterochromatin in the differentiation of sex chromosomes: An overview from Neotropical fishes. *J. Fish Biol.* **2012**, *80*, 2125–2139. [CrossRef]
57. Yano, C.F.; Bertollo, L.A.C.; Liehr, T.; Troy, W.P.; Cioffi, M.B. W chromosome dynamics in *Triportheus* species (Characiformes, Triportheidae)—An ongoing process narrated by repetitive sequences. *J. Hered.* **2016**, *107*, 342–348. [CrossRef]
58. Kejnovsky, E.; Hobza, R.; Cermak, T.; Kubat, Z.; Vyskot, B. The role of repetitive DNA in structure and evolution of sex chromosomes in plants. *Heredity* **2009**, *102*, 533–541. [CrossRef]
59. Ezaz, T.; Sarre, S.D.; O'Meally, D.; Marshall Graves, J.A.; Georges, A. Sex chromosome evolution in lizards: Independent origins and rapid transitions. *Cytogenet. Genome Res.* **2009**, *127*, 249–260. [CrossRef] [PubMed]
60. Castagnone-Sereno, P.; Danchin, E.G.; Deleury, E.; Guillemaud, T.; Malausa, T.; Abad, P. Genome-wide survey and analysis of microsatellites in nematodes, with a focus on the plant-parasitic species *Meloidogyne incognita*. *BMC Genom.* **2010**, *11*, 598. [CrossRef] [PubMed]
61. Kretschmer, R.; Furo, I.O.; Cioffi, M.B.; Gunski, R.J.; Garnero, A.D.V.; O'Brien, P.C.M.; Ferguson-Smith, M.A.; de Freitas, T.R.O.; de Oliveira, E.H.C. Extensive chromosomal fissions and repetitive DNA accumulation shaped the atypical karyotypes of two Ramphastidae (Aves: Piciformes) species. *Biol. J. Linn. Soc.* **2020**, *130*, 839–849. [CrossRef]
62. Johansson, U.S.; Fjeldså, J.; Bowie, R.C.K. Phylogenetic relationships within Passerida (Aves: Passeriformes): A review and a new molecular phylogeny based on three nuclear intron markers. *Mol. Phylogenetics Evol.* **2008**, *48*, 858–876. [CrossRef] [PubMed]
63. Hackett, S.J.; Kimball, R.T.; Reddy, S.; Bowie, R.C.K.; Braun, E.L.; Braun, M.J.; Chojnowski, J.L.; Cox, W.A.; Han, K.L.; Harshman, J.; et al. A phylogenomic study of birds reveals their evolutionary history. *Science* **2008**, *320*, 1763–1768. [CrossRef] [PubMed]
64. Jarvis, E.D.; Mirarab, S.; Aberer, A.J.; Li, B.; Houde, P.; Li, C.; Ho, S.Y.; Faircloth, B.C.; Nabholz, B.; Howard, J.T.; et al. Whole-genome analyses resolve early branches in the tree of life of modern birds. *Science* **2014**, *346*, 1320–1331. [CrossRef]
65. Prum, R.O.; Berv, J.S.; Dornburg, A.; Field, D.J.; Townsend, J.P.; Lemmon, E.M.; Lemmon, A.R. A comprehensive phylogeny of birds (Aves) using targeted next-generation DNA sequencing. *Nature* **2015**, *526*, 569–573. [CrossRef]
66. Romanov, M.N.; Farré, M.; Lithgow, P.E.; Fowler, K.E.; Skinner, B.M.; O'Connor, R.; Fonseka, G.; Backström, N.; Matsuda, Y.; Nishida, C.; et al. Reconstruction of gross avian genome structure, organization and evolution suggests that the chicken lineage most closely resembles the dinosaur avian ancestor. *BMC Genom.* **2014**, *15*, 1–18. [CrossRef]
67. Dobzhansky, T. *Genetics and the Origin of Species*; Columbia University Press: New York, NY, USA, 1937.
68. Krimbas, C.B.; Powell, J.R. *Drosophila Inversion Polymorphism*; CRC Press: Boca Raton, FL, USA, 1992.

69. Hoffmann, A.A.; Sgrò, C.M.; Weeks, A.R. Chromosomal inversion polymorphisms and adaptation. *Trends Ecol. Evol.* **2004**, *19*, 482–488. [CrossRef]
70. Kumar, S.; Stecher, G.; Suleski, M.; Hedges, S.B. TimeTree: A Resource for Timelines, Timetrees, and Divergence Times. *Mol. Biol. Evol.* **2017**, *34*, 1812–1819. [CrossRef] [PubMed]

Case Report

The Second Case of Non-Mosaic Trisomy of Chromosome 26 with Homologous Fusion 26q;26q in the Horse

Sharmila Ghosh [1,†], Josefina Kjöllerström [1,†], Laurie Metcalfe [2], Stephen Reed [2], Rytis Juras [1] and Terje Raudsepp [1,*]

1. Department of Veterinary Integrative Biosciences, Texas A&M University, College Station, TX 77843, USA; sharmila173@gmail.com (S.G.); jkjollerstrom@cvm.tamu.edu (J.K.); rjuras@cvm.tamu.edu (R.J.)
2. Rood & Riddle Equine Hospital, Lexington, KY 40580, USA; lmetcalfe@roodandriddle.com (L.M.); sreed@roodandriddle.com (S.R.)
* Correspondence: traudsepp@cvm.tamu.edu
† These authors contributed equally to this work.

Simple Summary: We present chromosome and DNA analysis of a normal Thoroughbred mare and her abnormal foal born with neurologic defects. We show that the foal has an abnormal karyotype with three copies of chromosome 26 (trisomy chr26), instead of the normal two. However, two of the three chr26 have fused, forming an unusual derivative chromosome. Chromosomes of the dam are normal, suggesting that the chromosome abnormality found in the foal happened during egg or sperm formation or after fertilization. Analysis of the foal and the dam with chr26 DNA markers indicates that the extra chr26 in the foal is likely of maternal origin and that the unusual derivative chromosome resulted from the fusion of two parental chr26. We demonstrate that although conventional karyotype analysis can accurately identify chromosome abnormalities, determining the mechanism and parental origin of these abnormalities requires DNA analysis. Most curiously, this is the second case of trisomy chr26 with unusual derivative chromosome in the horse, whereas all other equine trisomies have three separate copies of the chromosome involved. Because horse chr26 shares genetic similarity with human chr21, which trisomy causes Down syndrome, common features between trisomies of horse chr26 and human chr21 are discussed.

Abstract: We present cytogenetic and genotyping analysis of a Thoroughbred foal with congenital neurologic disorders and its phenotypically normal dam. We show that the foal has non-mosaic trisomy for chromosome 26 (ECA26) but normal 2n = 64 diploid number because two copies of ECA26 form a metacentric derivative chromosome der(26q;26q). The dam has normal 64,XX karyotype indicating that der(26q;26q) in the foal originates from errors in parental meiosis or post-fertilization events. Genotyping ECA26 microsatellites in the foal and its dam suggests that trisomy ECA26 is likely of maternal origin and that der(26q;26q) resulted from Robertsonian fusion. We demonstrate that conventional and molecular cytogenetic approaches can accurately identify aneuploidy with a derivative chromosome but determining the mechanism and parental origin of the rearrangement requires genotyping with chromosome-specific polymorphic markers. Most curiously, this is the second case of trisomy ECA26 with der(26q;26q) in the horse, whereas all other equine autosomal trisomies are 'traditional' with three separate chromosomes. We discuss possible ECA26 instability as a contributing factor for the aberration and likely ECA26-specific genetic effects on the clinical phenotype. Finally, because ECA26 shares evolutionary homology with human chromosome 21, which trisomy causes Down syndrome, cytogenetic, molecular, and phenotypic similarities between trisomies ECA26 and HSA21 are discussed.

Keywords: karyotyping; FISH; STR genotyping; parental origin; congenital abnormalities; neurologic disorders; Down syndrome

Citation: Ghosh, S.; Kjöllerström, J.; Metcalfe, L.; Reed, S.; Juras, R.; Raudsepp, T. The Second Case of Non-Mosaic Trisomy of Chromosome 26 with Homologous Fusion 26q;26q in the Horse. *Animals* **2022**, *12*, 803. https://doi.org/10.3390/ani12070803

Academic Editors: Pietro Parma and Leopoldo Iannuzzi

Received: 8 February 2022
Accepted: 18 March 2022
Published: 22 March 2022

Publisher's Note: MDPI stays neutral with regard to jurisdictional claims in published maps and institutional affiliations.

Copyright: © 2022 by the authors. Licensee MDPI, Basel, Switzerland. This article is an open access article distributed under the terms and conditions of the Creative Commons Attribution (CC BY) license (https://creativecommons.org/licenses/by/4.0/).

1. Introduction

Multiple forms of chromosome rearrangements have been reported in the domestic horse, *Equus caballus* (ECA) and most are associated with decreased fertility, embryonic or fetal loss, congenital and developmental disorders, causing significant economic loss to breeders and the equine industry [1,2]. The most commonly found chromosomal abnormalities in horses are X-monosomy and XY male-to-female sex reversal (also known as XY disorder of sex development or XY DSD) [1–3], which owe to the specific features of equine sex chromosome organization [2,4,5]. Rearrangements involving autosomes, however, are rare in horses and include mainly a few translocations and autosomal aneuploidies [2].

Aneuploidies cause genetic imbalance, due to which most of them are lethal [6], and the 14 reported live-born cases of autosomal trisomies involve only the six smallest equine autosomes—ECA23, 26, 27, 28, 30 and 31 [1,2,7]. Autosomal aneuploidies are equally rare in other domestic species. There are 16 reported cases of autosomal trisomies in cattle involving the 10 smallest autosomes, typically resulting in fetal death or postnatal culling by breeders due to congenital defects [8,9]. In the domestic pig, there are no reports of live-born animals with whole autosome aneuploidies [10], and all autosomal trisomies in dogs have exclusively been found in tumor cells [11]. Likewise, although aneuploidies occur in at least 5% of clinically recognized human (*Homo sapiens*, HSA) pregnancies and account for over 25% of spontaneous abortions, only trisomies of HSA13, 18 and 21 have been found in live born, of which only trisomy HSA21 survives to adulthood [12,13].

Extensive studies of human autosomal aneuploidies show that the majority are caused by errors in maternal meiosis I (MI) with advanced maternal age being a critical contributing factor, whereas only 5–10% of trisomies are caused by paternal errors [13]. At the same time, human data also show remarkable variation among trisomies regarding the parent and meiotic stage (MI or MII) of origin of the extra chromosome. For example, paternal errors account for nearly 50% of trisomy HSA2 but almost never for trisomy HSA16. Likewise, errors in maternal MI account for almost all cases of trisomy HSA16, whereas trisomy HSA18 is predominantly caused by errors in maternal MII, suggesting that the patterns of non-disjunction may have chromosome specific effects [13,14].

In rare occasions, trisomies of acrocentric autosomes are combined with Robertsonian fusion or isochromosome formation [15–17], so that despite of aneuploidy, the diploid chromosome number remains normal. For example, about 5–6% of cases with Down syndrome carry unbalanced heterologous or homologous fusions involving HSA21 [15,17]. The mechanism for heterologous fusions is Robertsonian translocation, of which the most common (82%) in Down syndrome patients is rob (14q;21q), with the remaining 8% represented by rob(13q;21q), rob(15q;21q) and rob(21q;22q) [17]. On the other hand, trisomy due to homologous fusion of (21q;21q) can result from different mechanisms—by isochromosome i(21q) formation or due to Robertsonian translocation rob(21q;21q). Since isochromosomes result from the duplication of a single chromosome arm [18], the duplicated parts are genetically identical and can be distinguished from homologous translocation by genotyping for allelic variation using chromosome specific polymorphic short tandem repeat (STR) markers [15,16,18,19].

In domestic animals, the only case of autosomal trisomy combined with centric fusion or isochromosome formation has been reported in horses for trisomy ECA26 [20,21]. The karyotype formula of the affected Thoroughbred mare was presented as 64,XX, −26,+t(26q;26q), but because polymorphic STR markers were not available for horses at that time, the researchers could not determine whether the abnormal chromosome (26q;26q) was an isochromosome or the result of a Robertsonian fusion.

In the present study, we report and characterize the second equine case of trisomy 26 involving homologous fusion 26q;26q. We will characterize the case using classical and molecular cytogenetic approaches and genotype the affected individual and its dam with ECA26 STR markers to determine the mechanism and likely parental origin of the aberration.

2. Material and Methods

2.1. Ethics Statement

Procurement of blood samples followed the United States Government Principles for the Utilization and Care of Vertebrate Animals Used in Testing, Research and Training. These protocols were approved as AUP and CRRC #2018-0342 CA at Texas A&M University.

2.2. Case Description and Sampling

A Thoroughbred foal (ID: H1063) was euthanized at the age of 5 months and 3 weeks due to stupors that gradually developed into ataxia, due to failure to thrive despite nursing well and being initially treated for possible neonatal mal-adjustment syndrome, and due to being inappropriate mentally. Although cervical radiographs did not provide an explanation for progressing ataxia, necropsy revealed axonal degeneration in brainstem and spinal cord suggestive of equine degenerative myeloencephalopathy. This was the first foal of a 5-year-old maiden Thoroughbred mare boarded on a large, well-managed farm. The sire had had several normal foals before. Peripheral blood samples in EDTA- and sodium heparin-containing vacutainers (VACUTAINERTM, Becton Dickinson) were obtained from the affected foal and its dam (ID: H1066) for cytogenetic and DNA analysis.

2.3. Cell Cultures and Chromosome Preparations

Metaphase chromosome spreads were prepared from peripheral blood lymphocytes following standard protocols [22]. Briefly, 1 mL of sodium heparin stabilized peripheral blood was grown for 72 h in 9 mL of culture medium RPMI-1640 supplemented with HEPES and Glutamax (Gibco), 30% fetal bovine serum (FBS; R&D Systems Inc., Minneapolis, MN, USA), 1X antibiotic-antimycotic (100×; Invitrogen, Waltham, MA, USA), and 15 µg/mL pokeweed mitogen (Sigma Aldrich, St. Louis, MO, USA). Lymphocyte cultures were harvested with demecolcine solution (10 µg/mL; Sigma Aldrich), treated with Optimal Hypotonic Solution (Rainbow Scientific, Windsor, CT, USA), and fixed in 3:1 methanol/acetic acid. The cells were dropped on clean, wet glass slides and checked under phase contrast microscope (×200) for quality.

2.4. Karyotyping and Cytogenetic Analysis

Chromosomes were stained by GTG-banding [23] for karyotyping. Karyotyping and chromosome analysis were performed with a motorized fluorescence microscope Axio Imager M2p (Zeiss) equipped with a high-resolution progressive scan CCD camera CoolCube 1 and Ikaros v5.3.18 software (MetaSystems GmbH, Altlußheim, Germany). Images of a minimum of 30 cells were captured and analyzed per individual. Horse chromosomes were identified and arranged into karyotypes according to the International System of Cytogenetic Nomenclature of the Domestic Horse [24] and chromosome aberrations were described following Human Cytogenomic, Nomenclature [25].

2.5. Fluorescence In Situ Hybridization (FISH)

The rearrangements identified by conventional cytogenetic analysis were validated by two-color FISH with ECA26-specific Bacterial Artificial Chromosome (BAC) clones (Table 1) from horse genomic BAC library CHORI-241 (https://bacpacresources.org/, last accessed 1 December 2021). The probes were labeled with biotin or digoxigenin by nick translation using Biotin or DIG Nick Translation Mix (Roche Diagnostics, Basel, Switzerland), following the manufacturer's protocol. Hybridization and signal detection followed standard protocols described elsewhere [22]. Biotin-labeled probes were detected with Alexa Fluor® 488 streptavidin conjugate (Molecular Probes, Life Technologies, Carlsbad, CA, USA) and digoxigenin-labeled probes with DyLight® 594 anti-digoxigenin conjugate (Vector Laboratories, Burlingame, CA, USA). Chromosomes were counterstained with 4′,6-diamidino-2-phenylindole (DAPI). At least 10 cells were captured and analyzed for each experiment using Isis v5.3.18 software (MetaSystems GmbH, Altlußheim, Germany).

Table 1. Information about ECA26 BAC clones used for FISH. Genomic location of BACs was retrieved from NCBI Genome (https://www.ncbi.nlm.nih.gov/genome/, last accessed 15 October 2021) and cytogenetic map information from [26].

CHORI-241 BAC Clone	BAC Location in EquCab3	Cytogenetic Location	Representative Genes
9N4	chr26:12,142,705–12,318,937	26q14	ROBO2
91H11	chr26:42,857,954–43,065,765	26q17	S100B

2.6. DNA Isolation, PCR Analysis and STR Genotyping

Genomic DNA was isolated from EDTA-stabilized blood with QIAamp DNA Blood Mini Kit (Qiagen, Hilden, Germany). Both horses were tested by PCR for the Y-linked *SRY* gene and X-linked androgen receptor (*AR*) gene as described earlier [27], followed by genotyping for the 15 autosomal STRs of the standard equine parentage panel [28], and an additional 24 STRs specific for ECA26 (Table 2). Genotyping was performed either with directly fluorescently labeled primers [29] or with three-primer nested PCR where the forward primer in each primer-pair had an M13-tail which was targeted by a fluorescently labeled universal M13 primer during PCR reactions [30]. Annealing temperature for all PCR reactions was 58 °C. The PCR products were resolved with an ABI PRISM 377 (Applied Biosystems, Foster City, CA, USA) and allele sizes were determined using GeneScan-500 LIZ Size Standard and GeneMapper® v4.1 (Applied Biosystems, Waltham, MA, USA).

Table 2. Information about ECA26 STR markers used for genotyping.

STR	Forward Primer: 5'-3'	Reverse Primer: 5'-3'	NCBI Accession or Reference
A-17 **	GTGGAGAGATAAAAGAAGATCC	GGCCACAAGGAATGAACACAC	X94446
COR071 **	CTTGGGCTACAACAGGGAATA	CTGCTATTTCAAACACTTGGA	AF142608
LEX044 *	TTGGGCTTCTTATCTTGTTAC	GGCCATATGATTTGCTTT	AF075646
NVHEQ070 **	GCTGGTCAAGTCACACTGTG	AACCTCACCCCAAGTTGTAT	AJ245765
TKY1155 *	AGCTCAGGGCGAATCTTACA	AAACCTGGGCATCTTCCTTT	AB104373
TKY275 *	TCTCAGTGGATATAACTAGC	GAGATGGATACAGATAGAAG	AB033926
TKY3385 *	TGACACCACCAGGGAAAAGT	CATGTTCCCTCACCTCTGGT	AB217328
TKY414 *	CCTGAAATCCGCTTCCATTA	ACCGGGTTATTTTGACATGG	AB103632
TKY488 *	TGTGTTTGTGTGCTATATACATGCTT	TGACATGAAGGCTGGACTTG	AB103706
TKY502 *	ACGGAAAACGTATGCCACTC	AGTGGGGACTTTGTTGAGGA	AB103720
TKY523 *	TGCACACCCATTCTAGCTCA	GTGGCTCACTCCTCGCTTAC	AB103741
TKY664 *	TACTGCCCTTGGCTGACTCT	CAGAACATGAACCCCTCCAG	AB103882
TKY766 *	ACTTTGCACCTGTGCAAAAAG	CTGATTCTTGGCATCTGGAAA	AB103984
TKY778 *	CTTAGATGGAGTCCTCCTAC	GGGTTCCTTTTACCTTCTCC	AB103996
TKY846 *	TCAAACCATCTGCTCAGAAG	AAATCCCAATCTGAGGGTAG	AB104064
TKY934 *	TTCCAGTGGTTAGGATGTAG	TTGAGCATAGTGATAGCATATG	AB104152
UM005 *	CCCTACCTGAAATGAGAATTG	GGCAAAAGATCAGGCCAT	AF195127
UMNe127 *	TTATAAATCACCACTGTTTACACAC	TCTTGAAGCAGGATGGGC	AY391298
UMNe153 *	GTGCTGGAGTGAGCTGACC	ATCCAAATCGGAGACCATATG	AF536265
UMNe188 *	GTTAACAAGGATTGTTTTGGGC	TGCGTTTCTGCTTCTCCC	AY391317
UMNe434 *	TCTGCTGTTGGCCATCATC	ACCTGCCTGCAAAACCTTC	[26]
UMNe542 *	TGAAAGAGACCATACACGATGC	CACGACTTAGAGACGTGTGAGC	AY735263

Table 2. Cont.

STR	Forward Primer: 5′-3′	Reverse Primer: 5′-3′	NCBI Accession or Reference
UMNe559 *	CTTCCCATTCTCTATCACCCC	CTGTTCTCCCAATTCTTTCTGG	[26]
UMNe588 *	CGCAGGTAGACTGTGTTAGGC	CAAGACTGGAAATTTTCAAGGG	[26]

* Forward primers had a M13 tail: TGTAAAACGACGGCCAGT ** Directly fluorescently labeled primers; Primer sequences were retrieved from [26,31].

3. Results

3.1. Chromosome Analysis

Cytogenetic analysis showed that the affected foal (H1063) had normal 2n = 64 diploid number, XY sex chromosomes, one copy of normal ECA26, and the karyotype contained a morphologically abnormal metacentric derivative chromosome (Figure 1A,B). Analysis of GTG-banding suggested that the derivative chromosome was composed of two copies of ECA26 likely fused at the centromeres. Molecular cytogenetic analysis by FISH with two ECA26 BAC clones, one corresponding to the proximal (BAC 9N4) and the other, to the distal (BAC 91H11) portion of the chromosome, confirmed that the derivative chromosome was the result of homologous centric fusion 26q;26q (Figure 1C). Thus, despite the normal diploid number, the foal carried trisomy ECA26 in all cells analyzed. However, by cytogenetic analysis alone, it was not possible to determine whether the derivative chromosome resulted from Robertsonian fusion rob(26q;26q) or from isochromosome formation i(26q).

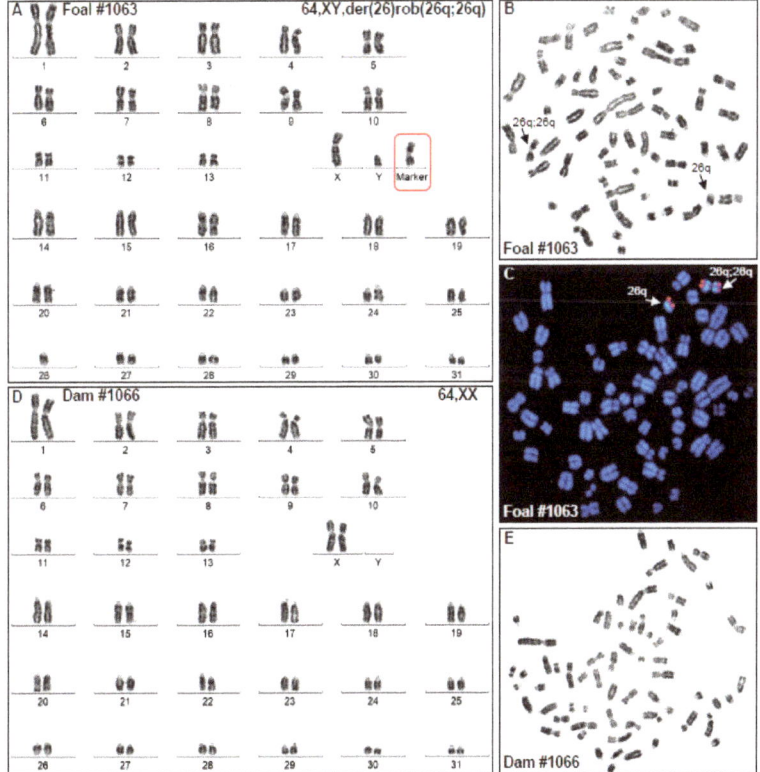

Figure 1. Cytogenetic analysis results. (**A**) GTG-banded karyotype of the affected foal H1063 showing 64,XY karyotype with a single ECA26 and a metacentric derivative chromosome with the arms corre-

sponding to ECA26q; (**B**) Metaphase spread corresponding to H1063 karyotype; arrows show the normal and derivative ECA26; (**C**) FISH results with ECA26 BAC clones (BAC 9N4 green; BAC 91H11 red) showing (arrows) the presence of a single ECA26 and a metacentric derivative chromosome 26q;26q; (**D**) GTG-banded karyotype of the dam (H1066) showing normal 64,XX female karyotype; (**E**) Metaphase spread corresponding to the karyotype of the dam (H1066).

Karyotype analysis of the dam (H1066) of the abnormal foal showed normal 64,XX female karyotype (Figure 1D,E) indicating that chromosomal abnormality of the foal must have originated from a parental meiotic error or a post-fertilization zygotic event.

As a standard part of cytogenetic analysis, both horses were tested by PCR for the *SRY* and *AR* genes and the results agreed with karyotype analysis and the phenotypic sex of the two horses: the XY foal H1063 was *SRY*-positive, the XX dam H1066 was *SRY*-negative, and both horses were positive for the X-linked control marker *AR*.

3.2. STR Genotyping: Parentage and the Origin of ECA26 Trisomy

Genotyping for 15 genome-wide autosomal STRs [28] qualified the cytogenetically normal Thoroughbred mare H1066 as the dam of the affected foal H1063. The two horses were also genotyped for 24 STR markers which were evenly distributed over ECA26, starting with UMNe588 as the most proximal marker and ending with TKY523 as the most distal one (Table 3). As expected, the STR markers showed the presence of one or two alleles in the cytogenetically normal dam H1066. However, five STRs had three alleles in the abnormal foal H1063 (Figure 2, Table 3), indicating that the metacentric derivative chromosome was the result of Robertsonian fusion rob(26q;26q) and not an isochromosome. The karyotype of the foal was designated as 64,XY,der(26),rob(26q;26q) [25].

Table 3. Genotyping results with ECA26 STRs. Markers are presented according to their linear order from centromere to telomere in ECA26; markers with three alleles in the foal are highlighted.

ECA26 Genomic Location, EquCab3	ECA26 STR	H1063: Alleles	H1066: Alleles
5,190,320–5,190,461	UMNe588	156	156
6,518,546–6,518,920	TKY934	158/160	158/160
7,006,025–7,006,186	UMNe559	173/175/177	173/175
8,845,111–8,845,452	TKY846	201/203	201
11,835,911–11,836,148	TKY766	104/110	110
19,109,482–19,110,003	TKY502	220	220
19,136,880–19,137,134	UMNe153	142/162	142/162
19,767,544–19,767,787	COR071	202/210	202/210
20,212,459–20,212,887	TKY275	142/158	142/158
20,367,221–20,367,742	LEX044	204/218	204/218
21,795,871–21,795,973	A-17	107/109	107/109
23,979,076–23,979,467	TKY778	226	226
24,637,783–24,638,172	TKY488	107/109	107/109
26,379,056–26,379,415	UMNe127	148	148

Table 3. Cont.

ECA26 Genomic Location, EquCab3	ECA26 STR	H1063: Alleles	H1066: Alleles
26,766,980–26,767,353	UM005	230/232/234	232/234
31,041,466–31,041,914	TKY1155	180/188/192	180/188
31,486,888–31,487,451	NVHEQ70	198/202/204	198/202
32,006,987–32,007,419	UMNe188	142/144	142/144
34,426,999–34,427,199	TKY3385	204	204
36,846,956–36,847,298	TKY664	271	271
37,488,847–37,489,215	UMNe542	270/276	270/276
38,794,949–38,795,212	UMNe434	284/286/288	284/288
39,259,334–39,259,638	TKY414	171/173	171/173
39,552,914–39,553,412	TKY523	162	162

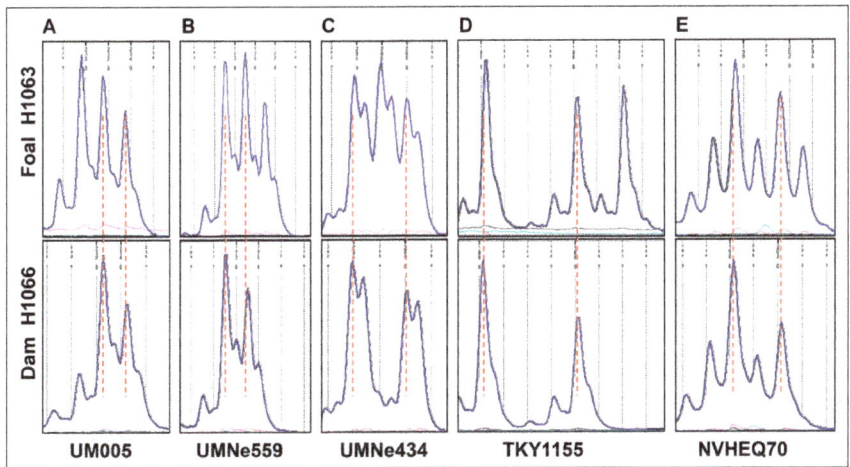

Figure 2. Genotyping results for ECA26 STRs UM005 (**A**), UMNe559 (**B**), UMNe434 (**C**), TKY1155 (**D**), and NVHEQ70 (**E**) showing the presence of three alleles in the foal H1063 (upper row) and two alleles in its dam H1066 (lower row). Note that for each STR, the two alleles present in the dam are shared with the foal. Allele size scales are aligned between the foal and the dam (vertical red dotted lines).

Further comparison of the genotyping patterns between the foal and the dam showed that in all 5 cases where the foal had 3 alleles, two of the alleles were identical with those of the dam (Figure 2). Additionally, of the 10 markers that were heterozygous both in the foal and the dam, the two horses shared the same alleles (Table 3). Based on these observations, and despite having no genotype information for the sire, it is very likely that the extra ECA26 in the foal was of maternal origin.

4. Discussion

Here, we characterized by chromosome analysis and STR genotyping an equine case of trisomy for chromosome 26 with homologous fusion 26q;26q (Figures 1 and 2, Table 2). Genotyping ECA26 STRs in the affected horse and its dam showed that the abnormal chromosome was the result of Robertsonian translocation and most likely of maternal origin. Since the dam of the affected foal had normal 64,XX karyotype (Figure 1D), the aneuploidy must have originated from maternal meiotic nondisjunction, though the following fusion could have taken place either in meiosis or post-fertilization.

It is certainly curious that this is the second case of trisomy ECA26 with a derivative chromosome 26q;26q in horses. The first case was described more than three decades ago [20,21], but because of uninformative blood typing, the mechanism (Robertsonian fusion or isochromosome) or parental origin of the aneuploidy remained unknown [21]. In our case, the presence of three alleles for 5 ECA26 STRs in the affected foal (Figure 2, Table 2) was a compelling piece of evidence that the derivative metacentric chromosome resulted from Robertsonian fusion. Furthermore, since all heterozygous STRs of the dam had the same two alleles also present in the affected foal (Table 2), we concluded that the extra chromosome ECA26 was likely of maternal origin. Though, complete evidence for the parental origin requires STR genotyping of the sire, whose samples were not available. Nevertheless, the findings underscore the importance of combining STR genotyping with cytogenetic analysis of possible isochromosomes or Robertsonian fusions. Isochromosome is formed by centromere mis-division of sister chromatids resulting in a bi-armed chromosome with identical genetic material in each arm [18,32]. Homologous Robertsonian fusions, on the other hand, result in genetically distinct arms preserving the heterozygosity from the parent from which the extra chromosome came from [16,18,33].

Another intriguing aspect of the present and the previous case [21] was that there have been no reports about ECA26 trisomy with three separate copies of the chromosome. This contrasts with other recurrently reported equine trisomies: all cases of cytogenetically studied trisomies of ECA27 (4 cases), ECA30 (5 cases), and ECA31 (2 cases) (reviewed by [2]) involve three separate chromosomes without homologous fusions. Furthermore, the trisomy ECA26 described in this study, is so far the only confirmed Robertsonian fusion in equine clinical cytogenetics [2], even though Robertsonian type rearrangements have been a normal part of equid and Perissodactyl karyotype evolution [34].

Can it be that ECA26 is more prone for centric fusion than other equine small acrocentric chromosomes? Chromosome-specific effects have been observed in humans where a small percentage of cases of Down and Patau syndrome with trisomy HSA21 and HSA13, respectively, have the extra chromosome in the form of Robertsonian fusion or an isochromosome [15–17,35,36]. In Down syndrome, there are even rare mosaic cases where one cell line carries HSA21 isochromosome and another, a Robertsonian fusion [17]. It is thought that some human chromosomes, such as HSA21, are inherently unstable and more prone to rearrangements [17] due to certain features of their sequence architecture (e.g., region-specific low copy number repeats) [18]. Based on our current knowledge of the horse genome [37], ECA26 does not stand out with any sequence peculiarities. Additionally, unlike HSA21 and other human acrocentric autosomes, ECA26 does not carry the satellite with multicopy rRNA genes that may contribute to instability [18]. On the other hand, and based on comparative chromosome painting [38] and gene mapping [26], ECA26 is more similar to HSA21 than to any other human chromosome because about 30 Mb (70%) of ECA26 shares evolutionary homology with the entire HSA21. However, the remaining 13 Mb (30%) of ECA26 is homologous to a part of HSA3 and this happens to be the pericentromeric/proximal portion of ECA26 which is involved in homologous fusion 26q;26q. Therefore, it is perhaps not relevant to expand the known instability of HSA21 [17] to ECA26 and it remains unclear whether the two cases of ECA26 trisomy with 26q;26q fusion were merely a coincidence or true reflections of presently unknown sequence properties of this horse autosome.

On the other hand, it is also possible that ECA26 instability and rearrangements are due to sequence variants segregating in certain horse breeds or families and not due to the genomic architecture of ECA26 per se. Indeed, the case described in this study and the one reported earlier [20,21], both occurred in Thoroughbreds. However, then again, two cases are too few for any conclusions.

Besides cytogenetics, there are several other shared features of interest between the two cases of trisomy ECA26 (this study; [20,21]). In both, the dams of the affected foals were young—5 years-old in this case and 3 years-old in the one described by Bowling et al. [21], thus excluding advanced maternal age as a contributing factor and rather supporting chromosome-specific effects. Additionally, both affected horses had gait deficits (ataxia), were not thriving, and had behavioral and mental issues. However, because the case presented in this study resulted in euthanasia at a young age but the horse described by Bowling et al. [21] lived many years, the basis of comparison is rather limited. It is, though, noteworthy that necropsy of the present case showed axonal degeneration in brainstem and spinal cord as seen in equine degenerative myeloencephalopathy (EDM) [39]. Although genetic basis for EDM is suspected but currently unknown [39], the present findings suggest that possible contribution of chromosome abnormalities/genetic imbalance should be considered. The fact that both cases were described as "inappropriate mentally" (this study) or "mentally dull" [21], and because of the homology between ECA26 and HSA21, there is a temptation to compare equine trisomy 26 with human Down syndrome. Indeed, there are some similarities: the horse described by Bowling et al. [21] lived many years and it is well-known that trisomy HSA21 is the only human autosomal trisomy surviving to adulthood [12,13]. Furthermore, at the age of 4, the mare with trisomy ECA26 gave birth to a chromosomally normal colt [21], and there are many cases of fertile women with Down syndrome in humans [40]. Despite this, drawing parallels between the two cases of ECA26 trisomy in horses with human Down syndrome should be taken with great caution. Firstly, genetic homology between ECA26 and HSA21 is not one-to-one since ECA26 is homologous also to part of HSA3 [26,38]. Secondly, stupors and ataxia which were the prevailing features of the two equine cases, are not the predominant characteristics of Down syndrome [41]. Most importantly, however, it is extremely narrow to compare the few phenotypic characteristics of two equine cases with the extensive research and clinical material available for Down syndrome since 1866 [41]. Furthermore, phenotypic features of the two equine cases share similarities with the phenotypes of other reported equine autosomal aneuploidies. For example, gait deficiencies, behavioral abnormalities and poor thriving have also been found in cases of trisomy ECA27 and ECA30 (reviewed by [2]), thus not being unique to trisomy ECA26. All in all, it is hard to tell which phenotypic features of trisomy ECA26 are the specific consequences of ECA26 overdose and which ones are due to general genomic imbalance.

5. Conclusions

We demonstrated that proper characterization of an autosomal (ECA26) trisomy with homologous fusion (26q;26q) and determining the mechanism and parental origin of the rearrangement, require the use of complementary approaches—cytogenetics and genotyping. To date, equine trisomy with homologous fusion has been unique to ECA26. However, to determine whether this is an ECA26-specific effect or just a coincidence, requires more cytogenetic cases and improved knowledge about the genomic architecture and functional annotation of ECA26. The latter is also needed to shed more light on the possible homology between trisomy ECA26 in the horse and the Down syndrome with trisomy HSA21 in humans.

Author Contributions: Conceptualization: T.R.; formal analysis: T.R., R.J., S.G. and J.K.; funding acquisition: T.R.; investigation: T.R., S.G., J.K., R.J., L.M. and S.R.; methodology: T.R., R.J., S.G. and J.K.; resources: T.R., R.J., L.M. and S.R.; supervision: T.R., R.J.; visualization: T.R., S.G. and J.K.; writing—original draft: T.R., J.K. and S.G.; writing—review and editing: T.R., S.G., J.K., R.J., L.M. and S.R. All authors have read and agreed to the published version of the manuscript.

Funding: Texas A&M Molecular Cytogenetics and Animal Genetics Laboratories.

Institutional Review Board Statement: Procurement of blood samples followed the United States Government Principles for the Utilization and Care of Vertebrate Animals Used in Testing, Research and Training. These protocols were approved as AUP and CRRC #2018-0342 CA at Texas A&M University.

Informed Consent Statement: Written informed consent has been obtained from the owners of the animals involved.

Data Availability Statement: Not applicable.

Acknowledgments: The authors thank Kathleen Paasch for conducting cervical radiography of the affected foal.

Conflicts of Interest: The authors declare no conflict of interest.

References

1. Lear, T.L.; Bailey, E. Equine clinical cytogenetics: The past and future. *Cytogenet. Genome Res.* **2008**, *120*, 42–49. [CrossRef] [PubMed]
2. Bugno-Poniewierska, M.; Raudsepp, T. Horse Clinical Cytogenetics: Recurrent Themes and Novel Findings. *Animals* **2021**, *11*, 831. [CrossRef]
3. Raudsepp, T.; Chowdhary, B. Chromosome Aberrations and Fertility Disorders in Domestic Animals. *Annu. Rev. Anim. Biosci.* **2016**, *4*, 15–43. [CrossRef] [PubMed]
4. Raudsepp, T.; Durkin, K.; Lear, T.L.; Das, P.J.; Avila, F.; Kachroo, P.; Chowdhary, B.P. Chowdhary. Molecular heterogeneity of XY sex reversal in horses. *Anim. Genet.* **2010**, *41* (Suppl. 2), 41–52. [CrossRef]
5. Janecka, J.E.; Davis, B.W.; Ghosh, S.; Paria, N.; Das, P.J.; Orlando, L.; Schubert, M.; Nielsen, M.K.; Stout, T.A.E.; Brashear, W.; et al. Horse Y chromosome assembly displays unique evolutionary features and putative stallion fertility genes. *Nat. Commun.* **2018**, *9*, 2945. [CrossRef] [PubMed]
6. Shilton, C.A.; Kahler, A.; Davis, B.W.; Crabtree, J.R.; Crowhurst, J.; McGladdery, A.J.; Wathes, D.C.; Raudsepp, T.; de Mestre, A.M. Whole genome analysis reveals aneuploidies in early pregnancy loss in the horse. *Sci. Rep.* **2020**, *10*, 13314. [CrossRef] [PubMed]
7. Brito, L.; Sertich, P.; Durkin, K.; Chowdhary, B.; Turner, R.; Greene, L.; McDonnell, S. Autosomic 27 Trisomy in a Standardbred Colt. *J. Equine Vet. Sci.* **2008**, *28*, 431–436. [CrossRef]
8. Holečková, B.; Schwarzbacherová, V.; Galdíková, M.; Koleničová, S.; Halušková, J.; Staničová, J.; Verebová, V.; Jutková, A. Chromosomal Aberrations in Cattle. *Genes* **2021**, *12*, 1330. [CrossRef]
9. Iannuzzi, A.; Parma, P.; Iannuzzi, L. Chromosome Abnormalities and Fertility in Domestic Bovids: A Review. *Animals* **2021**, *11*, 802. [CrossRef]
10. Donaldson, B.; Villagomez, D.A.F.; King, W.A. Classical, Molecular, and Genomic Cytogenetics of the Pig, a Clinical Perspective. *Animals* **2021**, *11*, 1257. [CrossRef]
11. Szczerbal, I.; Switonski, M. Clinical Cytogenetics of the Dog: A Review. *Animals* **2021**, *11*, 947. [CrossRef] [PubMed]
12. Chen, S.; Liu, D.; Zhang, J.; Li, S.; Zhang, L.; Fan, J.; Luo, Y.; Qian, Y.; Huang, H.; Liu, C.; et al. A copy number variation genotyping method for aneuploidy detection in spontaneous abortion specimens. *Prenat. Diagn.* **2017**, *37*, 176–183. [CrossRef] [PubMed]
13. Hassold, T.; Hunt, P. To err (meiotically) is human: The genesis of human aneuploidy. *Nat. Rev. Genet.* **2001**, *2*, 280–291. [CrossRef] [PubMed]
14. Hassold, T.; Maylor-Hagen, H.; Wood, A.; Gruhn, J.; Hoffmann, E.; Broman, K.W.; Hunt, P. Failure to recombine is a common feature of human oogenesis. *Am. J. Hum. Genet.* **2021**, *108*, 16–24. [CrossRef]
15. Hervé, B.; Quibel, T.; Taieb, S.; Ruiz, M.; Molina-Gomes, D.; Vialard, F. Are de novo rea(21;21) chromosomes really de novo? *Clin. Case Rep.* **2015**, *3*, 786–789. [CrossRef]
16. Antonarakis, S.E.; Adelsberger, P.A.; Petersen, M.B.; Binkert, F.; Schinzel, A.A. Analysis of DNA polymorphisms suggests that most de novo dup(21q) chromosomes in patients with Down syndrome are isochromosomes and not translocations. *Am. J. Hum. Genet.* **1990**, *47*, 968–972.
17. Bandyopadhyay, R.; McCaskill, C.; Knox-Du Bois, C.; Zhou, Y.; Berend, S.A.; Bijlsma, E.; Shaffer, L.G. Mosaicism in a patient with Down syndrome reveals post-fertilization formation of a Robertsonian translocation and isochromosome. *Am. J. Med. Genet. A* **2003**, *116*, 159–163. [CrossRef]
18. Shaffer, L.G.; Lupski, J.R. Molecular mechanisms for constitutional chromosomal rearrangements in humans. *Annu. Rev. Genet.* **2000**, *34*, 297–329. [CrossRef]
19. Israni, A.; Mandal, A. De Novo Robertsonian Translocation t(21; 21) in a Child with Down Syndrome. *J. Nepal Paediatr. Soc.* **2017**, *37*, 92–94. [CrossRef]
20. Bowling, A.T.; Millon, L.; Hughes, J.P. An update of chromosomal abnormalities in mares. *J. Reprod. Fertil. Suppl.* **1987**, *35*, 149–155.
21. Bowling, A.T.; Millon, L.V. Two autosomal trisomies in the horse: 64,XX, −26, +t(26q26q) and 65,XX, +30. *Genome* **1990**, *33*, 679–682. [CrossRef] [PubMed]
22. Raudsepp, T.; Chowdhary, B.P. FISH for mapping single copy genes. *Methods Mol. Biol.* **2008**, *422*, 31–49. [PubMed]
23. Seabright, M. A rapid banding technique for human chromosomes. *Lancet* **1971**, *2*, 971–972. [CrossRef]

24. Bowling, A.T.; Breen, M.; Chowdhary, B.P.; Hirota, K.; Lear, T.; Millon, L.V.; Ponce de Leon, F.A.; Raudsepp, T.; Stranzinger, G.; ISCNH. International system for cytogenetic nomenclature of the domestic horse. Report of the Third International Committee for the Standardization of the domestic horse karyotype, Davis, CA, USA, 1996. *Chromosome Res.* **1997**, *5*, 433–443. [CrossRef] [PubMed]
25. ISCNH. *An International System for Human Cytogenomic Nomenclature*; McGowan-Jordan, J., Simons, A., Schmid, M., Eds.; S. Karger: Basel, Switzerland, 2020; p. 163.
26. Raudsepp, T.; Gustafson-Seabury, A.; Durkin, K.; Wagner, M.L.; Goh, G.; Seabury, C.M.; Brinkmeyer-Langford, C.; Lee, E.J.; Agarwala, R.; Stallknecht-Rice, E.; et al. A 4103 marker integrated physical and comparative map of the horse genome. *Cytogenet. Genome Res.* **2008**, *122*, 28–36. [CrossRef]
27. Ghosh, S.; Carden, C.F.; Juras, R.; Mendoza, M.N.; Jevit, M.J.; Castaneda, C.; Phelps, O.; Dube, J.; Kelley, D.E.; Varner, D.D.; et al. Two Novel Cases of Autosomal Translocations in the Horse: Warmblood Family Segregating t(4;30) and a Cloned Arabian with a de novo t(12;25). *Cytogenet. Genome Res.* **2020**, *160*, 688–697. [CrossRef]
28. Khanshour, A.; Conant, E.; Juras, R.; Cothran, E.G. Microsatellite analysis of genetic diversity and population structure of Arabian horse populations. *J. Hered.* **2013**, *104*, 386–398. [CrossRef]
29. Juras, R.; Cothran, E.G.; Klimas, R. Genetic analysis of three Lithuanian native horse breeds. *Acta Agric. Scand. Sect. A Anim. Sci.* **2003**, *53*, 180–185. [CrossRef]
30. Schuelke, M. An economic method for the fluorescent labeling of PCR fragments. *Nat. Biotechnol.* **2000**, *18*, 233–234. [CrossRef]
31. Tozaki, T.; Penedo, M.C.; Oliveira, R.P.; Katz, J.P.; Millon, L.V.; Ward, T.; Pettigrew, D.C.; Brault, L.S.; Tomita, M.; Kurosawa, M.; et al. Isolation, characterization and chromosome assignment of 341 newly isolated equine TKY microsatellite markers. *Anim. Genet.* **2004**, *35*, 487–496. [CrossRef]
32. Shaffer, L.G.; Jackson-Cook, C.K.; Meyer, J.M.; Brown, J.A.; Spence, J.E. A molecular genetic approach to the identification of isochromosomes of chromosome 21. *Hum. Genet.* **1991**, *86*, 375–382. [CrossRef] [PubMed]
33. Harel, T.; Pehlivan, D.; Caskey, C.T.; Lupski, J.R. Chapter 1—Mendelian, Non-Mendelian, Multigenic Inheritance, and Epigenetics. In *Rosenberg's Molecular and Genetic Basis of Neurological and Psychiatric Disease*, 5th ed.; Rosenberg, R.N., Pascual, J.M., Eds.; Academic Press: Boston, MA, USA, 2015; pp. 3–27.
34. Trifonov, V.A.; Stanyon, R.; Nesterenko, A.I.; Fu, B.; Perelman, P.L.; O'Brien, P.C.; Stone, G.; Rubtsova, N.V.; Houck, M.L.; Robinson, T.J.; et al. Multidirectional cross-species painting illuminates the history of karyotypic evolution in Perissodactyla. *Chromosome Res.* **2008**, *16*, 89–107. [CrossRef]
35. Shaffer, L.G.; McCaskill, C.; Haller, V.; Brown, J.A.; Jackson-Cook, C.K. Further characterization of 19 cases of rea(21q21q) and delineation as isochromosomes or Robertsonian translocations in Down syndrome. *Am. J. Med. Genet.* **1993**, *47*, 1218–1222. [CrossRef] [PubMed]
36. Shaffer, L.G.; McCaskill, C.; Han, J.Y.; Choo, K.H.; Cutillo, D.M.; Donnenfeld, A.E.; Weiss, L.; Van Dyke, D.L. Molecular characterization of de novo secondary trisomy 13. *Am. J. Hum. Genet.* **1994**, *55*, 968–974.
37. Kalbfleisch, T.S.; Rice, E.S.; DePriest, M.S., Jr.; Walenz, B.P.; Hestand, M.S.; Vermeesch, J.R.; O'Connell, B.L.; Fiddes, I.T.; Vershinina, A.O.; Saremi, N.F.; et al. Improved reference genome for the domestic horse increases assembly contiguity and composition. *Commun. Biol.* **2018**, *1*, 197. [CrossRef] [PubMed]
38. Yang, F.; Fu, B.; O'Brien, P.C.M.; Nie, W.; Ryder, O.A.; Ferguson-Smith, M.A. Refined genome-wide comparative map of the domestic horse, donkey and human based on cross-species chromosome painting: Insight into the occasional fertility of mules. *Chromosome Res.* **2004**, *12*, 65–76. [CrossRef] [PubMed]
39. Burns, E.N.; Finno, C.J. Equine degenerative myeloencephalopathy: Prevalence, impact, and management. *Vet. Med.-Res. Rep.* **2018**, *9*, 63–67. [CrossRef]
40. Parizot, E.; Dard, R.; Janel, N.; Vialard, F. Down syndrome and infertility: What support should we provide? *J. Assist. Reprod. Genet.* **2019**, *36*, 1063–1067. [CrossRef]
41. Bull, M.J. Down Syndrome. *N. Engl. J. Med.* **2020**, *382*, 2344–2352. [CrossRef]

Article

Chromosome Instability in Pony of Esperia Breed Naturally Infected by Intestinal Strongylidae

Emanuele D'Anza [1,†], Francesco Buono [1,†], Sara Albarella [1,*], Elisa Castaldo [1], Mariagiulia Pugliano [1], Alessandra Iannuzzi [2], Ilaria Cascone [1], Edoardo Battista [3], Vincenzo Peretti [1] and Francesca Ciotola [1]

[1] Department of Veterinary Medicine and Animal Production, University of Naples Federico II, Via Delpino 1, 80137 Naples, Italy
[2] National Research Council (CNR), Institute of Animal Production System in Mediterranean Environment (ISPAAM), Piazzale E. Fermi, 1, 80055 Portici, Italy
[3] Independent Researcher, Via Rampa 6, 03038 Roccasecca, Italy
* Correspondence: sara.albarella@unina.it; Tel.: +39-0812536502
† These authors contributed equally to this work.

Simple Summary: Intestinal parasites are among the main causes of hidden economic losses in livestock farming. This study reports the results of chromosome instability analyses in Esperia ponies with different intestinal strongyles fecal egg counts. Interestingly, animals with higher fecal egg counts showed increased levels of chromosome instability. If this condition is confirmed in other horse breeds and livestock species, it will be important to understand the causes in order to implement therapeutic strategies for the management of intestinal parasites.

Abstract: The Pony of Esperia is an Italian autochthonous horse breed reared in the wild on the Aurunci and Ausoni Mountains. Currently, it is considered an endangered breed, as its population consists of 1623 animals. It is therefore essential to identify all aspects that can improve the management and economy of its breeding, favoring its diffusion. In this paper, the effects of intestinal strongyle infection on the chromosome stability of peripheral blood lymphocytes (PBLs) was evaluated through aneuploidy and chromosome aberration (gap, chromatid and chromosome breaks, and the number of abnormal cells) test. Statistical difference in the mean values of aneuploidy, cells with chromosome abnormalities, and chromosome and chromatid breaks were observed between ponies with high fecal egg counts (eggs per gram > 930) and those with undetectable intestinal strongylosis. The causes of this phenomenon and possible repercussions on the management of Pony of Esperia are discussed in the paper.

Keywords: Pony of Esperia; chromosome instability (CIN); intestinal strongylosis; eggs per gram (EPG); chromosome aberrations (CAs)

1. Introduction

The Pony of Esperia (Figure 1) is an autochthonous horse breed reared in the province of Frosinone (Lazio, Central Italy). It is characterized by a morello coat, sometimes with socks and head star, and a very thick mane and tail. The head is short and conical with a straight profile; the neck is proportionate and not excessively muscular; the robust shoulder is well-attached to the trunk; the withers is pronounced; the back is inclined; the chest is developed and muscular; the thorax is shallow; and the limbs are robust. It has a maximum withers height of 138 cm for males and 132 cm for females with a maximum live weight of 350 Kg. It originated in the area of the Aurunci and Ausoni Mountains and underwent crossbreeding along successive generations with Arabian horses of Nedjad origin. The selective pressure of the mountainous environment in which it developed determined its small size, the ability to live in hostile environments, and its rusticity. This horse was used for different purposes such as the transport of materials or people, riding, and mule production. The Pony of Esperia is currently bred as a saddle animal and for riding competitions.

Figure 1. (**a**) Pony of Esperia. (**b**) Grazing herd of Pony of Esperia.

This breed is reared in the wild and it stays in the pastures throughout the year. No food supplementations are administered, and pharmacological treatments are performed only when strictly necessary.

Currently, there are 1623 individuals enrolled in the studbook, so it is included in the breeds with limited diffusion to be safeguarded. For this purpose, it is necessary to improve our knowledge about their geno-morphofunctional characteristics and to highlight the conditions that make its farming more difficult and economically burdensome. The main breeding issues to which this breed is exposed are attacks by predators (which usually affect younger individuals or those in a precarious state of health), infectious diseases transmitted by other wild animals or transmitted from endo-ectoparasites. A study performed in 2006 on 230 individuals belonging to 33 families, showed the presence of polyparasitism in this breed. Intestinal Strongyles, *Parascaris* spp., *Oxyuris equi*, *Anoplocephala* spp., ticks, flies, and *Gasterophilus* spp. larvae have been found in all groups, suggesting that despite this breed being considered rustic and disease-resistant it needs parasitosis control plans [1].

Intestinal parasites are often responsible for delays in growth and the worsening of athletic performance, causing hidden economic losses in breeding [2]. Since they are often clinically asymptomatic, in livestock preventive therapies are carried out at specific times of the year, but this is not the case in Pony of Esperia breeding.

The aim of this study was to evaluate the effect of different degrees of natural intestinal strongyles infection on peripheral blood lymphocytes (PBLs) chromosome instability (CIN) in Pony of Esperia using aneuploidy and chromosome aberration (CA) tests [3]. Both tests have been used in humans and animals to evaluate the mutagenetic effects of environmental pollutants [4–6], drug and food supplements [7,8]; correlations among congenital malformations and chromosome stability [9–11], chromosome stability differences within different breeds of the same species [12], and the effects of micro- and macronutrient deficiencies [13].

PBLs are the ideal cells for CIN evaluation in an individual since they circulate throughout the body, thus being exposed to all possible risk conditions and thus representing an early and easy marker to analyze after in vitro cultivation [14].

CIN is a type of genomic instability with an increase in the numerical and structural alterations in chromosomes, and it is due to an increase in DNA damages, the malfunction of DNA damage repair mechanisms, or both [15]. The increase in CIN, as well as being a risk factor for the development of cancers, represents an index of altered homeostasis that negatively influences the well-being of the individual. Moreover, in livestock an increase in CIN has been negatively correlated to fertility and reproduction [16].

To the best authors knowledge, this is the first time that a relation between aneuploidy and CAs and the degree of intestinal parasitic infection was investigated in horse species.

2. Materials and Methods

2.1. Animals

For this study, fifty female Ponies of Esperia, aged between 3 and 20 years, were enrolled. All animals belonged to the same farm located in the province of Frosinone (Lazio Region) and were reared under the same conditions. All individuals were sampled twice (D0, before the treatment, and D14, 14 days after the treatment) for blood and feces to perform karyotype, aneuploidy, and CA tests in PBLs and fecal egg counts (FEC), respectively. At both sampling times, a clinical evaluation was performed, and the body condition scores (BCS) were determined using a five-point system (1 = poor, 2 = moderate, 3 = ideal, 4 = fat, and 5 = obese) [17] by the same investigator, and all animals were healthy, with a BCS = 3.

2.2. Coprological Analysis

Individual fecal samples were collected from all ponies involved in the study, and according to general recommendations proposed by Nielsen et al. [18], feces were taken directly from the rectum of each animal. Individual fecal egg counts (FECs) were performed for all ponies before the start of the trial (D0) and at 14 days post-treatment (D14) using a special modification of the McMaster method with a lower detection limit of 10 eggs per gram (EPG) using a Sheather's saturated sugar solution with a specific gravity of 1.250 as a flotation medium [19]. Based on the morphological identification [20], each egg was classified as belonging to intestinal strongyles, *Parascaris* spp., *Strongyloides westeri*, *Oxyuris equi*, or *Anoplocephala* spp.

2.3. Anthelmintic Treatment

After the first collection, the ponies were divided in two groups of 25 animals that were homogeneous per age and fecal egg count: in the treatment group (T group), animals were treated with fenbendazole (FBZ) at the horse dose rate (7.5 mg/kg BW, Panacur Oral Paste, MSD Animal Health, Walton, Milton Keynes, UK), and control group (C group) animals were left untreated. Two weeks after the treatment (day 14) the blood and feces of all animals were resampled to verify changes in aneuploidy, CA, and parasitic infection. To determine the anthelmintic efficacy of FBZ, the arithmetic mean (AM) of EPG was calculated 14 days post-treatment, and the percent efficacy (%) was considered in terms of a fecal egg count reduction (FECR) test using the formula: FECR = [(AM FEC PRE-TREATMENT − AM FEC POST-TREATMENT)/AM FEC PRE-TREATMENT] × 100, according to the American Association of Equine Practitioners (AAEP) guidelines [21]. The cut-off values used to interpret the results of the FECRT were the following: efficacy > 95%, suspected resistance 90–95%, and resistance < 90% [21]. Microsoft Office Excel 2010 software was used for data recording, and FEC reductions, expressed as percentages with 95% confidence intervals, were calculated using the RESO FECRT analysis program, version 4 [22], for Excel. The simultaneous finding of a lower confidence limit (LCL) below 90% [23] and a mean percentage of FECR below 90% [21] was indicative of resistance; if only one of these two criteria was present, resistance was suspected.

2.4. Cytogenetic Analyses

Cell cultures for chromosome isolation were set up as reported by Ciotola et al. [24]. Briefly, peripheral blood (1 mL) was cultured at 37.5 °C in RPMI medium and enriched with fetal calf serum (10%), L-glutamine (1%), and lectin (1.5%) for 48 h. Cells were harvested after colcemid (0.3 lg/mL) treatment for 1 h and given a hypotonic treatment (KCl 0.5%) and three fixations in methanol–acetic acid (3:1), the third occurring overnight. Three drops of cell suspension were air-dried on cleaned and wet slides that were stained a day later with acridine orange (0.1% in a phosphate buffer, pH 7.0) for 3 min, washed in tap and distilled water, and mounted in the same phosphate buffer. The slides were observed 24 h after staining or later (1 week). At least 10, 100, and 50 cells per animal were examined from slides of normal cultures to perform conventional karyotype, aneuploidy, and chromosome aberration (CA) tests, i.e., gaps, chromatid and chromosome breaks, and the percentage of abnormal cells (abnormal cells are those with at least one chromatid or chromosome break) (Figure 2) [8,25]. All metaphase plates were observed under a fluorescence Nikon Eclipse 80i microscope, captured with a Nikon Sight DS-5M digital camera, transferred to a PC, and later processed with an image analysis software by two cytogeneticists.

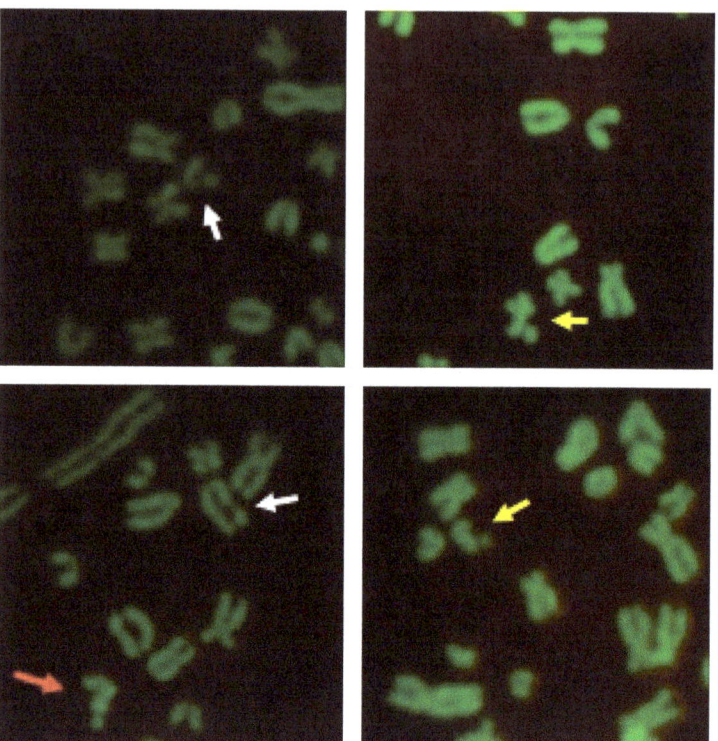

Figure 2. Details of chromosome metaphase plates of Pony of Esperia. The white arrow indicates a gap, the yellow arrow indicates a chromatid break, and the red arrow indicates a chromosome break.

2.5. Statistical Analyses

Microsoft Office Excel 2010 software was used for data recording then IBM SPSS for Windows software package version 22.0 (SPSS Inc., Chicago, IL, USA) was used for statistical analyses. The Esperia ponies were divided into five groups: group C + T included all the animals at day 0, groups C_{D0} and C_{D14} referred to the control group on day 0 and day 14, respectively, and groups T_{D0} and T_{D14} referred to the treated animals on day 0 and day 14, respectively. A statistical subanalysis was performed by dividing the groups C + T and T_{D0} according to the EPG and then according to age of the individuals (up to 6 years and equal or higher to 6 years). The parasitic burden ranges of the groups were chosen by making multiple comparisons to verify if there were ranges within which the CAs increased or decreased significantly. Age groups were established according to Wójcik and Smalec [25].

Student's t-test was used to compare the structural percentage and cells with the CA percentage in all groups. The independent-sample *t*-test (Mann–Whitney test) was used to compare the means of the quantitative variables in the groups [26]. A Spearman correlation was performed between FEC and CAs.

3. Results

3.1. Coprological Analysis and Anthelmintic Efficacy

Intestinal strongyle eggs were found in all tested animals. At the start of the study, the mean EPG count was 992.20 ± 443.75. The mean EPG values in the T and C groups were 1096.80 ± 424.31 and 887.60 ± 446.31, respectively, and no statistical differences were observed between the two group (t = 1.6986; p = 0.0959). For all animals, the dose was administered carefully, and no adverse reactions were observed in any of the treated ponies. FBZ was effective in reducing FECs at 14 days after treatment, showing an FECRT = 100%.

3.2. Cytogenetic Analyses

All investigated animals showed a normal karyotype, thus excluding the presence of congenital numerical and structural chromosome abnormalities that could alter CIN. Thus, the identified aneuploidy and chromosomal abnormalities observed in the metaphases of the analyzed horses did not cause birth defects.

With regards to aneuploidy and the CA assay: for seven animals at D0 and another seven animals at D14, it was not possible to analyze enough metaphases.

As it has been reported that chromosomal stability in horses is influenced by age [25], to verify if this parameter significantly influenced the values of aneuploidy and CAs in the studied population, the T_{D14} group (animals negative for intestinal parasites) was divided into two groups: $T_{D14} < 5$ (n = 9) and $T_{D14} > 6$ (n = 9), comprising, respectively, individuals aged less than or equal to 5 years and individuals older than or equal to 6 years. Statistically significant differences were observed for gaps and aneuploidy (Table 1).

When considering aneuploidy, it was decreased in all groups after the first collection in a statistically significant way, while all CA values were increased in group C_{D14} and decreased in T_{D14}.

When comparing groups divided according to EPG (Table 2), statistically significant differences were found for chromatid breaks, chromosome breaks, and CAs excluding gaps between groups $1T_{D0}$ and $2T_{D0}$, with all values being higher in $2T_{D0}$. In the comparison between $1T_{D0}$ and T_{D14}, a statistically significant difference was found only for gaps, while in the comparison between $2T_{D0}$ and T_{D14} all parameters showed a statistically significant differences. Statistically significant higher mean values for all CA parameters were also observed in group 3C + T when compared to T_{D14} and for chromatid breaks, chromosome breaks, and CAs excluding gaps in 3C + T when compared to groups $1T_{D0}$ and 1C + T.

When animals grouped according to EPG and age (up to 5 years and older than 6 years) were compared, a statistical difference was observed only between the aneuploidy percentage and gaps of animals with high EPG (930–1970).

Finally, the Spearman correlation test showed a negative correlation between EPG and gaps (r = −0.07; p < 0.01) and positive correlations between EPG and chromosome breaks (r = 0.05; p < 0.01) and EPG and CAs excluding gaps (r = 0.33; p < 0.05).

Table 1. Percentages of aneuploidy and abnormal cells and mean values and standard deviations of CA in groups C + T (all animals before the treatment), C_{D0} (untreated animals at T0), T_{D0} (treated animals before the treatment), C_{D14} (untreated animals at T14), T_{D14} (treated animals at T14), $T_{D14} < 6$ (treated animals at T14 up to 6 years), and $T_{D14} > 6$ (treated animals at T14 older than 6 years).

Group	N of Animals	Age Mean ± SD	Age Range	EPG Mean ± SD	EPG Range	Aneuploidy (2n ≠ 64) %	Abnormal Cells %	Gaps Mean ± SD	Chromatid Breaks Mean ± SD	Chromosome Breaks Mean ± SD	CAs Excluding Gaps Mean ± SD
C + T	43	7.63 ± 5.02	3–20	980.24 ± 471.75	260–1970	16.76 a,A	6.33	0.71 ± 0.93	0.06 ± 0.25 a	0.01 ± 0.11 a	0.07 ± 0.28 A
C_{D0}	20	8.00 ± 6.04	3–20	904.44 ± 536.34	260–1970	15.44 a,b,A	4.70	0.72 ± 0.98	0.04 ± 0.22 a,A	0.01 ± 0.10	0.05 ± 0.25 a
T_{D0}	23	7.45 ± 4.06	3–20	1039.57 ± 389.47	400–1970	17.90 a,A	7.74	0.70 ± 0.88	0.07 ± 0.27 B	0.01 ± 0.12	0.08 ± 0.30 A,b
C_{D14}	13	7.75 ± 6.25	3–20	1179.00 ± 801.82	110–2760	11.30 c,B	7.20	0.79 ± 0.91 A	0.07 ± 0.28 B	0.01 ± 0.12	0.08 ± 0.31 A,b
T_{D14}	23	7.45 ± 4.06	3–20	0	0	13.32 b,c	6.64	0.61 ± 0.74	0.04 ± 0.19 b,A	0.00 ± 0.07 b	0.04 ± 0.2 B
$T_{D14} < 6$	12	3.91 ± 0.95	3–5	0	0	11.25 c,B	6.31	0.77 ± 0.80 A	0.04 ± 0.18 b,A	0.00 ± 0.00 a	0.04 ± 0.18 B
$T_{D14} > 6$	11	10.25 ± 4.91	6–20	0	0	16.05 b,A	7.45	0.49 ± 0.66 B	0.04 ± 0.21 b,A	0.00 ± 0.06 a	0.05 ± 0.22 a

a,b,c = $p < 0.05$; A,B = $p < 0.005$.

Table 2. Percentages of aneuploidy and abnormal cells and mean values and standard deviations of chromosome abnormalities in ponies grouped according to EPG and age at D0 and compared with the T_{D14} group.

Group	N of Animals	Age Mean	Age Range	EPG Range	Aneuploidy (2n ≠ 64) %	Abnormal Cells %	Gaps Mean ± SD	Chromatid Breaks Mean ± SD	Chromosome Breaks Mean ± SD	CAs Excluding Gaps Mean ± SD
1T_{D0}	11	7.08 ± 4.32	3–16	400–800	17.39	5.69	0.87 ± 0.88 A	0.05 ± 0.23 A	0.00 ± 0.07 A	0.05 ± 0.31 A,a
2T_{D0}	9	8.23 ± 5.43	3–20	950–1970	14.47	10.22	0.81 ± 0.86 A	0.08 ± 0.29 B	0.03 ± 0.16 B	0.11 ± 0.33 B
1C + T	12	8.4 ± 6.12	3–20	260–640	18.62	5.20	0.74 ± 0.96	0.05 ± 0.23 A	0.00 ± 0.06 A	0.05 ± 0.24 A,a
2C + T	13	6.17 ± 3.24	3–12	690–930	12.26 a	6.92	0.76 ± 0.94	0.07 ± 0.27	0.01 ± 0.10	0.08 ± 0.30 b
3C + T	18	9.00 ± 6.14	3–20	950–1970	17.85	10.89	0.81 ± 0.88 A	0.08 ± 0.30 B	0.04 ± 0.19 B	0.12 ± 0.35 B
1–2C + T < 6	12	3.82 ± 0.94	3–5	260–930	19.13 b	10.00	0.76 ± 0.94	0.06 ± 0.27	0.01 ± 0.09	0.07 ± 0.29
1–2C + T > 6	11	10.24 ± 4.15	6–20	260–930	18.41	9.64	0.80 ± 0.95	0.06 ± 0.25	0.01 ± 0.08	0.07 ± 0.27
3C + T < 6	9	3.40 ± 0.80	3–5	950–1970	12.30 a	10.64	1.12 ± 0.98 A	0.06 ± 0.26	0.05 ± 0.21 B	0.11 ± 0.34 B
3C + T > 6	9	11.00 ± 4.79	6–20	950–1970	16.52	11.67	0.72 ± 0.83 B	0.08 ± 0.31 B	0.03 ± 0.17 B	0.11 ± 0.34 B
T_{D14}	23	7.45 ± 4.06	3–20	0	13.32	6.64	0.61 ± 0.74 B	0.04 ± 0.19 A	0.00 ± 0.07 A	0.04 ± 0.21 A,a

a,b = $p < 0.05$; A,B = $p < 0.005$.

4. Discussion

The Pony of Esperia is an endangered autochthonous breed reared in the Aurunci and Ausoni Italian Mountains. The implementation of safeguard plans for this breed is essential not only to avoid the loss of biodiversity but also to avoid the abandonment of marginal areas. For this purpose, it appears important to highlight the causes that can worsen the profitability of this breed and to verify the treatments that can improve its productivity.

One of the main causes of hidden economic losses in animal farming is parasitic infections of digestive tract. In fact, they can induce states of subclinical malnutrition due to the alteration of the intestinal absorption capacity as well as the actual subtraction of nutrients by parasites.

In this pony population, anthelmintic treatments are carried out without performing a coprological diagnosis once per year, and this anthelmintic treatment scheme is quite different from those reported in horses in Italy [27] but is very similar to that reported in the Italian donkey population [28]. Contrary to the diffusion of anthelmintic resistance worldwide [29], anthelmintic treatment with fenbendazole has been shown to be highly effective in this wild pony population, and for this reason, as reported for donkeys [30], ponies of Esperia can be considered an animal population in refugia.

With regards to the effect of intestinal strongyles on chromosome stability in PBLs, the aneuploidy and CA tests have provided interesting results.

After dividing group T_{D14} (animals with EPG = 0) into two groups based on age (up to 5 years and equal to or older than 6 years), a statistically significant difference was observed between aneuploidy and gap. It is interesting to note that, while the gaps were higher in younger ponies, a reduction was observed in the older ones in favor of the percentage of aneuploid cells that instead was higher in older animals. This result is congruent with what was observed by Wojcik and Smalec [25], who found that in horses the mean value of SCEs was significantly different between horses under 6 years of age and horses older than 6 years, indicating that the age of the investigated animals also represents a reference parameter for sampling when studying CIN by aneuploidy and CAs but only when considering gaps. In fact, abnormal cells percentages, chromatid and chromosome breaks, and CAs excluding gap are not affected by age.

The main finding is that the mean values of chromosome breaks and CAs excluding gaps significantly increase in ponies with high fecal egg counts. Another interesting result is that a statistically significant higher value of CAs was found in animals with EPG > 930 when compared to animals with EPG < 930. In this regard, it has been seen that, in equids, the number of parasites present could be decisive in evaluating the state of health of an individual, for which EPG < 500 represents a low infection, 500 < EPG < 1000 represents a moderate infection, and EPG > 1000 represents a high infection [31,32]. The only parameter in which an inconsistency was observed is aneuploidy, which was higher in subjects with 260 < EPG < 640 than in animals with 690 < EPG < 930. However, the differences were not statistically significant.

When animals were grouped by EPG and then by age, statistically significant differences were only observed in the mean number of gaps in individuals with high EPG (950–1970).

The data reported here show that intestinal parasites affect chromosomal stability as evaluated by aneuploidy and CAs. The studies published to date on the application of chromosomal stability tests in different animal species show correlations with age and breed, while correlations with any parasitic condition have never been considered. This study is therefore the first to report this effect, but the data are limited to aneuploidy and CA tests. It therefore appears necessary to verify whether other tests, such as SCE, micronuclei, the comet assay, and fragile sites, can also be affected by intestinal parasites or other types of endoparasitosis.

The CA test detects exposure to substances that cause DNA strand breaks or a deficiency of substances that are involved in DNA repair mechanisms but does not allow the identification of the clastogen class or micronutrient deficiency [33]. Thus, it is possible to hypothesize that increased CAs in animals with high EPG could be due to a subclinical deficiency of micro- and macronutrients linked to the action of parasites present in the intestine, which in addition to subtracting nutrients, cause alterations in the composition of the microbiota [34] (fundamental in these species to produce micro- and macronutrients) and to the structure of the intestinal barrier, with consequences for the absorption ability. Another possibility is that the alterations in the intestinal barrier of the microbiota and the inflammatory processes induced by parasites allow the passage of toxic substances into the circulation that would normally be eliminated with the feces and that have a negative effect on PBLs. Both situations may cause decreases in sport performance and fertility, which in both cases, become possible causes of reduced earnings with significant impacts on the profitability of the breeding of native breeds such as the Pony of Esperia.

These hypotheses, if confirmed through specific dosage analyses of micro- and macronutrients or of other substances at the serum level, would suggest the intervention, in subjects with EPG > 930, with appropriate food supplements. Moreover, it would be interesting to evaluate in other horse breeds and in other livestock species the impact on CIN of intestinal strongyle FEC, verifying the minimum EPG count at which it is appropriate to carry out pharmacological treatments, possibly associated with food supplement administration. These data would be useful for breeds raised in the wild, for the semi-extensive technique in which carrying out pharmacological treatments is not particularly easy, and for animals in intensive farming in which the improper use of pharmacological molecules causes a considerable environmental impact. Furthermore, in native breeds these data acquire further value since the low profitability of breeding involves an accurate evaluation of the cost/benefit ratios of any treatment that is selected.

5. Conclusions

According to the results observed in this study, it is possible to speculate that in the wildly reared Pony of Esperia breed: (1) an anthelmintic treatment with fenbendazole is highly effective, (2) a high FEC (>930 EPG) causes an increase in CAs that is reduced after only 14 days of drug treatment, and (3) in a cost–benefit assessment, the treatment of parasitosis could be useful in subjects with FEC > 930. Finally, an important aspect emerges that should be verified with future studies: individuals with FEC > 930 could benefit from food supplementation with micro- and macronutrients.

Author Contributions: Conceptualization, F.C. and A.I.; Methodology, S.A., F.B. and A.I.; Formal Analysis, E.D., E.C., I.C., M.P., E.B. and F.B.; Resources, F.C. and E.B.; Funding Acquisition, V.P. and F.C.; Data Curation, E.D., S.A., F.B. and I.C.; Writing—Original Draft Preparation, E.D., I.C., M.P., E.C. and E.B.; Writing—Review and Editing, S.A., F.B. and A.I.; Supervision, F.C. and V.P. All authors have read and agreed to the published version of the manuscript.

Funding: This research received no external funding.

Institutional Review Board Statement: This study was carried out following the recommendations of the European Council Directive (86/609/EEC) on the protection of animals. Ethical approval was not required in this study, as fecal and blood sampling were performed by the vet as part of a routine clinical visit.

Informed Consent Statement: Informed consent was obtained from the owner of the animals.

Data Availability Statement: The data presented in this study are available on request from the corresponding author.

Acknowledgments: The authors wish to thank Flavio Luongo for his technical support.

Conflicts of Interest: The authors declare no conflict of interest.

References

1. Rinaldi, L.; Musella, V.; Santaniello, A.; Santaniello, M.; Carbone, S.; Battista, E.; Cringoli, G. Mappe parassitologiche del pony di Esperia. In *Il Pony di Esperia*, 1st ed.; Ciociariaturismo: Frosinone, Italy, 2009; pp. 126–156.
2. Love, S.; Murphy, D.; Mellor, D. Pathogenicity of cyathostome infection. *Vet. Parasitol.* **1999**, *85*, 113–121. [CrossRef]
3. Peretti, V.; Ciotola, F.; Albarella, S.; Russo, V.; Di Meo, G.; Iannuzzi, L.; Roperto, F.; Barbieri, V. Chromosome fragility in cattle with chronic enzootic haematuria. *Mutagenesis* **2007**, *22*, 317–320. [CrossRef] [PubMed]
4. Di Meo, G.P.; Perucatti, A.; Genualdo, V.; Caputi-Jambrenghi, A.; Rasero, R.; Nebbia, C.; Iannuzzi, L. Chromosome fragility in dairy cows exposed to dioxins and dioxin-like PCBs. *Mutagenesis* **2011**, *26*, 269–272. [CrossRef]
5. Genualdo, V.; Perucatti, A.; Iannuzzi, A.; Di Meo, G.P.; Spagnuolo, S.M.; Caputi-Jambrenghi, A.; Coletta, A.; Vonghia, G.; Iannuzzi, L. Chromosome fragility in river buffalo cows exposed to dioxins. *J. Appl. Genet.* **2012**, *53*, 221–226. [CrossRef] [PubMed]
6. Perucatti, A.; Di Meo, G.P.; Albarella, S.; Ciotola, F.; Incarnato, D.; Caputi Jambreghi, A.; Peretti, V.; Vonghia, G.; Iannuzzi, L. Increased frequencies of both chromosome abnormalites and SCEs in two sheep flocks exposed to high dioxin levels during pasturages. *Mutagenesis* **2006**, *21*, 67–75. [CrossRef]
7. Ames, B.N. DNA damage from micronutrient deficiencies is likely to be a major cause of cancer. *Mutat. Res.* **2001**, *475*, 7–20. [CrossRef]
8. Savage, J.R. Classification and relationships of induced chromosomal structural changes. *J. Med. Genet.* **1975**, *13*, 103–122. [CrossRef]
9. Albarella, S.; Ciotola, F.; Dario, C.; Iannuzzi, L.; Barbieri, V.; Peretti, V. Chromosome instability in Mediterranean Italian buffaloes affected by limb malformation (transversal hemimelia). *Mutagenesis* **2009**, *24*, 471–474. [CrossRef]
10. Macrì, F.; Ciotola, F.; Rapisarda, G.; Lanteri, G.; Albarella, S.; Aiudi, G.; Liotta, L.; Marino, F. A rare case of simple syndactyly in a puppy. *J. Small Anim. Pract.* **2014**, *55*, 170–173. [CrossRef]
11. Peretti, V.; Ciotola, F.; Albarella, S.; Paciello, O.; Dario, C.; Barbieri, V.; Iannuzzi, L. XX/XY chimerism in cattle: Clinical and cytogenetic studies. *Sex. Dev.* **2008**, *2*, 24–30. [CrossRef]
12. Ciotola, F.; Albarella, S.; Scopino, G.; Carpino, S.; Monaco, F.; Peretti, V. Crossbreeding effect on genome stability in pig (*Sus scrofa scrofa*). *Folia Biol.* **2014**, *62*, 23–28. [CrossRef] [PubMed]
13. Ames, B.N. Micronutrient deficiencies: A major cause of DNA damage. *Ann. N. Y. Acad Sci.* **1999**, *889*, 87–106. [CrossRef] [PubMed]
14. Johannes, C.; Obe, G. Chromosomal aberration test in human lymphocytes. *Methods Mol. Biol.* **2013**, *1044*, 165–178. [CrossRef]
15. Siri, S.O.; Martino, J.; Gottifredi, V. Structural Chromosome Instability: Types, Origins, Consequences, and Therapeutic Opportunities. *Cancers* **2021**, *13*, 3056. [CrossRef] [PubMed]
16. Wójcik, E.; Sokól, A. Assessment of chromosome stability in boars. *PLoS ONE* **2020**, *15*, e0231928. [CrossRef]
17. Carroll, C.L.; Huntington, P.J. Body condition scoring and weight estimation of horses. *Equine Vet. J.* **1988**, *20*, 41–45. [CrossRef]
18. Nielsen, M.K.; Vidyashankar, A.; Andresen, U.; De Lisi, K.; Pilegaard, K.; Kaplan, R. Effects of fecal collection and storage factors on strongylid egg counts in horses. *Vet. Parasitol.* **2010**, *167*, 55–61. [CrossRef]
19. Lester, H.; Matthews, J.B. Faecal worm egg count analysis for targeting anthelmintic treatment in horse: Points to consider. *Equine Vet. J.* **2014**, *46*, 139–145. [CrossRef]
20. Zajac, A.M.; Conboy, G.A. *Veterinary Clinical Parasitology*, 8th ed.; Wiley Blackwell Press: Oxford, UK, 2011; 354p.
21. Nielsen, M.K.; Mittel, L.; Grice, A.; Erskine, M.; Graves, E.; Vaala, W.; Tully, R.C.; French, D.D.; Bowman, R.; Kaplan, R.M. American Association of Equine Practioners Parasite Control Guidelines. 2019. Available online: https://aaep.org/sites/default/files/Guidelines/AAEPParasiteControlGuidelines_0.pdf (accessed on 25 May 2022).
22. Wursthorn, L.; Martin, P. *Reso 2.0: Calculation for Fecal Egg Count Reduction Test (FECRT)*; Animal Health Research Laboratory—CSIRO: Canberra, Australia, 1990.
23. Tzelos, T.; Barbeito, J.; Nielsen, M.K.; Morgan, E.; Hodgkinson, J.; Matthews, J.B. Strongyle egg reappearance period after moxidectin treatment and its relationship with management factors in UK equine populations. *Vet. Parasitol.* **2017**, *237*, 70–76. [CrossRef]
24. Ciotola, F.; Albarella, S.; Pasolini, M.P.; Auletta, L.; Esposito, L.; Iannuzzi, L.; Peretti, V. Molecular and cytogenetic studies in a case of XX SRY-negative sex reversal in an Arabian horse. *Sex. Dev.* **2012**, *6*, 104–107. [CrossRef]
25. Wójcik, E.; Smalec, E. The effect of environmental factors on sister chromatid exchange incidence in domestic horse (*Equus caballus*) chromosomes. *Folia Biol.* **2013**, *61*, 199–204. [CrossRef] [PubMed]
26. Carrano, A.V.; Natarajan, A.T. International Commission for Protection Against Environmental Mutagens and Carcinogens. ICPEMC publication no. 14. Considerations for population monitoring using cytogenetic techniques. *Mutat. Res.* **1988**, *204*, 379–406. [CrossRef]
27. Veneziano, V.; Veronesi, F.; Buono, F.; Lia, P.R.; Manfredi, M.T.; Traversa, D.; Frangipane di Regalbono, A.; Papini, R.; Brianti, E.; Otranto, D.; et al. A national questionnaire survey of helminth control practices on horse in Italy—Preliminary results. In Proceedings of the 25th International Conference of the World Association for the Advancement of Veterinary Parasitology, Liverpool, UK, 16–20 August 2015; p. 366.
28. Buono, F.; Veronesi, F.; Pacifico, L.; Roncoroni, C.; Napoli, E.; Zanzani, S.A.; Mariani, U.; Neola, B.; Sgroi, G.; Piantedosi, D.; et al. Helminth infections in Italian donkeys: *Strongylus vulgaris* more common than *Dictyocaulus arnfieldi*. *J. Helminthol.* **2021**, *95*, e4. [CrossRef] [PubMed]

29. Peregrine, A.S.; Molento, M.B.; Kaplan, R.M.; Nielsen, M.K. Anthelmintic resistance in important parasites of horses: Does it really matter? *Vet. Parasitol.* **2014**, *201*, 1–8. [CrossRef]
30. Buono, F.; Roncoroni, C.; Pacifico, L.; Piantedosi, D.; Neola, B.; Barile, V.L.; Fagiolo, A.; Várady, M.; Veneziano, V. Cyathostominae Egg Reappearance Period After Treatment with Major Horse Anthelmintics in Donkeys. *J. Equine Vet. Sci.* **2018**, *65*, 6–11. [CrossRef]
31. Scala, A.; Tamponi, C.; Sanna, G.; Predieri, G.; Dessì, G.; Sedda, G.; Buono, F.; Cappai, M.G.; Veneziano, V.; Varcasia, A. Gastrointestinal Strongyles Egg Excretion in Relation to Age, Gender, and Management of Horses in Italy. *Animals* **2020**, *10*, 2283. [CrossRef]
32. Jota Baptista, C.; Sós, E.; Szabados, T.; Kerekes, V.; Madeira de Carvalho, L. Intestinal parasites in Przewalski's horses (*Equus ferus przewalskii*): A field survey at the Hortobágy National Park, Hungary. *J. Helminthol.* **2021**, *95*, e39. [CrossRef]
33. Mateuca, R.A.; Decordier, I.; Kirsch-Volders, M. Cytogenetic methods in human biomonitoring: Principles and uses. *Methods Mol. Biol.* **2012**, *817*, 305–334. [CrossRef]
34. Clark, A.; Sallé, G.; Ballan, V.; Reigner, F.; Meynadier, A.; Cortet, J.; Koch, C.; Riou, M.; Blanchard, A.; Mach, N. Strongyle Infection and Gut Microbiota: Profiling of Resistant and Susceptible Horses Over a Grazing Season. *Front. Physiol.* **2018**, *9*, 272. [CrossRef]

Article

Comparative Fluorescence In Situ Hybridization (FISH) Mapping of Twenty-Three Endogenous Jaagsiekte Sheep Retrovirus (enJSRVs) in Sheep (*Ovis aries*) and River Buffalo (*Bubalus bubalis*) Chromosomes

Angela Perucatti [1], Alessandra Iannuzzi [1,*], Alessia Armezzani [2], Massimo Palmarini [2] and Leopoldo Iannuzzi [1]

[1] National Research Council (CNR), Institute of Animal Production System on Mediterranean Environment (ISPAAM), Piazzale E. Fermi, 1, 80055 Portici, Italy
[2] MRC-University of Glasgow Centre for Virus Research, 464 Bearsden Road, Glasgow G61-1QH, UK
* Correspondence: alessandra.iannuzzi@cnr.it; Tel.: +39-32-8961-7073

Citation: Perucatti, A.; Iannuzzi, A.; Armezzani, A.; Palmarini, M.; Iannuzzi, L. Comparative Fluorescence In Situ Hybridization (FISH) Mapping of Twenty-Three Endogenous Jaagsiekte Sheep Retrovirus (enJSRVs) in Sheep (*Ovis aries*) and River Buffalo (*Bubalus bubalis*) Chromosomes. *Animals* **2022**, *12*, 2834. https://doi.org/10.3390/ani12202834

Academic Editor: Ettore Olmo

Received: 27 August 2022
Accepted: 17 October 2022
Published: 19 October 2022

Publisher's Note: MDPI stays neutral with regard to jurisdictional claims in published maps and institutional affiliations.

Copyright: © 2022 by the authors. Licensee MDPI, Basel, Switzerland. This article is an open access article distributed under the terms and conditions of the Creative Commons Attribution (CC BY) license (https:// creativecommons.org/licenses/by/ 4.0/).

Simple Summary: The genome of domestic sheep (*Ovis aries*) harbors at least twenty-seven copies of enJSRVs, endogenous retroviruses (ERVs) highly related to the exogenous and pathogenic Jaagsiekte sheep betaretrovirus (JSRV). Interestingly, some of these loci are insertionally polymorphic, that is they are present only in some individuals or populations of their host species. This differential distribution of enJSRVs has provided important insights into tracing host and viral evolution. In this study, we report the first comparative fluorescent in situ hybridization (FISH) mapping of genetically characterized enJSRVs in domestic sheep (2n = 54) and river buffalo (*Bubalus bubalis*, 2n = 50), and reveal a high conservation of enJSRVs chromosome localization between these two species.

Abstract: Endogenous retroviruses (ERVs) are the remnants of ancient infections of host germline cells, thus representing key tools to study host and viral evolution. Homologous ERV sequences often map at the same genomic locus of different species, indicating that retroviral integration occurred in the genomes of the common ancestors of those species. The genome of domestic sheep (*Ovis aries*) harbors at least twenty-seven copies of ERVs related to the exogenous and pathogenic Jaagsiekte sheep retrovirus (JSRVs), thus referred to as enJSRVs. Some of these loci are unequally distributed between breeds and individuals of the host species due to polymorphic insertions, thereby representing invaluable tools to trace the evolutionary dynamics of virus populations within their hosts. In this study, we extend the cytogenetic physical maps of sheep and river buffalo by performing fluorescent in situ hybridization (FISH) mapping of twenty-three genetically characterized enJSRVs. Additionally, we report the first comparative FISH mapping of enJSRVs in domestic sheep (2n = 54) and river buffalo (*Bubalus bubalis*, 2n = 50). Finally, we demonstrate that enJSRV loci are conserved in the homologous chromosomes and chromosome bands of both species. Altogether, our results support the hypothesis that enJSRVs were present in the genomes of both species before they differentiated within the *Bovidae* family.

Keywords: sheep; river buffalo; endogenous retrovirus; FISH-mapping; cytogenetic map

1. Introduction

Retroviruses possess the unique ability to integrate into the genome of infected cells. Occasionally, they can infect germline cells and give rise to endogenous retroviruses (ERVs), retroviral sequences transmitted vertically, in a Mendelian fashion, as part of the host genome. As such, ERVs represent fascinating tools to study both virus and host genome evolution [1]. ERVs have been found in all vertebrates studied to date, including fish, amphibians, birds, reptiles, and mammals [2]. Comparative genomic studies have shown that related species often share ERV families or specific ERV loci, and that, in many cases, homologous ERV sequences map at the same genomic locus in multiple species' genomes, indicating that retroviral integration events occurred in the genomes of the common ancestors of those species [3,4].

In 2013, Garcia-Etxebarria and Jugo traced the evolutionary history of bovine ERVs (BoERVs) by performing computational analyses on the genomes of several bovid species, including cattle, sheep, goat, and water buffalo [5]. Interestingly, they found twenty-six BoERV families in all the species studied, suggesting that most of these families could be present in all members of the *Bovidae* family. However, they could not detect four BoERV families (i.e., BoERV24, BoERV26, BoERV28, and BoERV29) in sheep or goat genomes, indicating that these families may be specific to the *Bovinae* subfamily. The authors hypothesize that the majority of the BoERV families invaded the genome of the common ancestor of the *Bovidae* family approximately 20 million years ago (MYA) and only later, between 12MYA and 20MYA, the ancestors of the BoERV24, BoERV26, BoERV28, and BoERV29 families might have been inserted into the genome of the *Bovinae* subfamily ancestor [5].

The domestic sheep (*Ovis aries*) harbors at least twenty-seven ERV loci related to the exogenous and pathogenic Jaagsiekte sheep retrovirus (JSRV), thus referred to as enJSRVs [6,7]. Interestingly, some enJSRV loci are insertionally polymorphic, that is they are present only in some individuals or populations of their host species. This differential distribution has provided important insights into tracing the evolutionary dynamics of virus populations within their hosts [8].

Along these lines, in previous studies, we used enJSRVs to (i) investigate the history of sheep domestication [9], (ii) explore the molecular mechanisms through which the most recent enJSRV—enJSRV-26—eludes the restriction exerted by enJS56A1 (which entered the sheep genome before and during the speciation within the genus *Ovis*) [10], or (iii) study the chromosome location of some enJSRV loci by fluorescent in situ hybridization (FISH) analyses of metaphase R-banded sheep chromosomes [9,10].

In the present work, we extend the cytogenetic physical maps of both sheep and river buffalo chromosomes by FISH mapping fifteen and twenty-three additional enJSRVs, respectively. In addition, we perform the first comparative FISH mapping of genetically characterized enJSRVs in domestic sheep and river buffalo (*Bubalus bubalis*).

2. Materials and Methods

Cell cultures. Peripheral blood lymphocytes of sheep (*Ovis aries*, OAR; 2n = 54) (four animals) and river buffalo (*Bubalus bubalis*, BBU; 2n = 50) (three animals) were cultured for 72 h in Roswell Park Memorial Institute (RPMI)-1640 culture medium enriched with 15% fetal calf serum (FCS), concanavalin A (15 μg/mL), penicillin/streptomycin (0.1 mL), L-glutamine (0.1 mL), and one drop of sterile sodium heparin to prevent coagulation. In order to obtain R-banding patterns and cause metaphase arrest, 7 h prior to harvesting, cells were labeled with BrdU (15 μg/mL) and Hoechst 33258 (30 μg/mL) and, 6 h later, they were treated with colcemid (0.1 μg/mL). After incubation in a hypotonic solution (KCl 0.075 M at 37.5 °C for 20 min), cells were fixed three times with 3:1 methanol–acetic acid (v/v) solution. Cell suspensions were then spread onto a slide, and stored at −20 °C.

Probes and FISH mapping. FISH analysis was performed using standard procedures [11,12]. Briefly, cells were pre-treated overnight at 50 °C, and subsequently stained with Hoechst 33258 (25 μg/mL) for 10 min. Slides were then exposed to UV light for 20 min, washed with distilled water, and air-dried. Hybridization, chromosome staining, signal detection and image processing were carried out as already described [11,12]. Slides were mounted in antifade mounting medium with propidium iodide to visualize both FITC-signals and RBPI-banding using two microscope filter combinations. Two images for each metaphase were acquired with both FITC signals and RBPI-banding. Next, FITC signals were superimposed on RBPI-banding to get a precise position of FITC-signals on chromosome bands. Thirty metaphases for each probe were examined. Chromosome identification and banding numbering system followed the standard nomenclature of both species [13,14]. The bacterial artificial chromosome (BAC) clones used for sheep cytogenetic mapping have been already characterized and described elsewhere [6]. Notably, each BAC clone used to obtain the sequences of the various enJSRV loci was subjected to Southern blot analyses to verify that it contained only one locus. The list of sheep BAC clones of the CHORI-243 library containing the 27 enJSRVs used in this study is reported in Table 1:

Table 1. List of sheep BAC clones containing the 27 enJSRVs used in the present study.

Provirus	BAC Clone
enJSRV-1	6g20
enJSRV-2	6j6
enJSRV-3	7i2
enJSRV-4	15c2
enJSRV-5	19l17
enJSRV-6	33k3
enJSRV-7	35o24
enJSRV-8	36d13
enJSRV-9	36p8
enJSRV-10	42k15
enJSRV-11	44o16
enJSRV-12	45l23
enJSRV-13	48b6
enJSRV-14	52d13
enJSRV-15	57m3
enJSRV-16	57m4
enJSRV-17	65c8
enJSRV-18	67f9
enJSRV-19	70l9
enJSRV-20	81j8
enJSRV-21	84d10
enJSRV-22	88g3
enJSRV-23	90k11
enJSRV-24	94c2
enJSRV-25	98k22
enJSRV-26	102b15
enJSRV-27	14c2

3. Results

In the present study, we conducted FISH mapping on sheep and river buffalo chromosomes (or chromosome arms) using twenty-seven sheep BAC clones (Table 1). As reported in Table 2, we obtained good hybridization signals only with twenty-three BAC clones, allowing us to map the corresponding twenty-three enJSRV loci. Interestingly, we localized these loci on twelve different chromosomes (or chromosome arms) of both species. As reported in Table 1, we found that, for all the probes used, hybridization signals and chromosome bands localized in the same homologous chromosome pairs of both sheep and river buffalo (Table 2). For example, we found that the enJSRV-1 and enJSRV-10 loci map onto homologous sheep and river buffalo R-banded chromosomes, as shown in Figure 1. Moreover, the BAC containing the enJSRV-1 map onto two different chromosomes in both sheep (chromosomes 6 and 18) and river buffalo (chromosomes 7 and 20) (Figure 1 and Table 2). In addition, the enJSRV-15, enJSRV-20, enJSRV-21, and enJSRV-27 display the same chromosome localization as the enJSRV-1 in both species (OAR6q13 and BBU7q21, respectively) (Table 2). Finally, the enJSRV-2 and enJSRV-6 map very close in both sheep (OAR1q45 and OARiq43, respectively) and river buffalo homologous chromosomes (BBU1q45 and BBU1q43, respectively). Notably, only the enJSRV-7 maps onto the centromeric regions of all autosomes [10], probably due to the presence of highly repetitive sequences in the BAC clones, the centromeric regions being highly rich in these sequences.

Table 2. Chromosomal localization of 23 enJSRVs in sheep (OAR) and river buffalo (BBU) chromosomes by FISH-mapping.

enJSRV	OAR	BBU
enJSRV-1	18q22	20q22
	6q13	7q13
enJSRV-2	1q45	1q45
enJSRV-3	14q24dist	18q24dist
enJSRV-4	2p21prox	3q21prox
enJSRV-6	1q41	1q41
enJSRV-7	centromeric	centromeric
enJSRV-8	3q21	4q21
enJSRV-10	14q24	18q24
enJSRV-11	1p13	6q13
enJSRV-12	19q24	21q24
enJSRV-13	14q24dist	18q24
enJSRV-14	3p24	12q34
enJSRV-15	6q13	7q21
enJSRV-16	10q24	13q24
enJSRV-17	19q24	21q24
enJSRV-18	11q17	3p22
enJSRV-19	1p23	6q25
enJSRV-20	6q13	7q13
enJSRV-21	6q13	7q13
enJSRV-22	15q23	16q23dist
enJSRV-24	18q24	20q24
enJSRV-26	2p25dist	3q25
enJSRV-27	6q13	7q13

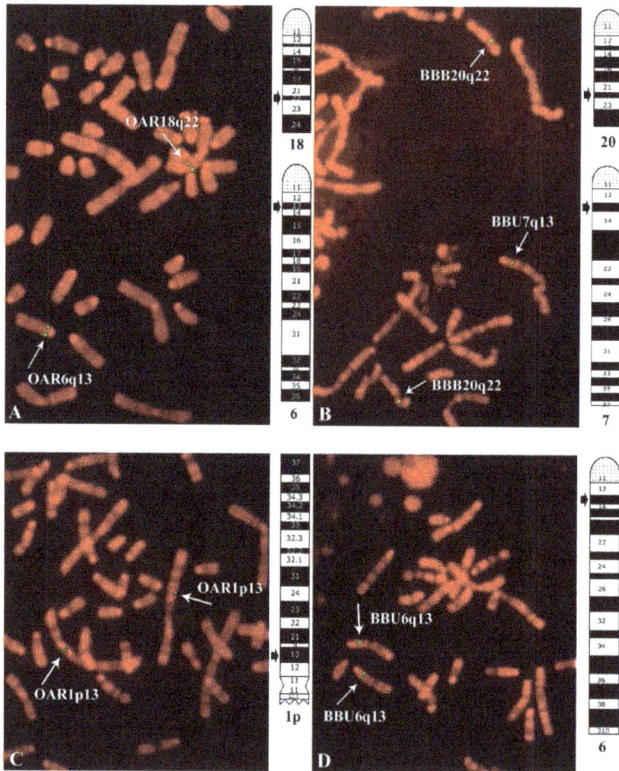

Figure 1. FISH mapping of enJSRV-1 (**A,B**) and enJSRV-10 (**C,D**) in sheep (OAR) and river buffalo (BBU) chromosomes. Two different images were taken, with hybridization FITC signal and with RBPI-banding. Subsequently, hybridization FITC signals were superimposed on RBPI-banding to get a precise localization of mapped loci in both species. Note that enJSRV-1 maps onto two different chromosomes in both species (**A,B**).

4. Discussion

FISH represents a very powerful cytogenetic technique for mapping a particular genomic sequence on a chromosome [15,16], and better anchoring of radiation hybrid (RH) and genomic maps to specific chromosome regions [17–20]. In more recent years, FISH has also been recognized as a reliable diagnostic and discovery tool to evaluate genetic anomalies, by studying chromosomal aberrations in both metaphase and interphase nuclei [reviewed in [21–25], and defects in chromosome segregation during meiosis [26,27]. In addition, implementations of FISH with whole-chromosome painting have led to the generation of detailed comparative maps to study chromosomal homologies and divergences between related and unrelated species [28–32]. Finally, FISH has also become instrumental in generating detailed comparative maps to study gene order, conserved chromosomal regions, and chromosomal rearrangements between related and unrelated species [33–36].

Along these lines, FISH analyses unveiled the phylogenetic relationships between the *Caprinae* subfamily and the earliest-diverging *Bovinae* subfamily, by showing two main chromosome events occurring at the autosomes 9 and 14, and the sex chromosomes (mainly the X-chromosome). More specifically, in previous studies carried out in our group, we demonstrated that a chromosome transposition has occurred from the proximal region of *Bovinae* chromosome 9 to the proximal region of *Caprinae* chromosome 14, and that at least four chromosome rearrangements (i.e., three transpositions and one inversion) differentiated the *Caprinae* from the *Bovinae* X-chromosomes reviewed in [37].

A comparative FISH mapping of enJSRVs has been reported previously in sheep and goat cell lines [38]. In this study, the authors only partially found enJSRVs localized on the same homologous chromosome band of the two species. In addition, they found enJSRV loci in seven and eight chromosomes of sheep and goats, respectively.

In the present work, we report the first comparative FISH mapping between two species belonging to the *Caprinae* (*Ovis aries*) and *Bovinae* (*Bubalus bubalis*) subfamilies of the *Bovidae* family by using well-identified and genetically characterized enJSRVs [5,6]. We show that hybridization signals of enJSRVs are found in at least twelve different chromosomes (or chromosome arms) of both species, and that all mapped loci are conserved in homologous chromosome regions and chromosome bands of these two species (Table 1). However, since BACs contain large genomic inserts, it is entirely possible that, besides enJSRVs, we also simultaneously mapped other genes and sequences present on such clones that share homology with some river buffalo chromosomal regions. Indeed, since some BoERV families (i.e., BoERV24, BoERV26, BoERV28, and BoERV29) are specific to the *Bovinae* family [5], the same could have occurred for some enJSRV which are present in the *Caprinae* subfamily (i.e., *Ovis aries*) but not in the *Bovinae* subfamily (i.e., *Bubabus bubalis*).

We observed that, in both species, the BAC clones containing enJSRV-1 map onto two different chromosomal locations (Table 2). Interestingly, we found that enJSRV-7 is the only locus mapping at the centromeric regions of both sheep [10] and buffalo chromosomes.

The same results were achieved by FISH mapping several ERV in sheep chromosomes which exhibited abundant centromeric to the dispersed distribution of various endoviruses, probably due to the abundance of genomic organization ERV-related repetitive elements which are particularly present at the centromeric regions of the chromosomes [39].

Our comparative FISH mapping in two different bovid species further confirms the high degree of chromosome (and chromosome arm) conservation among bovids reviewed in [37]. In addition, our study supports the hypothesis that enJSRVs were present in the genomes of their bovine ancestor before the differentiation of the *Caprinae* subfamily (including *Ovis aries*) from the most ancient *Bovinae* subfamily (including *Bubalus bubalis*) [37,40]. These results are in agreement with those published in a previous study [5] tracing BoERVs evolution in several species of the *Bovidae* family, including cattle, sheep, goats, and water buffalo [5]. Interestingly, these authors found that most of the BoERV families are present in all the species studied, supporting the hypothesis that BoERVs entered the genome of the common ancestor of the *Bovidae* family about 20 MYA or less. In addition, they detected higher BoERV copy numbers in cattle compared to other bovid species, suggesting that an additional expansion of retroviral copies might have occurred in the cattle genome [5].

Interestingly, we found five enJSRV loci (enJSRV-1, enJSRV-15, enJSRV-20, enJSRV-21, and enJSRV-27) on the same chromosome band (OAR6q13/BBU7q21) (Table 2). Previous FISH mapping conducted in our group revealed two type-one loci on this very same chromosome band: the pyroglutamylated RFamide peptide receptor (QRFPR) and the translocation-associated membrane protein 1-like 1 (TRAM1L1) [41]. The RFamide peptide family consists of several groups, including the neuropeptide FF group, the prolactin-releasing peptide group, the gonadotropin inhibitory hormone group, the kisspeptin group, and the pyroglutamylated RFamide peptide (26RFa/QRFP) group [42]. Interestingly, pyroglutamylated RFamide peptide 43 has been proven to be a putative modulator of testicular steroidogenesis, playing an important role in reproduction [43]. Notably, ERVs are key in placental morphogenesis and mammalian reproduction [44]. TRAM1L1 seems to be closely related to chronic widespread musculoskeletal pain (CWP), a common disorder affecting about 5–15% of the population, and one of the main symptoms of fibromyalgia, which has been shown to be associated with an altered gut microbiome [45]. By using the sheep genome reference sequence (https://www.ncbi.nlm.nih.gov/genome/gdv/browser/genome/?id=GCF_016772045.1, accessed on 12 May 2018), we identified sixteen genes included between QRFPR and TRAM1L (Supplementary Table S1). In this table we reported some of the functions of those genes (including QRFPR and TRAM1L), mostly involved in anti-tumor immune response probably to counteract the presence of several enJSRVs in these chromosomic regions. Indeed, genomic amplification within the 6q13 region was detected, and it was found that the number of enJSRV-6q13 is correlated to the number of protective mutations [46].

5. Conclusions

To our knowledge, this is the first comparative FISH mapping of sheep and river buffalo chromosomes using genetically characterized enJSRVs. Interestingly, our results reveal a high degree of conservation of enJSRVs localization in the homologous chromosomes and chromosome bands of both species. These findings support the hypothesis that enJSRVs entered the host genome before the differentiation of the *Caprinae* subfamily from the earliest-diverging *Bovinae* subfamily of the *Bovidae* family. Finally, the present study extends the current genetic physical maps of sheep and river buffalo by mapping, respectively, fifteen and twenty-three additional enJSRV loci on the chromosomes and chromosome arms of these two species.

Supplementary Materials: The following supporting information can be downloaded at: https://www.mdpi.com/article/10.3390/ani12202834/s1, Table S1: List of genes present in the ORA6q13 and included between QRFPR and TRAM1L1 (from: https://www.ncbi.nlm.nih.gov/genome/gdv/browser/genome/?id=GCF_016772045.1), accessed on 2 October 2022) their main functions and references.

Author Contributions: Conceptualization L.I. and M.P.; investigation, A.P. and A.A.; data curation, A.P. and A.I.; writing—original draft preparation L.I., A.A. and A.I.; writing—review and editing, A.I., L.I., A.P., A.A. and M.P.; supervision, L.I. and M.P.; funding acquisition L.I. and M.P. All authors have read and agreed to the published version of the manuscript.

Funding: This research was funded by the project PON1_486 GENOBU.

Institutional Review Board Statement: The study was approved by the CNR-ISPAAM Ethics Committee, protocol number 0000606 of 03/27/2015.

Informed Consent Statement: Not applicable.

Data Availability Statement: The data that support the findings of this study are available from the corresponding author, [AI], upon reasonable request.

Acknowledgments: We are grateful to Domenico Incarnato, CNR-ISPAAM, Portici, Italy for his excellent technical assistance.

Conflicts of Interest: The authors declare no conflict of interest.

References

1. Stoye, J.P. Studies of endogenous retroviruses reveal a continuing evolutionary saga. *Nat. Rev. Microbiol.* **2012**, *10*, 395–406. [CrossRef] [PubMed]
2. Johnson, W.E. Origins and evolutionary consequences of ancient endogenous retroviruses. *Nat. Rev. Microbiol.* **2019**, *17*, 355–370. [CrossRef] [PubMed]
3. Johnson, W.E. Endogenous Retroviruses. In *Encyclopedia of Virology*; Mahy, B.W.J., Van Regenmortel, M.H.V., Eds.; Elsevier: Oxford, UK, 2008; pp. 105–109.
4. Gifford, R.J.; Blomberg, J.; Coffin, J.M.; Fan, H.; Heidmann, T.; Mayer, J.; Stoye, J.; Tristem, M.; Johnson, W.E. Nomenclature for endogenous retrovirus (ERV) loci. *Retrovirology* **2018**, *15*, 59. [CrossRef] [PubMed]
5. Garcia-Etxebarria, K.; Jugo, B.M. Evolutionary history of bovine endogenous retroviruses in the Bovidae family. *BMC Evol. Biol.* **2013**, *13*, 256. [CrossRef] [PubMed]
6. Arnaud, F.; Caporale, M.; Varela, M.; Biek, R.; Chessa, B.; Alberti, A.; Golder, M.; Mura, M.; Zhang, Y.-P.; Yu, L.; et al. A paradigm for virus-host coevolution: Sequential counter-adaptations between endogenous and exogenous retroviruses. *PLoS Pathog.* **2007**, *3*, e170. [CrossRef] [PubMed]
7. Arnaud, F.; Varela, M.; Spencer, T.E.; Palmarini, M. Coevolution of endogenous betaretroviruses of sheep and their host. *Cell Mol. Life Sci.* **2008**, *65*, 3422–3432. [CrossRef] [PubMed]
8. Armezzani, A.; Varela, M.; Spencer, T.E.; Palmarini, M.; Arnaud, F. "Ménage à Trois": The evolutionary interplay between JSRV, enJSRVs and domestic sheep. *Viruses* **2014**, *6*, 4926–4945. [CrossRef]
9. Chessa, B.; Pereira, F.; Arnaud, F.; Amorim, A.; Goyache, F.; Mainland, I.; Kao, R.R.; Pemberton, J.M.; Beraldi, D.; Stear, M.J.; et al. Revealing the history of sheep domestication using retrovirus integrations. *Science* **2009**, *324*, 532–536. [CrossRef] [PubMed]
10. Armezzani, A.; Arnaud, F.; Caporale, M.; Di Meo, G.P.; Iannuzzi, L.; Murgia, C.; Palmarini, M. The signal peptide of a recently integrated endogenous sheep betaretrovirus envelope plays a major role in eluding gag-mediated late restriction. *J. Virol.* **2011**, *85*, 7118–7128. [CrossRef] [PubMed]
11. Iannuzzi, L.; Di Berardino, D. Tools of the trade: Diagnostics and research in domestic animal cytogenetics. *J. Appl. Genet.* **2008**, *49*, 357–366. [CrossRef]
12. Di Meo, G.P.; Perucatti, A.; Floriot, S.; Hayes, H.; Schibler, L.; Incarnato, D.; Di Berardino, D.; Williams, J.; Cribiu, E.; Eggen, A.; et al. An extended river buffalo (*Bubalus bubalus*, 2n = 50) cytogenetic map: Assignment of 68 autosomal loci by FISH-mapping and R-banding and comparison with human chromosomes. *Chromos. Res.* **2008**, *16*, 827–837. [CrossRef]
13. Cribiu, E.; Di Berardino, D.; Di Meo, G.; Eggen, A.; Gallagher, D.; Gustavsson, I.; Hayes, H.; Iannuzzi, L.; Popescu, C.; Rubes, J.; et al. International System for Chromosome Nomenclature of Domestic Bovids (ISCNDB 2000). *Cytogenet. Genome Res.* **2001**, *95*, 283–299. [CrossRef] [PubMed]
14. Iannuzzi, L. Report of the committee for standard karyotype of the river buffalo (*Bubalus bubalis* L., 2n = 50). *Cytogenet. Cell Genet.* **1994**, *67*, 102–113. [CrossRef] [PubMed]
15. Hayes, H.; Elduque, C.; Gautier, M.; Schibler, L.; Cribiu, E.; Eggen, A. Mapping of 195 genes in cattle and updated comparative map with man, mouse, rat and pig. *Cytogenet. Genome Res.* **2003**, *102*, 16–24. [CrossRef] [PubMed]
16. Iannuzzi, L.; Di Meo, G.P.; Gallagher, D.S.; Ryan, A.M.; Ferrara, L.; Womack, J.E. Chromosomal localization of omega and trophoblast interferon genes in goat and sheep by fluorescent in situ hybridization. *J. Hered.* **1993**, *84*, 301–304. [CrossRef]
17. Amaral, M.E.; Grant, J.R.; Riggs, P.K.; Stafuzza, N.B.; Filho, E.A.; Goldammer, T.; Weikard, R.; Brunner, R.M.; Kochan, K.J.; Greco, A.J.; et al. A first generation whole genome RH map of the river buffalo with comparison to domestic cattle. *BMC Genom.* **2008**, *9*, 631. [CrossRef]
18. Ianella, P.; Venancio, L.P.; Stafuzza, N.B.; Miziara, M.N.; Agarwala, R.; Schäffer, A.A.; Riggs, P.K.; Womack, J.E.; Amaral, M.E.J. First radiation hybrid map of the river buffalo X chromosome (BBUX) and comparison with BTAX. *Anim. Genet.* **2008**, *39*, 196–200. [CrossRef] [PubMed]
19. Goldammer, T.; Brunner, R.M.; Rebl, A.; Wu, C.H.; Nomura, K.; Hadfield, T.; Maddox, J.F.; Cockett, N.E. Cytogenetic anchoring of radiation hybrid and virtual maps of sheep chromosome X and comparison of X chromosomes in sheep, cattle, and human. *Chromos. Res.* **2009**, *17*, 497–506. [CrossRef] [PubMed]
20. De Lorenzi, L.; Parma, P. Identification of Some Errors in the Genome Assembly of *Bovidae* by FISH. *Cytogenet Genome Res.* **2020**, *160*, 85–93. [CrossRef] [PubMed]
21. Ducos, A.; Revay, T.; Kovacs, A.; Hidas, A.; Pinton, A.; Bonnet-Garnier, A.; Molteni, L.; Slota, E.; Switonski, M.; Arruga, M.; et al. Cytogenetic screening of livestock populations in Europe: An overview. *Cytogenet. Genome Res.* **2008**, *120*, 26–41. [CrossRef] [PubMed]
22. Iannuzzi, A.; Parma, P.; Iannuzzi, L. Chromosome Abnormalities and Fertility in Domestic Bovids: A Review. *Animals* **2021**, *11*, 802. [CrossRef] [PubMed]
23. Szczerbal, I.; Switonski, M. Clinical Cytogenetics of the Dog: A Review. *Animals* **2021**, *11*, 947. [CrossRef] [PubMed]
24. Iannuzzi, A.; Braun, M.; Genualdo, V.; Perucatti, A.; Reinartz, S.; Proios, I.; Heppelmann, M.; Rehage, J.; Hülskötter, K.; Beineke, A.; et al. Clinical, cytogenetic and molecular genetic characterization of a tandem fusion translocation in a male Holstein cattle with congenital hypospadias and a ventricular septal defect. *PLoS ONE* **2020**, *15*, e0227117. [CrossRef] [PubMed]

25. Donaldson, B.; Villagomez, D.; King, W.A. Classical, Molecular, and Genomic Cytogenetics of the Pig, a Clinical Perspective. *Animals* **2021**, *11*, 1257. [CrossRef] [PubMed]
26. Di Dio, C.; Longobardi, V.; Zullo, G.; Parma, P.; Pauciullo, A.; Perucatti, A.; Higgins, J.; Iannuzzi, A. Analysis of meiotic segregation by triple-color fish on both total and motile sperm fractions in a t(1p;18) river buffalo bull. *PLos ONE* **2020**, *15*, e0232592. [CrossRef] [PubMed]
27. Genualdo, V.; Turri, F.; Pizzi, F.; Castiglioni, B.; Marletta, D.; Iannuzzi, A. Sperm Nuclei Analysis and Nuclear Organization of a Fertile Boar-Pig Hybrid by 2D FISH on Both Total and Motile Sperm Fractions. *Animals* **2021**, *11*, 738. [CrossRef] [PubMed]
28. Hayes, H. Chromosome painting with human chromosome-specific DNA libraries reveals the extent and distribution of conserved segments in bovine chromosomes. *Cytogenet. Cell Genet.* **1995**, *71*, 168–174. [CrossRef]
29. Chowdhary, B.P.; Frönicke, L.; Gustavsson, I.; Scherthan, H. Comparative analysis of the cattle and human genomes: Detection of ZOO-FISH and gene mapping-based chromosomal homologies. *Mamm. Genome* **1996**, *7*, 297–302. [CrossRef] [PubMed]
30. Iannuzzi, L.; Di Meo, G.P.; Perucatti, A.; Bardaro, T. ZOO-FISH and R-banding reveal extensive conservation of human chromosome regions in euchromatic regions of river buffalo chromosomes. *Cytogenet. Cell Genet.* **1998**, *82*, 210–214. [CrossRef] [PubMed]
31. Iannuzzi, L.; Di Meo, G.P.; Perucatti, A.; Incarnato, D. Comparison of the human with the sheep genomes by use of human chromosome-specific painting probes. *Mamm. Genome* **1999**, *10*, 719–723. [CrossRef] [PubMed]
32. Iannuzzi, A.; Pereira, J.; Iannuzzi, C.; Fu, B.; Ferguson-Smith, M. Pooling strategy and chromosome painting characterize a living zebroid for the first time. *PLoS ONE* **2017**, *12*, e0180158. [CrossRef] [PubMed]
33. Iannuzzi, L.; Di Meo, G.P.; Perucatti, A.; Incarnato, D.; Schibler, L.; Cribiu, E.P. Comparative FISH-mapping of bovid X chromosomes reveals homologies and divergences between the subfamilies *Bovinae* and *Caprinae*. *Cytogen. Cell Genet.* **2000**, *89*, 171–176. [CrossRef] [PubMed]
34. Di Meo, G.P.; Perucatti, A.; Floriot, S.; Incarnato, D.; Rullo, R.; Jambrenghi, A.C.; Ferretti, L.; Vonghia, G.; Cribiu, E.; Eggen, A.; et al. Chromosome evolution and improved cytogenetic maps of the Y chromosome in cattle, zebu, river buffalo, sheep and goat. *Chromos. Res.* **2005**, *13*, 349–355. [CrossRef] [PubMed]
35. Di Meo, G.P.; Goldammer, T.; Perucatti, A.; Genualdo, V.; Iannuzzi, A.; Incarnato, D.; Rebl, A.; Di Berardino, D.; Iannuzzi, L. Extended cytogenetic maps of sheep chromosome 1 and their cattle and river buffalo homoeologues: Comparison with the OAR1 RH map and human chromosomes 2, 3, 21 and 1q. *Cytogenet. Genome Res.* **2011**, *133*, 16–24. [CrossRef]
36. Perucatti, A.; Genualdo, V.; Iannuzzi, A.; Rebl, A.; Di Berardino, D.; Goldammer, T.; Iannuzzi, L. Advanced comparative cytogenetic analysis of X chromosomes in river buffalo, cattle, sheep, and human. *Chromos. Res.* **2012**, *20*, 413–425. [CrossRef]
37. Iannuzzi, L.; King, A.W.; Di Berardino, D. Chromosome evolution in domestic bovids as revealed by chromosome banding and FISH-mapping techniques. *Cytogenet. Genome Res.* **2009**, *126*, 49–62. [CrossRef] [PubMed]
38. Carlson, J.; Lyon, M.; Bishop, J.; Vaiman, A.; Cribiu, E.; Mornex, J.-F.; Brown, S.; Knudson, D.; De Martini, J.; Leroux, C. Chromosomal Distribution of Endogenous Jaagsiekte Sheep Retrovirus Proviral Sequences in the Sheep Genome. *J. Virol.* **2003**, *77*, 9662–9668. [CrossRef]
39. Mustafa, S.I.; Schwarzacher, T.; Heslop-Harrison, J.S. The Nature and Chromosomal Landscape of Endogenous Retroviruses (ERVs) Integrated in the Sheep Nuclear Genome. *DNA* **2022**, *2*, 86–103. [CrossRef]
40. Balmus, G.; Trifonov, V.A.; Biltueva, L.S.; O'Brien, P.C.M.; Alkalaeva, E.S.; Fu, B.; Skidmore, J.A.; Allen, T.; Graphodatsky, A.S.; Yang, F.; et al. Cross-species chromosome painting among camel, cattle, pig and human: Further insights into the putative Cetartiodactyla ancestral karyotype. *Chromos. Res.* **2007**, *15*, 499–514. [CrossRef]
41. Goldammer, T.; Di Meo, G.P.; Lühken, G.; Drögemüller, C.; Wu, C.H.; Kijas, J.; Dalrymple, B.; Nicholas, F.; Maddox, J.; Iannuzzi, L.; et al. Molecular Cytogenetics and Gene Mapping in Sheep (*Ovis aries*, 2n = 54). *Cytogenet. Genome Res.* **2009**, *126*, 63–76. [CrossRef]
42. Ukena, K.; Osugi, T.; Leprince, J.; Vaudry, H.; Tsutsui, K. Molecular evolution and function of 26RFa/QRFP and its cognate receptor. *J. Mol. Endocrinol.* **2014**, *52*, T119–T131. [CrossRef] [PubMed]
43. Patel, S.K.; Singh, S.K. Pyroglutamylated RFamide peptide 43: A putative modulator of testicular steroidogenesis. *Andrology* **2020**, *8*, 1815–1823. [CrossRef]
44. Dunlap, K.A.; Palmarini, M.; Varela, M.; Burghardt, R.C.; Hayashi, K.; Farmer, J.L.; Spencer, T.E. Endogenous retroviruses regulate periimplantation placental growth and differentiation. *PNAS* **2006**, *103*, 14390–14395. [CrossRef] [PubMed]
45. Freidin, M.B.; Stalteri, M.A.; Wells, P.M.; Lachance, G.; Baleanu, A.-F.; Bowyer, R.C.E.; Kurilshikov, A.; Zhernakova, A.; Steves, C.J.; Williams, F.M.K. An association between chronic widespread pain and the gut microbiome. *Rheumatology* **2021**, *60*, 3727–3737. [CrossRef] [PubMed]
46. Cumer, T.; Pompanon, F.; Boyer, F. Old origin of a protective endogenous retrovirus (enJSRV) in the *Ovis* genus. *Heredity* **2019**, *122*, 187–194. [CrossRef] [PubMed]

Article

XX/XY Chimerism in Internal Genitalia of a Virilized Heifer

Izabela Szczerbal [1,†], Joanna Nowacka-Woszuk [1,†], Monika Stachowiak [1], Anna Lukomska [2], Kacper Konieczny [3], Natalia Tarnogrodzka [1], Jakub Wozniak [1] and Marek Switonski [1,*]

[1] Department of Genetics and Animal Breeding, Poznan University of Life Sciences, 60-637 Poznan, Poland
[2] Department of Preclinical Sciences and Infectious Diseases, Poznan University of Life Sciences, 60-637 Poznan, Poland
[3] Department of Internal Diseases and Diagnostics, Poznan University of Life Sciences, 60-637 Poznan, Poland
* Correspondence: marek.switonski@up.poznan.pl
† These authors contributed equally to this work.

Simple Summary: Freemartinism is the most common type of disorder of sex development (DSD) in heifers; it is caused by the formation of placental anastomoses between heterosexual twin fetuses and the transfer of masculine factors produced by the testes of the male co-twin to the female fetus. The abnormal development of external genitalia is commonly observed in such heifers, but it cannot be assumed that each heifer with ambiguous genitalia is an example of freemartinism. We genetically analyzed five DSD heifers, and four appeared to be freemartins, as revealed by the presence of XX/XY leukocyte chimerism. The fifth heifer had a normal XX sex chromosome complement and lacked the Y-chromosome-derived genes (*SRY*, *ZFY* and *AMELY*) in blood cells. This heifer was extensively studied through genetic, anatomical, and histological approaches. Postmortem anatomical and histological analysis showed the presence of normal ovaries, oviducts, and uterus, while three Y-linked genes (*SRY*, *ZFY*, and *AMELY*) were detected in DNA isolated from these organs. In conclusion, we suggest that among virilized heifers, there are, besides freemartins with XX/XY leukocyte chimerism, also cases with XX/XY chimerism in internal genitalia, the etiology of which remains unknown.

Abstract: Five DSD heifers underwent genetic analysis in the present study. We cytogenetically analyzed in vitro cultured leukocytes and searched for *SRY*, *AMELX/AMELY* and *ZFX/ZFY* genes in leukocytes and hair follicles, finding that four of the studied heifers were freemartins (XX/XY leukocyte chimerism). The fifth case had an underdeveloped vulva localized ventrally and cranially to the mammary gland, a normal female sex chromosome complement (60,XX) in the leukocytes, and a lack of Y-chromosome-derived genes in the leukocytes and hair follicles. Postmortem anatomical examination of this heifer revealed the presence of normal ovaries with follicles, uterus, and oviducts, but molecular detection of the *SRY*, *ZFX*, *ZFY*, *AMELX*, and *AMELY* genes in these organs indicated the presence of a cell line carrying the Y chromosome. Further analysis of twelve microsatellite markers revealed the presence of additional variants at six loci in DNA samples derived from the reproductive organs; XX/XY chimerism was thus suspected in these samples. On the basis of the detection of *AMELY* (Y-linked) versus *AMELX* (X-linked) and *SOX9* (autosomal) versus *AMELY* genes by droplet digital PCR (ddPCR), the Y/X and Y/autosome ratios were evaluated; they indicated the presence of XX and XY cell lines in the reproductive tissues. Our study showed that XX/XY chimerism can be present in the internal reproductive organs of the virilized heifers with a normal female set of sex chromosomes (60,XX) and a lack of Y-chromosome-derived genes in the leukocytes. The etiology of this phenomenon remains unknown.

Keywords: cattle; disorder of sex development; freemartinism; intersexuality; XX/XY chimerism; *SOX9*; *SRY*; *AMELX*; *AMELY*; *ZFX*; *ZFY*

Citation: Szczerbal, I.; Nowacka-Woszuk, J.; Stachowiak, M.; Lukomska, A.; Konieczny, K.; Tarnogrodzka, N.; Wozniak, J.; Switonski, M. XX/XY Chimerism in Internal Genitalia of a Virilized Heifer. *Animals* 2022, 12, 2932. https://doi.org/10.3390/ ani12212932

Academic Editors: Leopoldo Iannuzzi and Pietro Parma

Received: 14 September 2022
Accepted: 24 October 2022
Published: 26 October 2022

Publisher's Note: MDPI stays neutral with regard to jurisdictional claims in published maps and institutional affiliations.

Copyright: © 2022 by the authors. Licensee MDPI, Basel, Switzerland. This article is an open access article distributed under the terms and conditions of the Creative Commons Attribution (CC BY) license (https:// creativecommons.org/licenses/by/ 4.0/).

1. Introduction

Freemartinism is the most common type of disorder of sex development (DSD) in cattle. It is caused by a transfer of masculine factors from a male fetus to a co-twin female fetus through placental anastomoses. This abnormality is classified as a sex chromosome DSD, and its diagnosis is mainly based on cytogenetic or molecular detection of XX/XY leukocyte chimerism [1–3]. Other forms of sex chromosome DSD, such as sex chromosome aneuploidies, have rarely been reported in cattle, while there have been no reports of gene mutations responsible for the DSD phenotype in individuals with the normal complement of sex chromosomes—i.e., XX DSD or XY DSD [4].

The identification of the mechanisms causing DSD phenotype is an important issue from the point of view of breeding. Some DSDs have a de novo origin (e.g., freemartinism, X monosomy, and XXY syndrome) and are not heritable, as the affected animals are sterile. On the contrary, carriers of gene mutations responsible for XX DSD or XY DSD can easily spread the mutation in populations. It is important to point out that distinguishing between heritable and non-heritable DSDs, based on the appearance of external genitalia, is not possible.

In domestic animals, heritable forms of XX DSD are quite common, but it is associated with the presence of ovotestis or testis. Until now, the causative mutation affecting the expression of the *FOXL2* gene involved in ovarian development has been identified only in goats [5,6]. In pigs, the XX DSD is associated with DNA variants in a region harboring the *SOX9* gene, which plays a crucial role in development of the testes [7,8]; in dogs, it is associated with variants near *SOX9* or *PADI6* [9–12]. Upstream DNA variants of *SOX9* are also known to cause of XX DSD in humans [13].

In cattle, three cases of XX DSD have been reported, and in all these cases, sequences derived from the Y chromosome were detected in the urogenital organs [14], leukocytes [15], blood cells, ovaries, and lymph nodes [16]. However, the *SRY* gene has only been detected in two reports [14,16]. Interestingly, in some DSD heifers, mosaicism with the presence of a triploid cell line carrying the Y chromosome (60,XX/90,XXY) has also been observed (summarized in [17]).

In this study, we analyzed five DSD heifers with ambiguous external genitalia, including a case with extensive virilization. This case was the main subject of molecular analysis due to the presence of a normal set of female sex chromosomes (60,XX) in leukocytes.

2. Material and Methods

2.1. Animals

Five heifers (four Holstein Friesians and one Limousin × Simmental crossbred) were subjected to genetic analysis on the request of breeders or veterinarians due to the presence of ambiguous external genitalia (Table 1). These heifers were not related and originated from four farms located in western or central Poland. In four of the heifers, enlarged clitoris or extended anus–vulva distance was observed (Figure 1a–d). The most extensive virilization was observed in the fifth case (#7514), with a rudimentary vulva being ventrally located near mammary gland (Figure 1e,f).

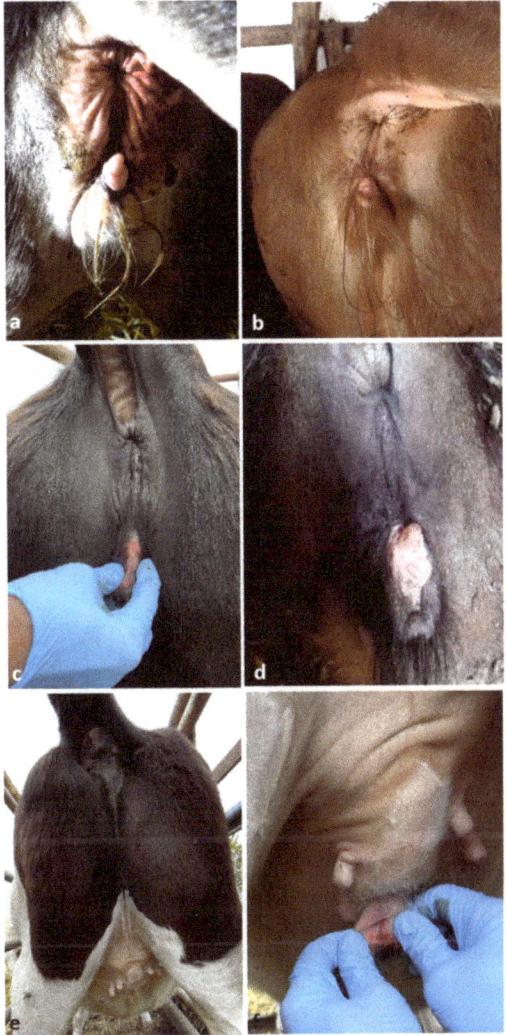

Figure 1. Virilized external genitalia of studied cases. (**a**) Case #7497. (**b**) Case #7502. (**c**) Case #7515. (**d**) Case #7518. (**e**,**f**) Case #7514.

Table 1. Phenotypes and classification of DSD cases.

Lab No. (Breed *)	External Genitalia	Sex Chromosomes in Leukocytes	AMELY/AMELX Copy Number Ratio in Blood Cells	Detection of SRY and ZFY Genes	DSD Classification
7497 (HF)	enlarged clitoris	not analyzed	0.074	present in blood cells, absent in hair follicles	freemartinism
7502 (L × S)	enlarged clitoris	XX [71%]/XY [29%] leukocyte chimerism	0.317	present in blood cells, absent in hair follicles	freemartinism

Table 1. Cont.

Lab No. (Breed *)	External Genitalia	Sex Chromosomes in Leukocytes	AMELY/AMELX Copy Number Ratio in Blood Cells	Detection of SRY and ZFY Genes	DSD Classification
7515 (HF)	extended anus–vulva distance	XX [98%]/XY [2%] leukocyte chimerism	0.0277	present in blood cells, absent in hair follicles	freemartinism
7518 (HF)	enlarged clitoris, extended anus–vulva distance	XX [25%]/XY [75%] leukocyte chimerism	0.367	present in blood cells, absent in hair follicles	Freemartinism
7514 (HF)	rudimentary vulva localized ventrally, near mammary gland	XX	AMELY not detected	present in internal genitalia, absent in blood and hair follicles	XX DSD, with XX/XY chimerism in internal genitalia

* HF: Holstein Friesian; L × S: Limousin × Simmental crossbred.

2.2. Histological Studies

Samples of the uterus (approx. 4 cm), oviducts (approx. 1.5 cm), and gonads (approx. 3 and 4 cm) collected postmortem were fixed in neutral buffered 10% formalin solution and used for preparation of paraffin sections (3 µm), which were stained with hematoxylin and eosin (H&E). Microscopic observations were carried out under an Axio Lab.A1 microscope (Carl Zeiss, Oberkochen, Germany) equipped an ERc5s digital camera (Carl Zeiss, Oberkochen, Germany) and analyzed with the use of Zen 2.3 software (blue edition; Carl Zeiss Microscopy, 2011).

2.3. Cytogenetic Analysis

Blood samples were collected in heparinized tubes for establishing short-term (48 h) in vitro leukocyte cultures. The cells were cultured in RPMI-1640 medium (Sigma-Aldrich, St. Louis, MO, USA) supplemented with 15% (v/v) fetal calf serum and 1% (v/v) penicillin/streptomycin and phytohemagglutinin at 37 °C in a humidified atmosphere of 5% CO_2. A standard cell culture harvesting procedure was used, including colcemid, hypotonic and fixative treatments. Chromosomes were analyzed using Giemsa staining and C- and G-banding techniques (applied to case #7514), according to methods reviewed by Iannuzzi and Di Berardino [18]. Bovine sex chromosomes were identified based on their biarmed morphology (a large submetacentric X and a small metacentric Y), contrasting with the one-arm morphology of all autosomes, lack of centromeric C band and the characteristic G banding pattern (case #7514). One hundred metaphase Giemsa-stained spreads were analyzed for each case. In addition, twenty C-banded and G-banded spreads derived from DSD heifer #7514 were also evaluated. The slides were examined with an epifluorescence Nikon E600 Eclipse microscope (Melville, NY, USA) equipped with a cooled CCD digital camera (Melville, NY, USA) and Lucia software (Laboratory Imaging, Prague, Czech Republic).

2.4. Molecular Detection of X-Linked and Y-Linked Genes

DNA was isolated from blood using DNA Blood Mini kit (A&A Biotechnology, Gdansk. Poland) and from hair follicles using Sherlock AX kit (A&A Biotechnology, Gdansk, Poland). The SRY gene fragment covering the whole coding sequence (851 bp) was amplified by PCR using the primers shown in Supplementary Table S1, and its presence was verified using agarose gel electrophoresis. The X-linked and Y-linked (ZFX and ZFY, respectively) genes were amplified (448 bp) by PCR (Supplementary Table S1) and distinguished by restriction enzyme (BsmI) digestion at 37 °C for 4 h following agarose gel electrophoresis (448 bp for ZFY; 391 and 57 bp for ZFX). Moreover, PCR detection of the Y-chromosome-derived genes was also performed on DNA samples isolated from the ovaries, uterus, and oviduct (Genomic Mini kit, A&A Biotechnology, Gdansk). All PCR primers were designed using

Primer3 (http://www.bioinformatics.nl/cgi-bin/primer3plus/primer3plus.cgi; accessed on 10 August 2009), and all details (primer sequences, annealing temperatures and the amplicon lengths) are shown in Supplementary Table S1.

2.5. Analysis of SOX9 and AMELY/AMELX Copy Number

Droplet digital PCR (ddPCR) was used to detect the *AMELX* (X-linked) and *AMELY* (Y-linked) genes, with a fluorescent ratio of *AMELY/AMELX* amplicons below 1.0 confirming the presence of XX/XY chimerism, following the procedure described by Szczerbal et al. [3]. Moreover, ddPCR was also used to estimate the copy number of the *SOX9* gene by taking the copy number of the *F2* autosomal gene as a reference [19]. To establish the amplicon ratio of the Y-derived gene (*AMELY*) to the autosomal gene (*SOX9*), an additional reaction was performed with these genes. The procedure described by Nowacka-Woszuk et al. [11] was followed. Briefly, the reaction mixture contained 10 μL of 2×ddPCR Supermix for Probes (Bio-Rad, Hercules, CA, USA), 1 μL of 20× primers/FAM probe, 1 μL of 20× primers/HEX probe, and 1 μL of the *Bsu*I and *Hae*III restriction enzymes for the *AMELX* and *AMELY* genes and the *SOX9* and *F2* genes, respectively. The PCR mixtures were partitioned into approximately 20,000 droplets using a QX200 droplet generator (Bio-Rad, Hercules, CA, USA). PCR was run using following conditions: denaturation at 95 °C for 10 min; 40 cycles at 94 °C for 30 s, and at 55 °C (for *AMELX* and *AMELY*), 57 °C (for *SOX9* and *F2*) and 56 °C (for *AMELY* and *SOX9*) for 60 s (ramp rate 2 °C/s); 98 °C for 10 min, and 10 °C until reading time. The droplets were analyzed on a QX200 droplet reader (Bio-Rad, Hercules, CA, USA). The concentration of the genes was calculated by Poisson distribution using Quantasoft software (Bio-Rad, Hercules, CA, USA). The primer and probe sequences are shown in Supplementary Table S1.

2.6. Genotyping of Selected Tissues Using Microsatellite Markers

The genotyping of DSD heifer #7514 was performed using microsatellite (short tandem repeats—STR) markers in DNA samples collected from the blood, hair follicles, ovaries, uterus, and oviduct. Altogether, twelve markers (BM1818, BM1824, BM2113, ETH3, ETH10, ETH225, INRA23, SPS115, TGLA53, TGLA122, TGLA126 and TGLA227) recommended by the International Society of Animal Genetics (ISAG) for parentage testing and genetic profiling were analyzed by a certified laboratory at the Institute of Animal Production (Balice, Poland). Briefly, the analysis was based on the amplified fragment length polymorphisms (AFLP) method, where all markers were amplified in a single multiplex using TypeIt Microsatellite PCR Kit (Qiagen, Hilden, Germany). The amplicons were separated by capillary electrophoresis on Genetic Analyzer 3500 xL (Applied Biosystems, Waltham, MA, USA) with the use of POP-7 polymer (Thermofisher Scientific, Waltham, MA, USA). The length of amplicon was determined using GeneMapper Software 5 (Applied Biosystems, Waltham, MA, USA).

3. Results

The microscopic evaluation of cytogenetic slides obtained from in vitro cultured leukocytes could be performed for the four DSD heifers (#7502, #7514, #7515 and #7518), and the molecular detection of the Y-derived sequences could be performed for all DSD heifers (#7497, #7502, #7514, #7515 and #7518).

A normal XX sex chromosome complement, analyzed by Giemsa staining, as well as by C- and G-banding (Figure 2) was observed in one of the heifers (#7514), and XX/XY leukocyte chimerism was detected in the other three (#7502, #7515, and #7518) (data not shown). The proportion of XX and XY metaphase spreads varied from XX [98%]/XY [2%] to XX [25%]/XY [75%] (Table 1). In the next step, Y-derived genes (*SRY* and *ZFY*) were not detected in case #7514. On the other hand, both genes were found in blood cells, though not in DNA samples isolated from hair follicles, in the other four cases (#7497, #7502, #7515 and #7518) (Supplementary Figure S1). In addition, the presence of the chimerism was confirmed by estimating the Y/X copy number through ddPCR, based on the number of amplicons derived from the *AMELY* (Y-derived) and *AMELX* (X-linked) genes (Supplementary Figure S2). The results were concordant with those of the cytogenetic analysis. On the basis of these results, case #7514 was tentatively classified as an XX (*SRY*-negative) DSD, while the remaining four cases appeared to be typical freemartins (Table 1).

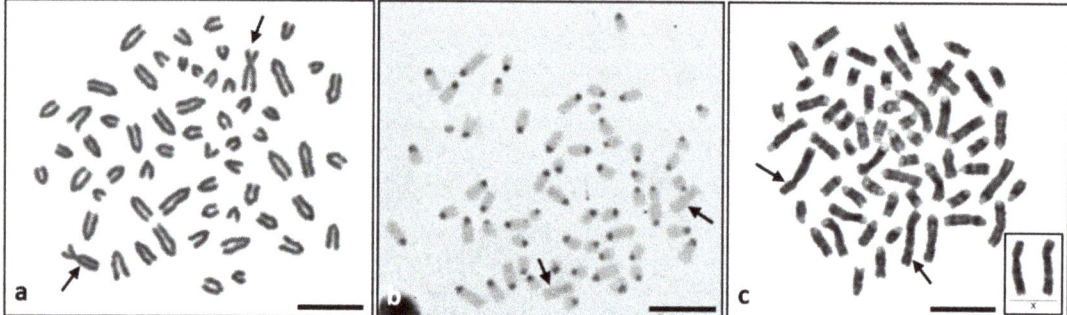

Figure 2. Representative metaphase spreads derived from in vitro leukocyte culture of DSD heifer #7514: (**a**) Giemsa staining—submetacentric X chromosomes are indicated by arrows, (**b**) C-banding—X chromosomes with no centromeric positive C band block are indicated by arrows, (**c**) G-banding—X chromosomes with normal patterns are indicated by arrows and enlarged in a right down corner. Scale bar = 10 µm.

Further analysis focused on DSD heifer #7514. Postmortem anatomical examination of the genitourinary system revealed normal female internal genitalia and virilized external genital organs. The uterine horns and cervix were of normal structure, shape, and consistency (Figure 3a). The ovaries were also normal in structure and shape but contained only a few follicles and corpora lutea. Both oviducts were complete and normal in size. The vagina, despite its normal structure in the cranial part, was dilated in the caudal part. Vulva, vestibule, cervix, uterine body and uterine horns were connected to each other and unobstructed. The absence of a vulval cleft in the perineal area was noted. The urinary bladder and ureter were of normal structure and shape. The urethral orifice was in its normal position in the vagina, and the urethra was connected to the bladder. There was a hypoplastic penile-like structure with a penile retractor muscle connected to the vestibule of the vagina; inside this, there was a virilized urethra with a secondary external orifice located on the ventral body aspect, cranial to the udder. This was in the form of a vulval cleft-like structure or preputial-like structure. The male external genitalia were absent from the inguinal area.

Histological analysis of gonads, oviducts, and uterus showed them to have normal structure (Figure 3b–d). In the ovaries were observed follicles, including a Graafian follicle, as well as corpora lutea and corpora albicans. No structures resembling testicular organization were found.

We focused in the first step of the molecular study on elucidating the background of the observed phenotype in DSD heifer #7514. PCR revealed the presence of Y-chromosome-derived genes (*SRY* and *ZFY*) in the gonads, oviduct, and uterus (Figure 4). This observation indicated the presence of another cell line or lines.

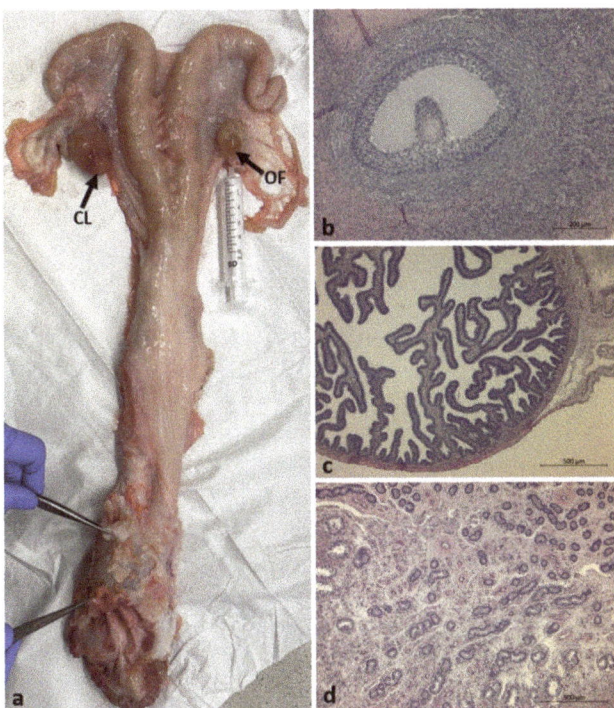

Figure 3. Anatomical and histological analysis of the DSD heifer #7514: (**a**) internal genitalia—uterus with oviducts and ovaries; (**b**) Graafian follicle with an oocyte surrounded by granulosa cells, scale bar = 200 µm; (**c**) cross section of the oviduct: folded mucosa and thin muscularis, scale bar = 500 µm; (**d**) uterine mucosa with small endometrial glands in the basal layer and tubular glands in the functional layer, scale bar = 500 µm. Corpus luteum (CL) and ovarian follicle (OF) are indicated by arrows.

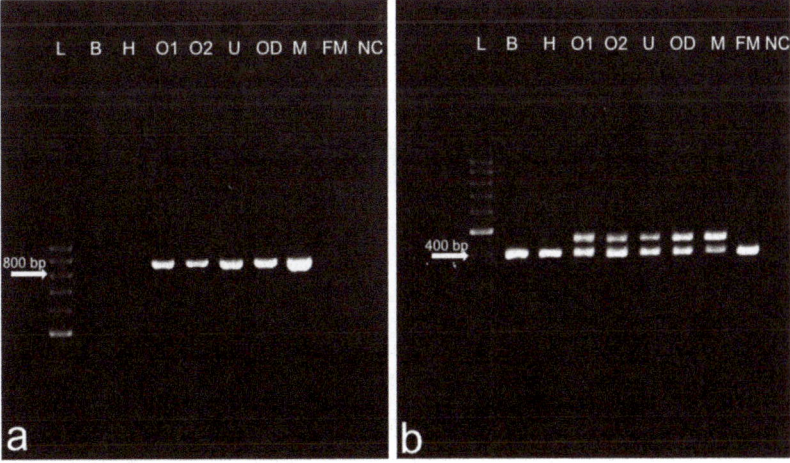

Figure 4. Detection of *SRY* (**a**) and *ZFY* (**b**) genes in DSD heifer #7514. L: GeneRuler DNA ladder; B: blood; H: hair follicles; O1: ovary 1; O2: ovary 2; U: uterus; OD: oviduct; M: control male; FM: control female; NC: negative control (no DNA).

We thus genotyped DNA samples isolated from blood, hair follicles, ovaries, oviduct, and uterus at twelve microsatellite loci, as is commonly done in parentage testing. Additional variants were observed at six loci (ETH3, ETH10, ETH225, SPS115, TGLA53, and TGLA227) in the internal genitalia, while only one or two variants were found in blood and hair follicles (Figure 5; Supplementary Table S2).

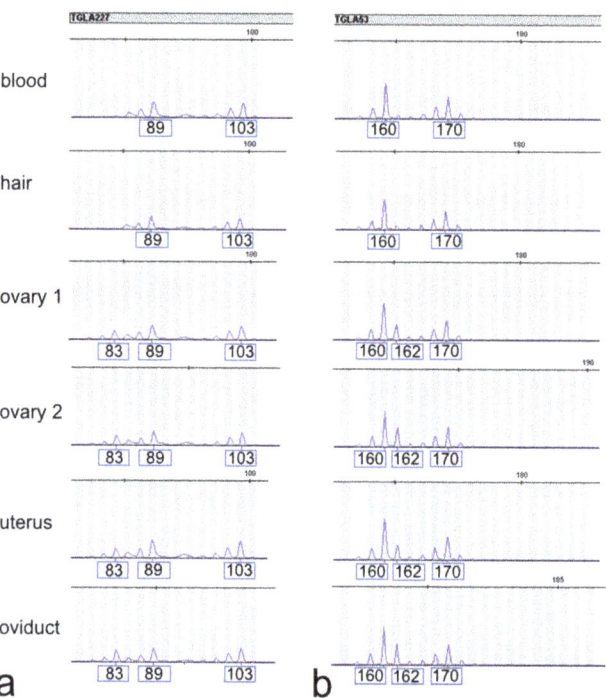

Figure 5. Genotypes for selected microsatellite markers in different tissues of DSD heifer #7514: TGLA227 (**a**) and TGLA53 (**b**). Three size variants for TGLA227 (83, 89 and 103 bp) and TGLA53 (160, 162 and 170) in internal genitalia, instead of two variants observed in blood cells and hair follicles, are visible.

This result, indicating the presence of chimerism in the internal genitalia, was followed by molecular detection of the number of copies of sex chromosomes and autosomes. Firstly, ddPCR was used to estimate the number of copies of the *SOX9* gene, since an elevated number usually affects gonadal development in females (ovotestis or testis instead of ovaries) and causes virilization. In all studied tissues, the copy number of *SOX9* was normal (two copies), as it was also observed in the studied freemartins (Supplementary Figure S3). Next, ddPCR was used to estimate the Y/X ratio, on the basis of the amplicon numbers of the *AMELY* (Y-linked) and *AMELX* (X-linked) genes. The expected Y/X ratio for a normal male cell line carrying X and Y chromosomes is 1.0, while for a normal female line (XX), it is 0. In our case, the Y/X ratio was low (<0.3, Figure 6), thus confirming the presence of two cell lines (XX and XY). In addition, the *AMELY/SOX9* ratio was lowered than the expected 0.5 for a single XY cell line (Supplementary Figure S4). This also indicated the presence of another cell line carrying the Y chromosome in this heifer.

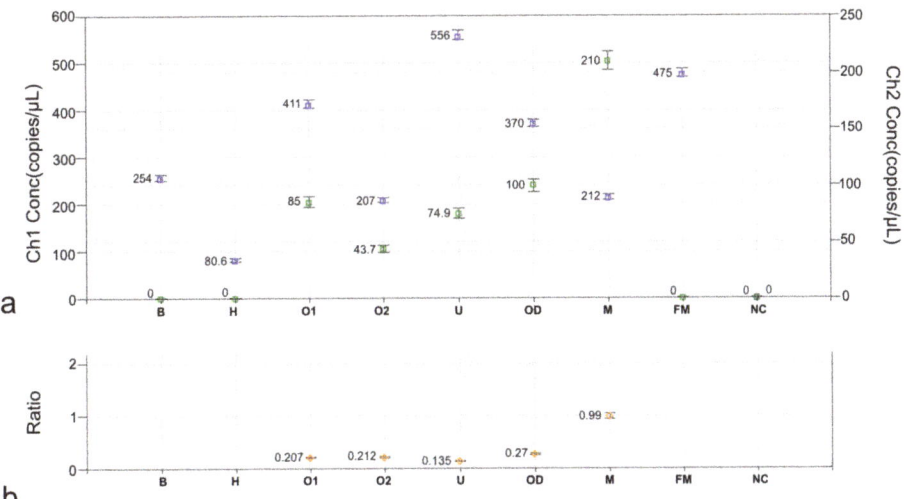

Figure 6. Estimation of the Y/X ratio by ddPCR based on the proportion of *AMELY* (Ch2) and *AMELX* (Ch1) amplicons in the DSD heifer #7514. (**a**) Amplification signals from chromosome X (blue color) and Y (green color). (**b**) Y/X ratio is presented. B: blood; H: hair follicles; O1: ovary 1; O2: ovary 2; U: uterus; OD: oviduct; M: control male; FM: control female; NC: negative control (no DNA).

4. Discussion

The incidence, consequences, and background of bovine freemartinism have been frequently reported on Esteves et al. [2]. It is well-known that such heifers have underdeveloped internal genitalia and that their external genitalia are often virilized. Heifers born as co-twins to males are usually culled due to the high risk of freemartinism (>90%). However, some freemartins are born as singletons due to early fetal death of the male co-twin [20]. Where virilized genitalia are observed in such heifers, distinguishing between nonheritable freemartinism and other types of DSD requires genetic analysis. Unfortunately, knowledge of the mechanisms responsible for DSD phenotype in heifers with a normal set of XX chromosomes is scarce.

To our best knowledge, there have only been three reports of XX DSD heifers, and in all these cases, Y-chromosome-derived sequences were detected [14–16]. In the heifer reported by Takagi et al. [14], a vulval orifice-like structure localized ventrally and cranially to the mammary gland, as well as normal internal genitalia, including ovaries with follicles and corpora lutea, were observed. Interestingly, DSD case #7514 in the present study had a very similar phenotype, and in both cases three Y-linked genes (*SRY*, *ZFY* and *AMELY*) were detected in the internal genitalia, though not in blood cells.

A different XX DSD heifer phenotype was reported by Payan-Carreira et al. [15], who observed rudimentary external genitalia with a small clitoris-like structure, bilateral streak gonads, a normal uterus, a long vagina, and the urethral orifice at the normal location. Fluorescent in situ hybridization (FISH) with a genomic degenerate oligonucleotide-primed (DOP)-PCR probe derived from the heifer revealed the presence of Y chromosome sequences in both X chromosomes, though the *SRY* gene was not detected by PCR.

The third XX DSD case reported in the literature was a female with a normal uterus, ovaries, and mammary gland, but also with a prepuce, normally urinating penis, and scrotum [16]. Molecular analysis revealed the presence of the *SRY* gene in several tissues, including the blood, ovaries, and lymph nodes.

The external genitalia of the DSD heifer (case # 7514) described here were extensively virilized, though a normal uterus and ovaries with follicles were observed. Since earlier reports of XX DSD heifers indicated the presence of Y-derived sequences in some organs, we also searched for three Y-derived genes (*SRY*, *ZFY* and *AMELY*); we detected them in internal genitalia. Our study thus confirmed that the presence of Y-chromosome-derived genes plays a crucial role in the virilization of XX DSD heifers with female internal genitalia, including ovaries.

It is well-known that the presence of a functional *SRY* gene triggers the development of undifferentiated fetal gonads of mammals toward testis, while ovaries develop when functional SRY transcription factor is not expressed [21]. Thus, the detection of the *SRY* gene in ovaries is a very unusual situation. On the other hand, there are reports suggesting that the XX/XY chimerism can be present in both blood cells and gonadal tissue of bulls originating from heterosexual twins [22,23]. There has also been a report concerning an XY (*SRY*-positive) DSD heifer, in which the *SRY* gene was detected in the blood and ovaries with follicles and a large corpus luteum, despite XX/XY chimerism (freemartinism) being excluded though analysis of in vitro cultured leukocytes and genotyping of eighteen microsatellite markers in DNA isolated from blood [24]. In the present case, we also observed a normal set of female sex chromosomes in leukocytes, and the analysis of microsatellites in DNA isolated from blood cells excluded the presence of the chimerism. In fact, it was in the internal genitalia that the chimerism was detected, as it was revealed by the microsatellite genotyping and ddPCR of sex-linked genes.

5. Conclusions

Our study has shown the presence of XX/XY chimerism in internal genitalia, including ovaries, in a DSD heifer with a normal set of female sex chromosomes (XX) in the leukocytes. Since chimersism was not observed in leukocytes, we could exclude a classical form of freemartinism. The most intriguing issue of the presence of the *SRY* gene in normally developed ovaries and the mechanisms responsible for the migration of XY cells to internal female genitalia requires further studies to be elucidated.

Supplementary Materials: The following supporting information can be downloaded at: https://www.mdpi.com/article/10.3390/ani12212932/s1. Supplementary Table S1. PCR conditions applied in molecular analysis of the selected genes. Supplementary Table S2. STR markers analyzed in DSD heifer #7514. Supplementary Figure S1. Detection of *SRY* (a) and *ZFY* (b) genes in freemartin cases. L: GeneRuler DNA ladder; B: blood; H: hair follicles; M: control male; FM: control female; NC: negative control (no DNA). Supplementary Figure S2. Estimation of the Y/X ratio by ddPCR based on the proportion of *AMELY* and *AMELX* amplicons in freemartin cases. (a) amplification signals from chromosome X (blue color) and Y (green color). (b) Y/X ratio is presented. B: blood; H: hair follicles; M: control male; FM: control female; NC: negative control (no DNA). Supplementary Figure S3. *SOX9* copy number by ddPCR for freemartin cases (a) and DSD heifer #7514 (b). B: blood; H: hair follicles; O1: ovary 1; O2: ovary 2; U: uterus; OD: oviduct; FM: control female; NC: negative control (no DNA). Supplementary Figure S4. Estimation of the Y/autosome ratio by ddPCR based on the proportion of *AMELY* and *SOX9* amplicons in the DSD heifer #7514. B: blood; H: hair follicles; O1: ovary 1: O2: ovary 2; U: uterus; OD: oviduct; M: control male; FM: control female; NC: negative control (no DNA).

Author Contributions: Each author made substantial contributions to this project. I.S. performed cytogenetic studies and contributed to molecular studies and preparation of the manuscript. J.N.-W. performed molecular studies and contributed to the preparation of the manuscript. M.S. (Monika Stachowiak) performed molecular studies. A.L. performed histological studies. K.K. performed anatomical studies and contributed to the collection of the samples. N.T. contributed to cytogenetic studies and the collection of the samples. J.W. contributed to molecular studies and the collection of the samples. M.S. (Marek Switonski) designed and supervised the study, prepared the manuscript and contributed to the collection of the samples. All authors have read and agreed to the published version of the manuscript.

Funding: The study was financed from the statutory funds of the Department of Genetics and Animal Breeding (No. 506.534.05.00), Faculty of Veterinary Medicine and Animal Science, Poznan University of Life Sciences, Poznan, Poland.

Institutional Review Board Statement: The study was conducted according to the guidelines of the Declaration of Helsinki and approved by the local Bioethical Commission for Animal Care and Use in Poznan, Poland (certificate no 3/2019).

Informed Consent Statement: The blood samples were collected by veterinarians with the consent of heifers owners for reporting the obtained scientific results.

Data Availability Statement: The data presented in this study are available on request from the corresponding author.

Acknowledgments: We thank veterinarian Konstancja Balcer for blood sample collection.

Conflicts of Interest: The authors declare that they have no conflict of interest.

References

1. Padula, A.M. The Freemartin Syndrome: An Update. *Anim. Reprod. Sci.* **2005**, *87*, 93–109. [CrossRef]
2. Esteves, A.; Bage, R.; Payan-Carreira, R. Freemartinism in Cattle. In *Ruminants: Anatomy, Behavior and Diseases*; Mendes, R.E., Ed.; Nova Science Publishers Inc.: Hauppauge, NY, USA, 2012; pp. 99–120.
3. Szczerbal, I.; Nowacka-Woszuk, J.; Albarella, S.; Switonski, M. Technical Note: Droplet Digital PCR as a New Molecular Method for a Simple and Reliable Diagnosis of Freemartinism in Cattle. *J. Dairy Sci.* **2019**, *102*, 10100–10104. [CrossRef]
4. Iannuzzi, A.; Parma, P.; Iannuzzi, L. Chromosome Abnormalities and Fertility in Domestic Bovids: A Review. *Animals* **2021**, *11*, 802. [CrossRef]
5. Pailhoux, E.; Vigier, B.; Chaffaux, S.; Servel, N.; Taourit, S.; Furet, J.-P.; Fellous, M.; Grosclaude, F.; Cribiu, E.P.; Cotinot, C.; et al. A 11.7-Kb Deletion Triggers Intersexuality and Polledness in Goats. *Nat. Genet.* **2001**, *29*, 453–458. [CrossRef]
6. Simon, R.; Lischer, H.E.L.; Pieńkowska-Schelling, A.; Keller, I.; Häfliger, I.M.; Letko, A.; Schelling, C.; Lühken, G.; Drögemüller, C. New Genomic Features of the Polled Intersex Syndrome Variant in Goats Unraveled by Long-read Whole-genome Sequencing. *Anim. Genet.* **2020**, *51*, 439–448. [CrossRef] [PubMed]
7. Rousseau, S.; Iannuccelli, N.; Mercat, M.-J.; Naylies, C.; Thouly, J.-C.; Servin, B.; Milan, D.; Pailhoux, E.; Riquet, J. A Genome-Wide Association Study Points out the Causal Implication of SOX9 in the Sex-Reversal Phenotype in XX Pigs. *PLoS ONE* **2013**, *8*, e79882. [CrossRef] [PubMed]
8. Stachowiak, M.; Szczerbal, I.; Nowacka-Woszuk, J.; Jackowiak, H.; Sledzinski, P.; Iskrzak, P.; Dzimira, S.; Switonski, M. Polymorphisms in the SOX9 Region and Testicular Disorder of Sex Development (38,XX; SRY -Negative) in Pigs. *Livest. Sci.* **2017**, *203*, 48–53. [CrossRef]
9. Rossi, E.; Radi, O.; De Lorenzi, L.; Vetro, A.; Groppetti, D.; Bigliardi, E.; Luvoni, G.C.; Rota, A.; Camerino, G.; Zuffardi, O.; et al. Sox9 Duplications Are a Relevant Cause of Sry-Negative XX Sex Reversal Dogs. *PLoS ONE* **2014**, *9*, e101244. [CrossRef]
10. Marcinkowska-Swojak, M.; Szczerbal, I.; Pausch, H.; Nowacka-Woszuk, J.; Flisikowski, K.; Dzimira, S.; Nizanski, W.; Payan-Carreira, R.; Fries, R.; Kozlowski, P.; et al. Copy Number Variation in the Region Harboring SOX9 Gene in Dogs with Testicular/Ovotesticular Disorder of Sex Development (78,XX; SRY-Negative). *Sci. Rep.* **2015**, *5*, 14696. [CrossRef]
11. Nowacka-Woszuk, J.; Szczerbal, I.; Stachowiak, M.; Szydlowski, M.; Nizanski, W.; Dzimira, S.; Maslak, A.; Payan-Carreira, R.; Wydooghe, E.; Nowak, T.; et al. Association between Polymorphisms in the SOX9 Region and Canine Disorder of Sex Development (78,XX; SRY-Negative) Revisited in a Multibreed Case-Control Study. *PLoS ONE* **2019**, *14*, e0218565. [CrossRef]
12. Nowacka-Woszuk, J.; Stachowiak, M.; Szczerbal, I.; Szydlowski, M.; Szabelska-Beresewicz, A.; Zyprych-Walczak, J.; Krzeminska, P.; Nowak, T.; Lukomska, A.; Ligocka, Z.; et al. Whole Genome Sequencing Identifies a Missense Polymorphism in PADI6 Associated with Testicular/Ovotesticular XX Disorder of Sex Development in Dogs. *Genomics* **2022**, *114*, 110389. [CrossRef]
13. Croft, B.; Ohnesorg, T.; Sinclair, A.H. The Role of Copy Number Variants in Disorders of Sex Development. *Sex. Dev. Genet. Mol. Biol. Evol. Endocrinol. Embryol. Pathol. Sex Determ. Differ.* **2018**, *12*, 19–29. [CrossRef]
14. Takagi, M.; Yamagishi, N.; Oboshi, K.; Kageyama, S.; Hirayama, H.; Minamihashi, A.; Sasaki, M.; Wijayagunawardane, M.P.B. A Female Pseudohermaphrodite Holstein Heifer with Gonadal Mosaicism. *Theriogenology* **2005**, *63*, 60–71. [CrossRef]
15. Payan-Carreira, R.; Pires, M.A.; Quaresma, M.; Chaves, R.; Adega, F.; Guedes Pinto, H.; Colaço, B.; Villar, V. A Complex Intersex Condition in a Holstein Calf. *Anim. Reprod. Sci.* **2008**, *103*, 154–163. [CrossRef]
16. Bresciani, C.; Parma, P.; De Lorenzi, L.; Di Ianni, F.; Bertocchi, M.; Bertani, V.; Cantoni, A.M.; Parmigiani, E. A Clinical Case of an *SRY*-Positive Intersex/Hermaphrodite Holstein Cattle. *Sex. Dev.* **2015**, *9*, 229–238. [CrossRef]
17. Szczerbal, I.; Komosa, M.; Nowacka-Woszuk, J.; Uzar, T.; Houszka, M.; Semrau, J.; Musial, M.; Barczykowski, M.; Lukomska, A.; Switonski, M. A Disorder of Sex Development in a Holstein–Friesian Heifer with a Rare Mosaicism (60,XX/90,XXY): A Genetic, Anatomical, and Histological Study. *Animals* **2021**, *11*, 285. [CrossRef]
18. Iannuzzi, L.; Di Berardino, D. Tools of the Trade: Diagnostics and Research in Domestic Animal Cytogenetics. *J. Appl. Genet.* **2008**, *49*, 357–366. [CrossRef]
19. Floren, C.; Wiedemann, I.; Brenig, B.; Schütz, E.; Beck, J. Species Identification and Quantification in Meat and Meat Products Using Droplet Digital PCR (DdPCR). *Food Chem.* **2015**, *173*, 1054–1058. [CrossRef]

20. Szczerbal, I.; Kociucka, B.; Nowacka-Woszuk, J.; Lach, Z.; Jaskowski, J.M.; Switonski, M. A High Incidence of Leukocyte Chimerism (60,XX/60,XY) in Single Born Heifers Culled Due to Underdevelopment of Internal Reproductive Tracts . *Czech J. Anim. Sci.* **2014**, *59*, 445–449. [CrossRef]
21. Lamothe, S.; Bernard, V.; Christin-Maitre, S. Gonad Differentiation toward Ovary. *Ann. Endocrinol.* **2020**, *81*, 83–88. [CrossRef]
22. Stranzlnger, G.; Dolf, G.; Fries, R.; Stocker, H. Some Rare Cases of Chimerism in Twin Cattle and Their Proposed Use in Determining Germinal Cell Migration. *J. Hered.* **1981**, *72*, 360–362. [CrossRef] [PubMed]
23. Rejduch, B.; Słota, E.; Gustavsson, I. 60,XY/60,XX Chimerism in the Germ Cell Line of Mature Bulls Born in Heterosexual Twinning. *Theriogenology* **2000**, *54*, 621–627. [CrossRef]
24. De Lorenzi, L.; Arrighi, S.; Rossi, E.; Grignani, P.; Previderè, C.; Bonacina, S.; Cremonesi, F.; Parma, P. XY (**SRY**-Positive) Ovarian Disorder of Sex Development in Cattle. *Sex. Dev.* **2018**, *12*, 196–203. [CrossRef] [PubMed]

Article

Supernumerary Marker Chromosome Identified in Asian Elephant (*Elephas maximus*)

Halina Cernohorska [1,*], Svatava Kubickova [1], Petra Musilova [1], Miluse Vozdova [1], Roman Vodicka [2] and Jiri Rubes [1]

[1] Department of Genetics and Reproductive Biotechnologies, Veterinary Research Institute, 62100 Brno, Czech Republic
[2] Zoo Praha, 17100 Praha, Czech Republic
* Correspondence: halina.cernohorska@vri.cz; Tel.: +420-533331425

Simple Summary: Supernumerary marker chromosomes, as they are known in the human population, are usually small chromosomes that differ morphologically and structurally from the standard ones and in many cases are formed by genetically inert heterochromatin. Similar features were observed for a supernumerary chromosome discovered in two Asian elephants, a mother and her male offspring. In this study, we present its detailed analysis using several molecular cytogenetic techniques including laser microdissection and fluorescence in situ hybridization that allowed identification of this marker chromosome. Based on our findings, we propose the most possible mechanism for the origin of the marker studied. We extended our investigation and showed that the distribution of nucleolar organizer regions on the chromosomes of Asian and savanna elephants may be related to the distribution of heterochromatin. Supernumerary chromosomes or, in other words, additional or extra chromosomes added to typical human or animal karyotypes, have recently gained the attention of scientists as model systems for the study of chromosome evolution, which may include the chromosome marker described here.

Abstract: We identified a small, supernumerary marker chromosome (sSMC) in two phenotypically normal Asian elephants (*Elephas maximus*): a female (2n = 57,XX,+mar) and her male offspring (2n = 57,XY,+mar). sSMCs are defined as structurally abnormal chromosomes that cannot be identified by conventional banding analysis since they are usually small and often lack distinct banding patterns. Although current molecular techniques can reveal their origin, the mechanism of their formation is not yet fully understood. We determined the origin of the marker using a suite of conventional and molecular cytogenetic approaches that included (a) G- and C-banding, (b) AgNOR staining, (c) preparation of a DNA clone using laser microdissection of the marker chromosome, (d) FISH with commercially available human painting and telomeric probes, and (e) FISH with centromeric DNA derived from the centromeric regions of a marker-free Asian elephant. Moreover, we present new information on the location and number of NORs in Asian and savanna elephants. We show that the metacentric marker was composed of heterochromatin with NORs at the terminal ends, originating most likely from the heterochromatic region of chromosome 27. In this context, we discuss the possible mechanism of marker formation. We also discuss the similarities between sSMCs and B chromosomes and whether the marker chromosome presented here could evolve into a B chromosome in the future.

Keywords: small supernumerary marker chromosome; sSMC; laser microdissection; FISH; karyotype; heterochromatin; NOR; Asian elephant; savanna elephant

Citation: Cernohorska, H.; Kubickova, S.; Musilova, P.; Vozdova, M.; Vodicka, R.; Rubes, J. Supernumerary Marker Chromosome Identified in Asian Elephant (*Elephas maximus*). *Animals* 2023, 13, 701. https://doi.org/10.3390/ani13040701

Academic Editors: Clive J. C. Phillips and Leopoldo Iannuzzi

Received: 20 December 2022
Revised: 14 February 2023
Accepted: 14 February 2023
Published: 17 February 2023

Copyright: © 2023 by the authors. Licensee MDPI, Basel, Switzerland. This article is an open access article distributed under the terms and conditions of the Creative Commons Attribution (CC BY) license (https:// creativecommons.org/licenses/by/ 4.0/).

1. Introduction

The occurrence of a small supernumerary marker chromosome (sSMC) in the human karyotype is relatively rare and its identification is always difficult using standard cytogenetic methods [1]. This is because sSMCs represent a heterogeneous group of derivative chromosomes in terms of their chromosomal origin and shape as well as their clinical consequences [2]. It is estimated that in the human population, approximately 0.044% of newborn children are sSMC carriers [3]. About half of sSMCs are represented by heterochromatic markers that are usually harmless to their carriers. Most of them are derived from short arms and pericentric regions of acrocentric chromosomes, in which the most implicated acrocentrics are HSA15 (https://cs-tl.de/DB/CA/sSMC/0-Start.html; (accessed on 14 January 2023) [4]. The risk of an abnormal phenotype for the carrier depends on factors such as the size of the marker, genetic content, and level of mosaicism [5]. Approximately one third of the published cases correlate with specific clinical signs and symptoms, such as Emanuel, Pallister-Killian, Turner, or cat eye syndromes, while two-thirds of cases have not been associated with clinical abnormalities [5,6]. Because of the wide variety of marker chromosomes in the human population, it remains difficult to correlate a particular sSMC with a particular phenotype, especially in de novo cases [7]. Recently, it has been suggested that approximately 77% are de novo mutations, while 23% are inherited either maternally (16%) or paternally (7%) [8]. In most of the familial cases, there is no discernible increased risk of offspring abnormalities if one parent has the same marker and their phenotype is normal [9]. Familial sSMCs are preferentially maternally transmitted [10–15], suggesting either reduced fertility in male carriers or that the marker is excluded in spermatogenesis [14].

Even though numerous studies have been published on human cytogenetics, the presence of sSMCs in animals has not been reported to our knowledge. We identified a small metacentric marker chromosome in two phenotypically normal Asian elephants, which is undoubtedly a karyotypic novelty within elephants. Living elephantids (*Elephantidae* family) include three species: (a) two species of the genus *Loxodonta*, the savanna elephant (*Loxodonta africana*) and forest elephant (*Loxodonta cyclotis*), which are restricted to Africa, and (b) one species of the genus *Elephas*, the Asian or Indian elephant (*Elephas maximus*), which is endemic to Asia [16]. The latter species is of considerable economic significance in many Asian countries. Chromosomal data based on G- and C- banding and comparative FISH available for *E. maximus* and *L. africana* show a high level of chromosome band homology [17–19]. Their karyotypes possess 56 chromosomes and differ only in the amount and distribution of C-band positive heterochromatin [19].

In the present study, we report the outcome of a detailed molecular cytogenetic dissection of the marker chromosome and its identification. Moreover, we provide new information about the location and number of NORs and distribution of heterochromatin in Asian and savanna elephants. We hypothesize that the sSMC identified in this study might have some features that could contribute to its future development into a B chromosome.

2. Materials and Methods

2.1. The Asian Elephant Family

The female Asian elephant originating from the Pinnawala Elephant Orphanage, Sri Lanka Island, was imported to Prague Zoo (Prague, Czech Republic). In the Prague Zoo, she gave birth to two calves, a male and a female, who were sired by two different males. The pedigree chart is presented in Figure 1. All members of the elephant family were cytogenetically examined except the male who sired the male offspring because he is currently kept in a Zoo in Switzerland and his karyotype is not available.

2.2. Samples and Banding Techniques

Peripheral blood samples were collected from four Asian elephants (*E. maximus*, EMA) held in the Prague Zoo: two females and two males. A blood sample was also taken from the female savanna elephant (*L. africana*, LAF) held in the Dvur Kralove Zoo. Blood samples were collected by zoo veterinarians for the purpose of preventive examinations or other medical procedures and an aliquot of the blood was used for cytogenetic studies. Metaphase spreads were prepared using culture protocols described by Cernohorska et al. [20]. Conventional protocols for G- and C-banding and AgNOR staining followed Seabright [21],

Sumner [22], and Goodpasture and Bloom [23], respectively. The G-banded karyotype of the Asian elephant was arranged according to Yang et al. [19].

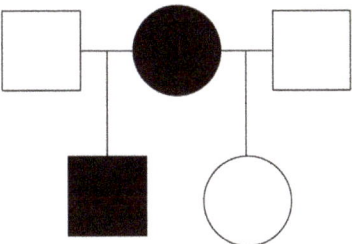

Figure 1. Pedigree chart for the Asian elephant family. Circles represent females and squares represent male individuals. sSMC carriers are marked in black.

2.3. DNA Probes and Fluorescence In Situ Hybridization (FISH)

2.3.1. Preparation of the EMAM1 Clone

We used the PALM Microlaser system (Carl Zeiss MicroImaging GmbH, Munich, Germany) to collect 20 copies of the marker chromosome. DNA of the collected chromosomes was amplified by degenerate oligonucleotide primed polymerase chain reaction (DOP-PCR), labeled during the secondary PCR with Orange-dUTP (Abbott, IL, USA) as described by Kubickova et al. [24] and checked by FISH. Amplification products derived from the marker were cloned into a pDrive vector (Qiagen, Hilden, Germany). The clones were screened by DOT-BLOT hybridization [25], fluorescently labeled by Orange-dUTP, and checked for specificity by FISH. Plasmid DNA of the selected clone was subsequently isolated and sequenced by Sanger sequencing. The clone comprised repetitive DNA but was not long enough to represent a basic repeat unit (BRU). Therefore, primers amplifying the 5′- and 3′- flanking regions were designed and inverse PCR was performed on the genomic DNA [20]. The amplification products representing the BRU obtained by PCR were cloned and the plasmid DNA was isolated, fluorescently labeled by Orange-dUTP, and used in the FISH analysis. The BRU clone was named EMAM1 clone, sequenced, and deposited in GenBank under accession number OP918028.

2.3.2. Preparation of the LAFM1 Clone

Primers selected for inverse PCR in *E. maximus* were used on *L. africana* genomic DNA to obtain the BRU. The amplification products were cloned and the plasmid DNA was isolated, labeled by Orange-dUTP, and checked for specificity by FISH (see the procedure described above). One clone was chosen based on fluorescence intensity and sequenced. The BRU was named LAFM1 clone and deposited in GenBank under accession number OP918029. The sequences of both clones were compared using BLAST2 software and screened for interspersed repeats using RepeatMasker (https://www.repeatmasker.org/cgi-bin/WEBRepeatMasker; (accessed on 20 May 2019).

2.3.3. Preparation of the Centromeric Probe

For generation of the centromeric probe, the DNA template was taken from the centromeric regions of the selected marker-free Asian elephant chromosomes by laser microdissection. The pooled DNA was amplified by DOP-PCR, labeled during the secondary PCR with Orange-d UTP, and checked by FISH [24].

2.3.4. Telomere-Specific Probe

A commercially available Telomere PNA/FITC probe (DAKO A/S, Glostrup, Denmark) was used for FISH following the manufacturer's recommendations.

2.3.5. FISH

FISH procedures for chromosome painting and specific probes followed previously described protocols [20,26]. Hybridization signals were examined using Zeiss Axio imager.Z2 fluorescence microscope with appropriate fluorescent filters; images were captured by a CoolCube CCD camera (MetaSystems, Altlussheim, Germany) and analyzed by ISIS (MetaSystems).

2.3.6. Identification of the NOR-Bearing Chromosomes in *E. maximus* and *L. africana*

NOR-bearing chromosomes were identified by FISH using human whole chromosome painting probes. The chromosome correspondence between human and elephants (Asian, EMA and African, LAF) was inferred from the comparative chromosome map established by Yang et al. [19]. A subset of Green- or Orange-labeled commercially available human chromosome-specific probes (MetaSystems, Altlussheim, Germany) were applied to both *E. maximus* and *L. africana* chromosomes following the hybridization protocol described by Yang et al. [19]. After hybridization, digital images were captured and the slides were subsequently treated with AgNOR staining [23]. The obtained images were compared to the FISH results to identify the NOR bearing chromosomes.

3. Results

Chromosome G-banding revealed a supernumerary chromosome in the elephant mother (2n = 57,XX,+mar) (Figure 2) and her son (2n = 57,XY,+mar). The marker was present in all metaphases examined in both animals (we examined 100 cells per animal). The diploid chromosome number of both the daughter and her father was 2n = 56, which was in accordance with the normal Asian elephant karyotype [17,27,28]. The marker chromosome was identified as small, metacentric (Figures 2 and 3), and C-band positive (Figure 3) with NOR sites at the terminal ends (Figure 4a). Hybridization with the telomeric probe showed (a) strong signals at the terminal ends of all *E. maximus* chromosomes including the marker, (b) weak signals at the centromeric region of biarmed chromosomes, and (c) a strong signal in the central constriction of the biarmed marker (Figure 5a). Hybridization with the centromeric probe showed signals at the centromeric regions of most *E. maximus* chromosomes. The centromeric regions of the biarmed autosomes including the marker were not painted (data not shown).

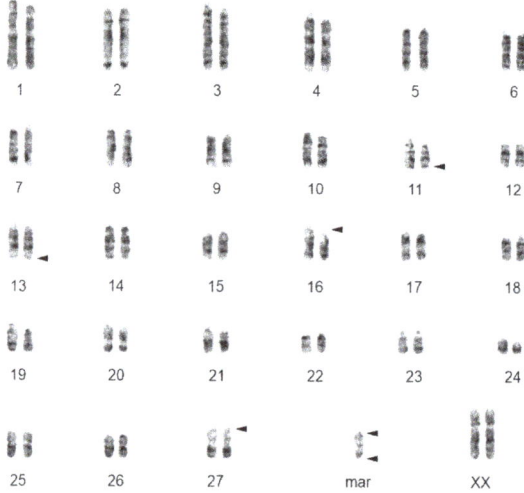

Figure 2. G-banded karyotype of *Elephas maximus* (2n = 57,XX,+mar). The chromosomes were arranged according to Yang et al. [19]. The arrowheads show the NOR positions.

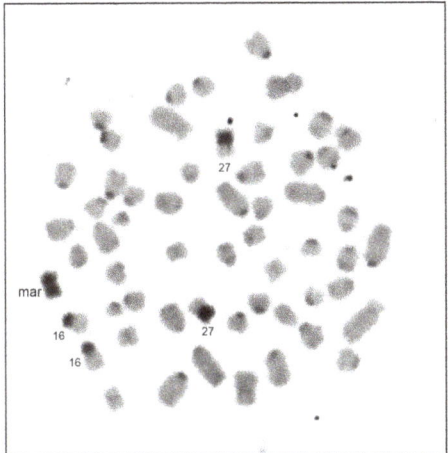

Figure 3. C-banded chromosomes of *E. maximus* (2n = 57,XX,+mar).

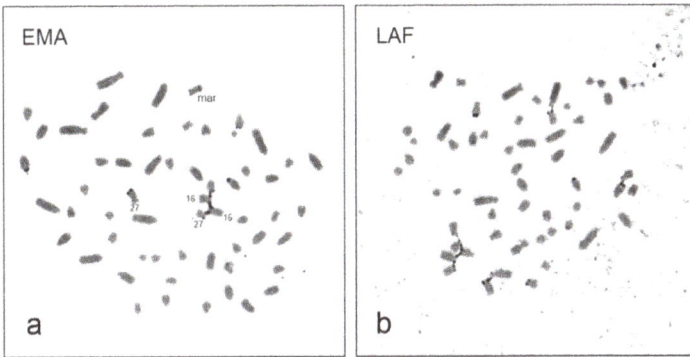

Figure 4. NOR positions in (**a**) *E. maximus* (2n = 57,XY,+mar) and (**b**) *L. africana* (2n = 56,XX).

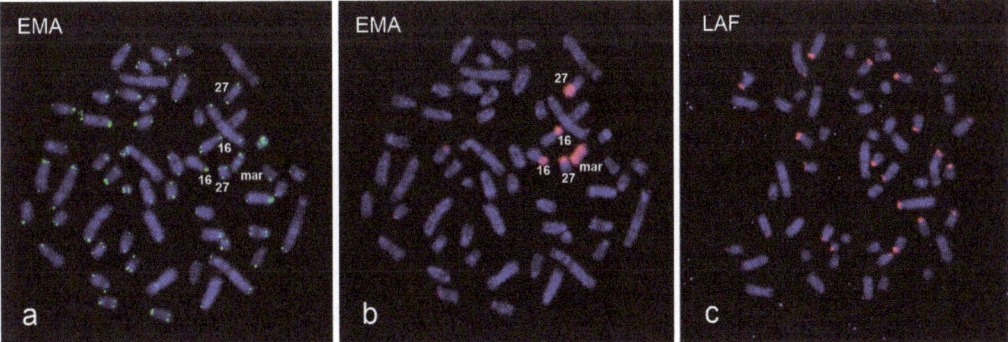

Figure 5. (**a**) FISH of a telomeric probe (green) to the *E. maximus*, EMA (2n = 57,XY,+mar). (**b**) The same metaphase spread hybridized with the EMAM1 probe (red). (**c**) FISH of the LAFM1 probe (red) to the *L. africana*, LAF chromosomes (2n = 56,XX). The chromosomes are counterstained with DAPI (blue).

3.1. Identification of the Marker Origin

In order to identify the origin of the marker chromosome, we applied several conventional and molecular cytogenetic methods. FISH with the EMAM1 probe prepared by microdissection of the marker revealed strong signals at the marker and p-arms of two small autosomal pairs, one metacentric and one submetacentric, which appeared almost entirely heterochromatic upon C-banding. Weaker signals were observed at the centromeric regions of several other autosomes (Figures 3 and 5b). We identified the two autosomal pairs with heterochromatic p-arms using human (*Homo sapiens*, HSA) painting probes. The q-arm of the small metacentric pair was painted by the HSA2 probe and corresponded to the EMA27 chromosome on the comparative map. The q-arm of the small submetacentric pair was painted by the HSA13 probe corresponding to EMA16 (Figure 6). The subsequent silver-staining revealed that the terminal ends of the marker and heterochromatic p-arms of EMA27 and EMA16 possessed NORs (Figure 4a). Both FISH and NOR staining results indicated that the marker may have originated either from the EMA27 or EMA16 chromosomes. Based on the amount of heterochromatin included into the marker, it seems reasonable to suggest that the marker originated from p-arms of EMA27 rather than EMA16 (Figures 3 and 5b). On closer inspection, variation in the amount of heterochromatin in EMA27 chromosomes was found in all of the Asian elephants examined. The metacentric shape of the marker suggested the isochromosome nature of the sSMC.

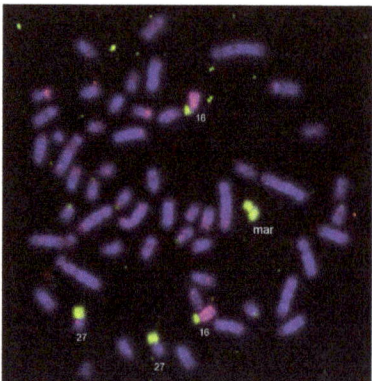

Figure 6. Co-hybridization of the EMAM1 (green) and HSA13 probes (red) to metaphase chromosomes of *E. maximus* (2n = 57,XX,+mar). The HSA13 probe shows signals on the EMA16 chromosome. The chromosomes are counterstained with DAPI (blue).

3.2. Identification of the NOR-Bearing Chromosomes Using Human Painting Probes and AgNOR Staining

3.2.1. *E. maximus*

The NOR positions determined in both marker carrier and marker-free Asian elephants were on four autosomal pairs and their number ranged from 5 to 8 in the metaphases examined (Figure 4a). The NORs, all terminal, were located on the heterochromatic p-arms of EMA16 and EMA27 (see above) and q-arms of two acrocentric autosomal pairs corresponding to HSA17 (EMA11) and HSA18 (EMA13).

3.2.2. *L. africana*

The NOR sites detected in the savanna elephant were located on nine terminal regions of eight autosomal pairs and their number varied from 11–16 in the metaphases examined (Figure 4b). The NORs, all terminal, were identified on: (a) the minute heterochromatic p-arms of six autosomal pairs corresponding to HSA1 (LAF2), HSA5 (LAF3), HSA4 (LAF5), HSA2 (LAF6), HSA13 (LAF16), and HSA2 (LAF27); (b) the q-arms of the autosomal pair corresponding to HSA17 (LAF11), and (c) both p- and q-arms of the autosomal pair corresponding to HSA18 (LAF13).

3.3. Comparison of the EMAM1 and LAFM1 Clones

The sequence homology of the BRU clones obtained from the Asian (EMAM1, 2619 kb in length) and savanna (LAFM1, 2629 kb in length) elephants was high (94%) and both clones lacked interspersed repeats (i.e., SINE, LINE or LTR elements), as revealed by RepeatMasker. We designated both sequences as satellite DNA based of the fact that the organization of repeat units in head-to-tail (or tandem) fashion permitted inverse PCR amplification [29]. The hybridization results with the EMAM1 probe to the *E. maximus* (Figure 5b) chromosomes are mentioned above. Hybridization with the LAFM1 probe to the *L. africana* chromosomes resulted in positive signals in the short p-arms of about half of the autosomes, whereas other chromosomes were unlabeled. (Figure 5c).

4. Discussion

Detection of an sSMC in phenotypically normal individuals is almost always an unexpected result in cytogenetic analysis and several molecular cytogenetic techniques are usually needed for their characterization. Using laser microdissection of the marker with subsequent reverse FISH that permitted the identification of the marker origin, we provide the first finding of an sSMC in wild mammals, to our knowledge. In the elephant family, the male offspring inherited the supernumerary chromosome from his mother while her daughter did not. Our observation fit the recently outlined fact that in humans, familial sSMCs are predominantly transmitted through the maternal line and familial marker chromosomes are usually harmless to their carriers [6]. Our marker is metacentric and only one central constriction was apparent. Since the supernumerary chromosome is mitotically stable, it presumably contains a functional centromere, even though the centromeric sequences were not detected with the centromeric probe, as with other biarmed chromosomes. The mirror-image shape of the marker indicates that the marker could have arisen in the same manner as isochromosomes [30]. The most possible explanation for the formation of our marker is that it might have originated in any ancestor during meiosis with an initial break in the (peri)centromeric region of EMA27 followed by horizontal separation of the p- and q-arms giving rise to isochromosomes i(27p) and i(27q). The separated heterochromatic p-arms, i(27p), later formed the marker chromosome (Figure 7). The strongest support for this explanation comes from the FISH results using a telomeric probe, which showed stronger fluorescence in the centromeric region of the marker in comparison to EMA27 (terminal telomeres are not considered here). As a consequence of the joining of the sister heterochromatic arms, the telomeric sequences located in the pericentromeric region appeared very close to one another, amplifying the signal in the center (Figure 5a). The nondisjunction of EMA27p during the first meiotic division subsequently resulted in maturation of an abnormal gamete, leading to the abnormal zygote with the heterochromatic sSMC after fertilization (Figure 7).

4.1. NORs

The marker chromosome detected in our study was heterochromatic with NORs at the terminal ends. In order to identify the origin of the marker, we determined chromosomes bearing both heterochromatin and NORs using a combination of AgNOR staining and FISH with human painting probes. Previous studies have used AgNOR staining on chromosomes of only *E. maximus* [31,32] without the identification of NOR-bearing chromosomes. Here, we present this information for *E. maximus* and also include the data for *L. africana*. Although the karyotypes of both species are largely conserved [19], the location and number of NORs show differences. Four NOR-bearing sites found in *E. maximus* (11q, 13q, 16p, and 27p) were shared by *L. africana*. Five other NOR sites identified in *L. africana* (2p, 3p, 5p, 6p, and 13p) were not found in *E. maximus*. In both species, NOR sites seem to be preferentially associated with heterochromatin, suggesting that the expansion of heterochromatic regions in *L. africana* might be related to the expansion of NORs and vice versa; the reduction of heterochromatic regions in *E. maximus* might be related to the reduction of NOR sites in the species.

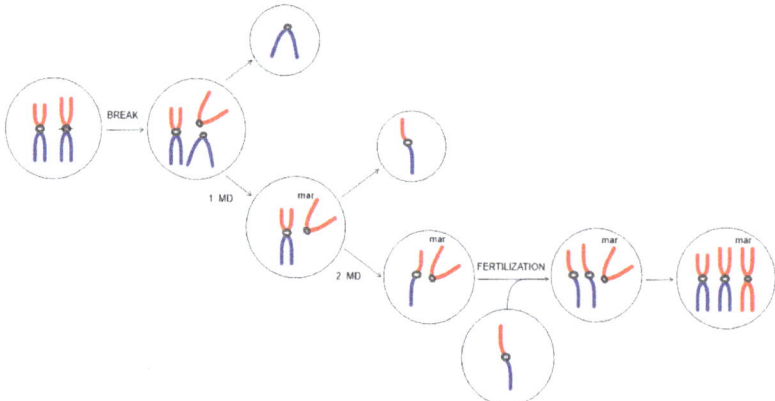

Figure 7. Schematic reconstruction showing the most possible marker formation during meiosis. The initial break in the (peri)centromeric region of one of the EMA27 homologs led to the horizontal separation of the p- and q-arms producing two isochromosomes, i(27p) and i(27q). One product of the misdivision was the sSMC formed by heterochromatic p-arms, i(27p). The other product was formed by the q-arms of the chromosome, i(27q). During the first meiotic division (MD), the marker along with the normal EMA27 chromosome segregated into one daughter cell. During the second MD, the sister chromatids of the normal EMA27 were released and segregated from one another. The marker chromosome along with one of the normal EMA27 chromatids segregated into the daughter cell, giving rise to an abnormal gamete, and leading to the abnormal zygote with the heterochromatic sSMC after fertilization.

4.2. Can We Consider the Marker Chromosome Identified in Two Asian Elephants as a Kind of Proto-B Chromosome?

There is some recent discussion in the literature about possible similarities between supernumerary marker chromosomes described in human and B chromosomes, which are enigmatic elements in eukaryotic karyotypes [33,34]. Both represent additional material to the main karyotype and may consist of heterochromatin and/or euchromatin. They are generally small in size and often lack specific phenotypic effects on the organisms that carry them. They are predominantly maternally transmitted and may be prone to mitotic instability [1,3,6,7,34,35]. It is estimated that B chromosomes occur in approximately 15% of eukaryotic species, the vast majority of which have been discovered in plants [36]. Other species in which B chromosomes have evolved include fungi, insects, helminth parasites, crustaceans, fish, amphibians, reptiles, birds, and mammals [35]. For the last group, up to 85 species carrying B chromosomes have been listed [37]. We reviewed the available literature in an attempt to determine whether the marker chromosome identified in our study in Asian elephants could evolve into a B chromosome in the future. It is believed that the best candidates for future B chromosomes in humans are (i) genetically inert (heterochromatic) supernumerary chromosomes that might manage to drive in either sex [34], (ii) sSMCs on which only their own DNA hybridizes [38], and (iii) acrocentric-derived inverted duplication sSMC with normal phenotypes [1,3,4]. Due to its heterochromatic nature, phenotypic inertness, and ability to be transmitted from parent to offspring, the marker chromosome identified in the two elephants in the current study can be included among the proposed candidates. In the recent review by Vujošević et al. ([37] and references in the article), the authors summarized that a typical B chromosome in mammals is seen as supernumerary, heterochromatic, smaller, and morphologically different from chromosomes of the standard set and does not evoke visible phenotypic effects. The size of the most common B chromosomes in mammals corresponds to the size of the smallest chromosome in the genome (occurring in 52 species, 65%), with metacentric and submetacentric shapes being more common than acrocentric shapes in this group. B chromosomes usually contain various repetitive sequences originally derived from autosomes, among which ribosomal and telomeric sequences have also been identified. Based on its morphological and molecular structure (i.e., small metacentric, heterochromatic chromosome with telomeric and rDNA sequences) and regardless of possible fertility disorders of their carriers, it seems that the marker chromosome identified in the current study has the potential to evolve into a B chromosome in the future. Additional rDNA sequences and/or other repeats or sequences on the marker may give some selective advantage to the carrier and thus may spread in a population of Asian elephants. In the future, it would be useful to collect DNA from the marker chromosome by microdissection and use it as a template for sequencing [39], as knowing the DNA content in the marker could help us determine how these chromosomes are formed.

5. Conclusions

We describe here the finding of a small supernumerary marker chromosome in a female elephant and her male offspring. Both animals were phenotypically normal, as is the case for most human carriers of markers containing heterochromatin. The fertility of the female did not appear to be affected by marker carriage, as she gave birth to two healthy offspring. Currently, we do not have any information about the fertility of the male offspring because he is not yet sexually mature. However, before being included in the captive breeding population, we recommend that his examination should entail microscopic and cytogenetic evaluation of the semen sample. Supernumerary marker chromosomes, especially B chromosomes, have recently gained the attention of scientists as model systems for the study of chromosome evolution, so it would be interesting to follow the fate of both elephants and their offspring in the future given that they carry a unique supernumerary chromosome that is not detrimental to their health and fitness.

Author Contributions: Conceptualization, H.C., S.K. and P.M.; investigation, H.C., S.K., P.M. and M.V.; methodology, H.C., S.K., P.M., M.V. and R.V.; supervision, J.R.; writing—original draft, H.C. and P.M., and writing—review and editing, H.C., P.M., S.K. and J.R. All authors have read and agreed to the published version of the manuscript.

Funding: The research was funded by the institutional support from the Ministry of Agriculture of the Czech Republic (MZE RO0523).

Institutional Review Board Statement: All procedures in this study were performed in accordance with the ethical standards of the Veterinary Research Institute (Brno, Czech Republic), which complies with the Czech and European Union Legislation for the protection of animals used for scientific purposes. According to these regulations, ethics approval was not required since the biological material (blood) was collected from live animals by zoo veterinarians for the purpose of preventive examinations or other medical procedures and an aliquot of the blood was used for the cytogenetic studies. All collaborating Zoos had licenses issued by the Ministry of the Environment of the Czech Republic (Act No 162/2003 Coll.).

Informed Consent Statement: Not applicable.

Data Availability Statement: All data is contained within the manuscript. The FISH images are available from the authors upon request.

Acknowledgments: The authors are grateful to Jiri Vahala, VMD (Dvur Kralove Zoo, Dvur Kralove n. L., Czech Republic) for providing samples for the analyses.

Conflicts of Interest: The authors declare no conflict of interest.

Abbreviations

sSMC	small supernumerary marker chromosome
NOR	nucleolar organizer region
FISH	fluorescence in situ hybridization
EMA	*Elephas maximus*
LAF	*Loxodonta africana*
DOP-PCR	degenerate oligonucleotide primed polymerase chain reaction
BRU	basic repeat unit
PNA	peptide nucleic acid

References

1. Liehr, T.; Claussen, U.; Starke, H. Small supernumerary marker chromosomes (sSMC) in humans. *Cytogenet. Genome. Res.* **2004**, *107*, 55–67. [CrossRef] [PubMed]
2. Trifonov, V.; Fluri, S.; Binkert, F.; Nandini, A.; Anderson, J.; Rodriguez, L.; Gross, M.; Kosyakova, N.; Mkrtchyan, H.; Ewers, E.; et al. Complex rearranged small supernumerary marker chromosomes (sSMC), three new cases; evidence for an underestimated entity? *Mol. Cytogenet.* **2008**, *15*, 1–6. [CrossRef]
3. Liehr, T.; Weise, A. Frequency of small supernumerary marker chromosomes in prenatal, newborn, developmentally retarded and infertility diagnostics. *Int. J. Mol. Med.* **2007**, *19*, 719–731. [CrossRef] [PubMed]
4. Liehr, T. The sSMC Homepage. Available online: https://cs-tl.de/DB/CA/sSMC/0-Start.html (accessed on 14 January 2023).
5. Jafari-Ghahfarokhi, H.; Moradi-Chaleshtori, M.; Liehr, T.; Hashemzadeh-Chaleshtori, M.; Teimori, H.; Ghasemi-Dehkordi, P. Small supernumerary marker chromosomes and their correlation with specific syndromes. *Adv. Biomed. Res.* **2015**, *4*, 140.
6. Liehr, T. Familial small supernumerary marker chromosomes are predominantly inherited via the maternal line. *Genet. Med.* **2006**, *8*, 459–462. [CrossRef] [PubMed]
7. Paoloni-Giacobino, A.; Morrisand, M.A.; Dahoun, S.P. Prenatal supernumerary r(16) chromosome characterized by multiprobe-FISH with normal pregnancy outcome. *Prenat. Diagn.* **1998**, *18*, 751–756. [CrossRef]
8. Zhou, L.; Zheng, Z.; Wu, L.; Xu, C.; Wu, H.; Xu, X.; Tang, S. Molecular delineation of small supernumerary marker chromosomes using a single nucleotide polymorphism array. *Mol. Cytogenet.* **2020**, *27*, 13–19. [CrossRef]
9. Brondum-Nielsen, K.; Mikkelsen, M. A 10-year survey, 1980–1990, of prenatally diagnosed small supernumerary marker chromosomes, identified by FISH analysis. Outcome and follow-up of 14 cases diagnosed in a series of 12,699 prenatal samples. *Prenat. Diagn.* **1995**, *15*, 615–619. [CrossRef]
10. Buckton, K.E.; Spowart, G.; Newton, M.S.; Evans, H.J. Forty four probands with an additional "marker" chromosome. *Hum. Genet.* **1985**, *69*, 353–370. [CrossRef]
11. De Braekeleer, M.; Dao, T.N. Cytogenetic studies in male infertility: A review. *Hum. Reprod.* **1991**, *6*, 245–250. [CrossRef]
12. Morel, F.; Douet-Guilbert, N.; Le Bris, M.J.; Amice, V.; Le Martelot, M.T.; Roche, S.; Valéri, A.; Derrien, V.; Amice, J.; De Braekeleer, M. Chromosomal abnormalities in couples undergoing intracytoplasmic sperm injection. A study of 370 couples and review of the literature. *Int. J. Androl.* **2004**, *27*, 178–182. [CrossRef]
13. Douet-Guilbert, N.; Marical, H.; Pinson, L.; Herry, A.; Le Bris, M.J.; Morel, F.; De Braekeleer, M. Characterisation of supernumerary chromosomal markers: A study of 13 cases. *Cytogenet. Genome Res.* **2007**, *116*, 18–23. [CrossRef] [PubMed]
14. Wang, W.; Hu, Y.; Zhu, H.; Li, J.; Zhu, R.; Wang, Y.-P. A case of an infertile male with a small supernumerary marker chromosome negative for M-FISH and containing only heterochromatin. *J. Assist. Reprod. Genet.* **2009**, *26*, 291–295. [CrossRef] [PubMed]
15. Oracova, E.; Musilova, P.; Kopecna, O.; Rybar, R.; Vozdova, M.; Vesela, K.; Rubes, J. Sperm and Embryo Analysis in a Carrier of Supernumerary inv dup(15) Marker Chromosome. *J. Androl.* **2009**, *30*, 233–239. [CrossRef] [PubMed]
16. Palkopoulou, E.; Lipson, M.; Mallick, S.; Nielsen, S.; Rohland, N.; Baleka, S.; Karpinski, E.; Ivancevic, A.M.; To, T.-H.; Kortschak, R.D.; et al. A comprehensive genomic history of extinct and living elephants. *Proc. Natl. Acad. Sci. USA* **2018**, *13*, 115. [CrossRef] [PubMed]
17. Houck, M.L.; Kumamoto, A.T.; Gallagher, D.S.; Benirschke, K. Comparative cytogenetics of the African elephant (*Loxodonta africana*) and Asiatic elephant (*Elephas maximus*). *Cytogenet. Cell Genet.* **2001**, *93*, 249–252. [CrossRef] [PubMed]
18. Frönicke, L.; Wienberg, J.; Stone, G.; Adams, L.; Stanyon, R. Towards the delineation of the ancestral eutherian genome organization: Comparative genome maps of human and the African elephant (*Loxodonta africana*) generated by chromosome painting. *Proc. Biol. Sci.* **2003**, *270*, 1331–1340. [CrossRef] [PubMed]
19. Yang, F.; Alkalaeva, E.Z.; Perelman, P.L.; Pardini, A.T.; Harrison, W.R.; O'Brien, P.C.M.; Fu, B.; Graphodatsky, A.S.; Ferguson-Smith, M.A.; Robinson, T.J. Reciprocal chromosome painting among human, aardvark, and elephant (superorder Afrotheria) reveals the likely eutherian ancestral karyotype. *Proc. Natl. Acad. Sci. USA* **2003**, *100*, 1062–1066. [CrossRef]
20. Cernohorska, H.; Kubickova, S.; Vahala, J.; Rubes, J. Molecular insights into X; BTA5 chromosome rearrangements in the tribe Antilopini (Bovidae). Cytogenet. *Genome Res.* **2012**, *136*, 188–198.
21. Seabright, M. A rapid banding technique for human chromosomes. *Lancet* **1971**, *2*, 971–972. [CrossRef]
22. Sumner, A.T. A simple technique for demonstrating centric heterochromatin. *Exp. Cell Res.* **1972**, *75*, 304–306. [CrossRef]

23. Goodpasture, C.; Bloom, S.E. Visualization of nucleolar organizer in mammalian chromosomes using silver staining. *Chromosoma* **1975**, *53*, 37–50. [CrossRef]
24. Kubickova, S.; Cernohorska, H.; Musilova, P.; Rubes, J. The use of laser microdissection for the preparation of chromosomespecific painting probes in farm animals. *Chromosome Res.* **2002**, *10*, 571–577. [CrossRef] [PubMed]
25. Pauciullo, A.; Kubickova, S.; Cernohorska, H.; Petrova, K.; Di Berardino, D.; Ramunno, L.; Rubes, J. Isolation and physical localization of new chromosome specific repeats in farm animals. *Vet. Med. Czech.* **2006**, *51*, 224–231. [CrossRef]
26. Cernohorska, H.; Kubickova, S.; Kopecna, O.; Vozdova, M.; Matthee, C.A.; Robinson, T.J.; Rubes, J. Nanger, Eudorcas, Gazella, and Antilope form a well-supported chromosomal clade within Antilopini (Bovidae, Cetartiodactyla). *Chromosoma* **2015**, *124*, 235–247. [CrossRef] [PubMed]
27. Hungerford, D.A.; Sharat Chandra, H.; Snyder, R.L.; Ulmer, F.A., Jr. Chromosomes of three elephants, two Asian (*Elephas maximus*) and one African (*Loxodonta africana*). *Cytogenetics* **1966**, *5*, 243–246. [CrossRef]
28. Norberg, H.S. The chromosomes of the Indian female elephant (*Elephas indicus* syn. *E. maximus* L.). *Hereditas* **1969**, *63*, 279–281. [CrossRef]
29. Kopecna, O.; Kubickova, S.; Cernohorska, H.; Cabelova, K.; Vahala, J.; Martinkova, N.; Rubes, J. Tribe-specific satellite DNA in non-domestic Bovidae. *Chromosome Res.* **2014**, *22*, 277–291. [CrossRef]
30. Barra, V.; Fachinetti, D. The dark side of centromeres: Types, causes and consequences of structural abnormalities implicating centromeric DNA. *Nat. Commun.* **2018**, *9*, 4340. [CrossRef]
31. Hartl, G.B.; Kurt, F.; Hemmer, W.; Nadlinger, K. Electrophoretic and chromosomal variation in captive Asian elephants (*Elephas maximus*). *Zoo. Biol.* **1995**, *14*, 87–95. [CrossRef]
32. Rattanayuvakorn, S.; Tanomtong, A.; Phimphan, S.; Sangpakdee, W.; Pinmongkhonkul, S.; Phintong, K. Karyological Study of Tusker and Tuskless Male Asian Elephant (*Elephas maximus*) by Conventional, GTG-, and Ag-NOR Banding Techniques. *CYTOLOGIA* **2017**, *82*, 349–354. [CrossRef]
33. Liehr, T.; Mrasek, K.; Kosyakova, N.; Ogilvie, C.M.; Vermeesch, J.; Trifonov, V.; Rubtsov, N. Small supernumerary marker chromosomes (sSMC) in humans; are there B chromosomes hidden among them. *Mol. Cytogenet.* **2008**, *1*, 12. [CrossRef] [PubMed]
34. Fuster, C.; Rigola, M.A.; Egozcue, J. Human supernumeraries: Are they B chromosomes? *Cytogenet. Genome Res.* **2004**, *106*, 165–172. [CrossRef] [PubMed]
35. Camacho, J.P.M. B chromosomes in the eukaryote genome. *Cytogenet. Genome Res.* **2004**, *106*, 147–410. [CrossRef]
36. Jones, R.N.; Viegas, W.; Houben, A. A century of B chromosomes in plants: So what? *Ann. Bot.* **2008**, *101*, 767–775. [CrossRef]
37. Vujošević, M.; Rajičić, M.; Jelena Blagojević, J. B chromosomes in populations of mammals revisited. *Genes* **2018**, *9*, 487. [CrossRef]
38. Mackie Ogilvie, C.; Harrison, R.H.; Horsley, S.W.; Hodgson, S.V.; Kearney, L. A mitotically stable marker chromosome negative for whole chromosome libraries, centromere probes and chromosome specific telomere regions: A novel class of supernumerary marker chromosome? *Cytogenet Cell Genet.* **2001**, *92*, 69–73. [CrossRef]
39. Seifertova, E.; Zimmerman, L.B.; Gilchrist, M.J.; Macha, J.; Kubickova, S.; Cernohorska, H.; Zarsky, V.; Owens, N.D.L.; Sesay, A.K.; Tlapakova, T.; et al. Efficient high-throughput sequencing of a laser microdissected chromosome arm. *BMC Genom.* **2013**, *14*, 357. [CrossRef]

Disclaimer/Publisher's Note: The statements, opinions and data contained in all publications are solely those of the individual author(s) and contributor(s) and not of MDPI and/or the editor(s). MDPI and/or the editor(s) disclaim responsibility for any injury to people or property resulting from any ideas, methods, instructions or products referred to in the content.

Article

Prevalence of Sex-Related Chromosomal Abnormalities in a Large Cohort of Spanish Purebred Horses

Sebastián Demyda-Peyrás [1,2,*], Nora Laseca [3], Gabriel Anaya [3], Barbara Kij-Mitka [4], Antonio Molina [3], Ayelén Karlau [1,2] and Mercedes Valera [5]

1. Departamento de Producción Animal, Facultad de Ciencias Veterinarias, Universidad Nacional de La Plata, La Plata 1900, Argentina
2. Consejo Nacional de Investigaciones Científicas y Técnicas (CONICET LA PLATA), La Plata 1900, Argentina
3. Departamento de Genética, Universidad de Córdoba, 14071 Córdoba, Spain
4. Department of Animal Reproduction, Anatomy and Genomics, University of Agriculture in Krakow, Mickiewicza 24/28, 30-059 Krakow, Poland
5. Departamento Agronomía, Escuela Técnica Superior de Ingeniería Agromómica, Universidad de Sevilla, Ctra Utrera Km 1, 41013 Sevilla, Spain
* Correspondence: sdemyda@fcv.unlp.edu.ar

Simple Summary: Horses are well known for the increased number of individuals carrying chromosomal abnormalities related to the sex pair, which have been identified as a major cause of idiopathic infertility. However, large-scale populational studies evaluating the occurrence of these chromosomal aberrations in commercial or wild populations are extremely scarce. We, therefore, performed a cytogenetic analysis on a large dataset of 25,237 individuals, gathered over a period of 24 months, using a two-step genomic-based diagnostic methodology. We first screened the entire population, analyzing the results of short tandem repeats parentage testing to determine individuals showing abnormal results. Thereafter, the positive samples, together with the individuals showing morphological abnormalities in the reproductive tract, were reanalyzed using a single nucleotide polymorphism (SNP)-based procedure to determine the occurrence of chromosomal abnormalities. Our results showed that the overall prevalence of individuals carrying chromosomal alterations was close to 0.05%, with blood chimerism and 64,XY sex-reversed mares the most common type of aberrations detected. In addition, one case of Turner and one of Klinefelter syndrome, as well as a small number of individuals carrying complex karyotypes, were also detected. However, these results should be taken with caution since the occurrence of X chromosome monosomy, a sex-related chromosomal aberration commonly reported in mares, cannot be screened using the methodology employed in this study. To our knowledge, this is the largest study performed aimed at determining the prevalence of the most important chromosomal abnormalities in the domestic horse.

Abstract: Chromosomal abnormalities are largely associated with fertility impairments in the domestic horse. To date, over 600 cases of individuals carrying abnormal chromosome complements have been reported, making the domestic horse the species with the highest prevalence. However, studies analyzing the prevalence of chromosomal diseases in whole populations are scarce. We, therefore, employed a two-step molecular tool to screen and diagnose chromosomal abnormalities in a large population of 25,237 Pura Raza Español horses. Individuals were first screened using short tandem repeats parentage testing results and phenotypic evaluations. Those animals showing results suggesting chromosomal abnormalities were re-tested using a single nucleotide polymorphism (SNP)-based diagnostic methodology to accurately determine the chromosomal complements. Thirteen individuals showed a positive screening, all of which were diagnosed as chromosomally abnormal, including five 64,XY mares with sex development disorders (DSD) and four cases of blood chimerism (two male/female and two female/female cases). In addition, we detected one Turner and one Klinefelter syndrome and two individuals carrying complex karyotypes. The overall prevalence in the entire population was ~0.05%, with the prevalence of 64,XY DSD and blood chimerism ~0.02% and ~0.016%, respectively. However, the overall results should be taken with caution since the individuals carrying Turner syndrome (in full (63,X) or mosaic (mos 63,X/64,XX) forms) cannot be

Citation: Demyda-Peyrás, S.; Laseca, N.; Anaya, G.; Kij-Mitka, B.; Molina, A.; Karlau, A.; Valera, M. Prevalence of Sex-Related Chromosomal Abnormalities in a Large Cohort of Spanish Purebred Horses. *Animals* **2023**, *13*, 539. https://doi.org/10.3390/ani13030539

Academic Editors: Leopoldo Iannuzzi and Pietro Parma

Received: 2 January 2023
Revised: 27 January 2023
Accepted: 28 January 2023
Published: 3 February 2023

Copyright: © 2023 by the authors. Licensee MDPI, Basel, Switzerland. This article is an open access article distributed under the terms and conditions of the Creative Commons Attribution (CC BY) license (https:// creativecommons.org/licenses/by/ 4.0/).

detected due to limitations in the methodology employed. Finally, the lack of agreement between populational studies performed using karyotyping or molecular methods is discussed. To our knowledge, this is the largest populational study performed evaluating the prevalence of the most common chromosomal abnormalities in the domestic horse.

Keywords: chromosomal abnormalities; horse; genomic; prevalence; SNP-array; cytogenetics

1. Introduction

Chromosomal abnormalities related to the sex pair are a common genetic disease in domestic horses, *Equus caballus*. This knowledge is not new [1] but was demonstrated 30 years ago by Power [2], who compiled the karyotyping results of nearly 400 cases showing chromosomal aberrations. More recently, Bugno-Poniewierska and Raudsepp [3] established that chromosomal disorders are among the most common non-infectious causes of subfertility, infertility, and congenital defects in the species, accounting for ~30% of horses with reproductive or developmental problems. However, cytogenetic analysis in such individuals is not a common practice and is far from being a systematic practice in large populations and/or commercial herds [4]. This lack of testing may make it difficult to determine the prevalence of chromosomal abnormalities in horses, especially since some individuals carrying chromosomal aberrations may be phenotypically normal, thus avoiding detection [5].

The most comprehensive and largest cytogenetic evaluation of a horse population existing to date was performed 15 years ago by Bugno, et al. [6], who reported a prevalence of chromosomal abnormalities related to the sex pair close to 2% by karyotyping 500 horses selected randomly. In contrast, Kakoi, et al. [7] conducted a large-scale analysis of 17,471 newborn light-breed foals in Japan using high-throughput molecular methods (instead of karyotyping) and found a much lower prevalence of chromosomal abnormalities, approximately 0.01%. Similarly, Anaya, et al. [8] used molecular methods to detect blood chimerism in 21,097 Pura Raza Español (PRE) horses, reporting a prevalence of 0.01%, which is 20 times lower than the previous existing reports. Given the discrepancy existing between results obtained using different methodologies, any additional data produced will help us to determine a more accurate rate of chromosomal abnormalities in the domestic horse.

Thirty years ago, Bowling, et al. [9] demonstrated the usefulness of short tandem repeats (STR) genotyping to detect chimerism in horses. More recently, this approach was further employed to detect the same type of chromosomal abnormality in an American Bashkir Curly [10] and several Pura Raza Español horses [8,11]. However, the standardized STR panel employed for parentage testing in the species includes only one ECAX marker and none located in the ECAY. Therefore, most of the aberrations associated with reproductive impairments, such as sex reversions (DSD) or ECAX monosomy [5], cannot be detected. To solve this issue, Kakoi, et al. [7] developed an extended STR panel with better coverage of sex chromosomes, detecting 17 individuals with abnormal complements in a large population of Japanese horses. However, more recently, our laboratory validated a novel, more accurate methodology for chromosome testing, based on the analysis of the information provided by single nucleotide polymorphism (SNP) genotyping arrays, in the PRE breed [12]. This method, which can detect almost any type of chromosomal abnormalities, was integrated as an auxiliary tool of the PRE breeding program in 2021.

The Pura Raza Español horse is one of the oldest and most important horse breeds bred in Europe [13], with more than 250,000 active individuals bred in over 60 countries in the present day [14]. Its studbook was created in 1912, and since then, it has been managed by the Real Asociación Nacional de Criadores de Caballos de Pura Raza Española (ANCCE), following a closed enrolment policy. For this reason, to be included in the PRE studbook, all individuals need to perform mandatory DNA testing to confirm the parentage assignment, as well as a phenotypic characterization to avoid the enrolment of individuals with morphological variations forbidden by the PRE breeding program bylaws. Since 2021, all PRE individuals showing any reproductive abnormality in the mandatory pre-enrolment phenotypic assessment or those whose STR-based parentage test results showed any abnormal or incongruent results (more than 2 alleles per loci or incompatibility between genotypes and phenotypic sex) are being flagged as presumptive carriers of chromosomal abnormalities, and submitted for further investigation using SNP genotyping. Two years later, almost 25,237 horses have now been screened in one of the largest cytogenetic studies conducted on the species.

Here, we present the results of the screening for sex-related chromosomal alterations in a large population of PRE horses, with the aim of establishing a more accurate estimation of the prevalence of chromosomal abnormalities in the species.

2. Materials and Methods

2.1. Animal Samples

All the individuals analyzed in this study are included in the screening program for chromosomal abnormalities of the Real Asociación Nacional de Criadores de Caballo Pura Raza Español (ANCCE) studbook. All of them provided blood samples collected by the ANCCE official veterinary services, according to the breeding association standard protocol for parentage testing, before their enrollment in the studbook. During the last 24 months, 25,237 individuals were evaluated (12,569 in 2021 and 12,668 in 2022).

2.2. Genotyping and Chromosomal Analysis

DNA was obtained from biological samples (either whole blood or hair bulbs) using regulation extraction kits from Qiagen (Madrid, Spain). Thereafter, the samples were first genotyped using the 17 STR panel for determining parentage in horses proposed by the International Society for Animal Genetics (ISAG) in a multiplexed determination using a set of commercially-available fluorescent-labeled primers (StockMarks kit for horses, PE Applied Biosystems, Foster City, CA, USA). In the same reaction, we determined the presence or absence of ECAX and ECAY gene-specific *amelogenin* markers using a slight modification of the PCR reaction proposed by Hasegawa, et al. [15]. Finally, the genotyping and allele calling was performed by capillary electrophoresis using an Applied Biosystems 3130xl DNA sequencer (ANCCE, Spain).

Any individuals who showed abnormal genotyping results (more than 3 alleles in different loci or those showing a discrepancy between the phenotypic and genotypic sex in the ECAX specific marker (LEX003) or AMELX and AMELY fragments, according to Anaya-Calvo [16]) or which showed phenotypic abnormalities in the external reproductive organs were further genotyped using a medium density SNP array chip (Equine GGP 70K, Neogen Inc, Scotland, UK). Finally, the chromosomal complements of these animals were determined according to the methodology validated in the PRE by Pirosanto, et al. [12]. In addition, these individuals were reinspected phenotypically by an official ANCCE veterinarian in situ to confirm the phenotypic sex and to determine the presence or absence of phenotypic abnormalities in the reproductive tract.

3. Results

Thirteen individuals (0.051% of the total population analyzed) were submitted for chromosomal analysis during the 24-month period (Table 1). Among them, three were submitted due to the existence of morphological abnormalities in the external gonads, four showed three or more alleles in a single locus in the parentage testing STR panel, and six showed incongruences between phenotypic and genotypic sex. In all the cases, the presence of chromosomal abnormalities was confirmed by the SNP genotyping (Table 1).

Table 1. Individuals carrying chromosomal abnormalities during the 24-month period in the PRE breed.

Individual	Year	Phenotypic Sex	Parentage STR	SNP Genotyping
1	2021	Intersex	Female	63,X
2	2021	Male	Multiallellic	chi 64,XX,64,XY
3	2021	Female	Multiallellic	chi 64,XX,64,XX
4	2021	Female	Multiallellic	chi 64,XX,64,XY
5	2021	Female	Sex incongruence	64,XY
6	2021	Female	Sex incongruence	64,XY
7	2021	Intersex	Normal female	mos 63,X/64,XX
8	2022	Female	Sex incongruence	64,XY
9	2022	Male	Sex incongruence	65,XXY
10	2022	Female	Multiallellic	chi 64,XX,64,XX
11	2022	Intersex	Male	mos 63,X/64,XY
12	2022	Female	Sex incongruence	64,XY
13	2022	Female	Sex incongruence	64,XY

A short tandem repeat (STR) based parentage test was performed according to Demyda-Peyras, et al. [11]. Single nucleotide polymorphism (SNP) genotyping analyses were performed according to Pirosanto, et al. [12]. N.d.: not detected.

The results show that 64,XY DSD sex reversal mare (see Figure 1) was the most common syndrome detected (five cases, 0.02%), followed by blood chimerism (four cases, 0.015%). Interestingly, two of them were male/female chimeras, whereas the remaining two were female/female chimeras (see Figure 2).

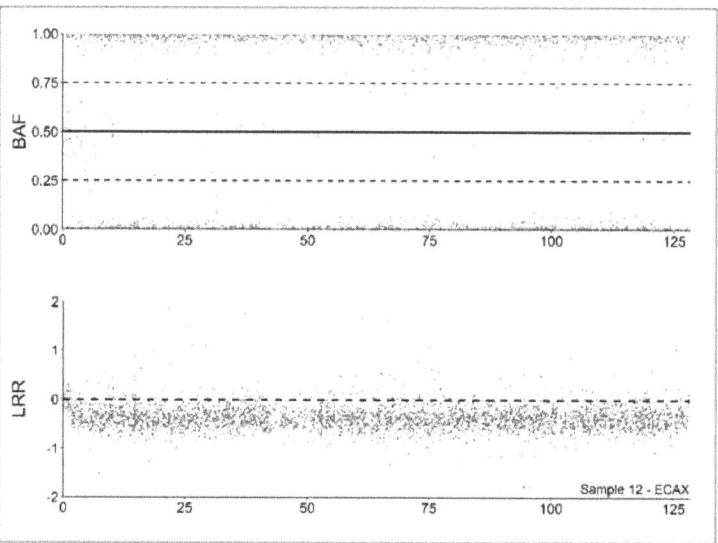

Figure 1. Copy number alterations (CNA) analysis profile of a 64,XY DSD mare. Analysis of the B allele frequency (BAF) and Log R ratio (LRR) values from a 64,XY DSD individual according to [10]. BAF values (upper part) depicting hemizygous markers (close to 0 or 1) in the non-pseudoautosomal region (XPAR, green). Conversely, the XPAR (in yellow) is mostly heterozygous (values close to 0.5). LRR values (lower part) are close to 0 in XPAR (in yellow), depicting diploidy. On the contrary, values in the non-XPAR region (in purple) are close to −0.5, depicting monosomy.

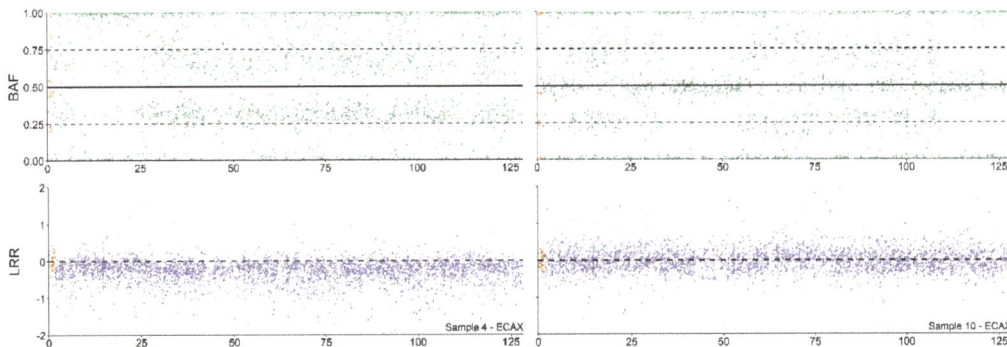

Figure 2. Copy number alterations (CNA) analysis profile of 64,XY/64,XX and 64,XX/64,XX blood chimeras. Analysis of the B allele frequency (BAF) and Log R ratio (LRR) values of the ECAX from 64,XY/64,XX (Sample 4, left part) and 64,XX/64,XX individuals (Sample 10, shown on the right of the figure) according to Pirosanto, et al. [12].

In addition, we detected a Klinefelter horse (65,XXY), a Turner's mare (63,X) and two mosaicisms (63,X/64,XX and 63,X/64,XY). Finally, the overall prevalence of individuals carrying chromosomal abnormalities in the PRE during the last 24 months was 0.051% (13/25,237). However, it should be mentioned that individuals carrying 63,X ECAX monosomy or low-level chimerisms could be screened using the diagnostic pipeline employed in this study, and therefore, the overall prevalence is most likely underestimated.

4. Discussion

It is well established that the domestic horse is the domestic species with the highest number of individuals carrying chromosomal alterations [17], most of which are associated with reproductive failures and abnormal phenotypes [5]. This knowledge is mostly built on the compelling number of cases reported to date, such as the ~400 horses compiled by Power [2] 30 years ago or the 214 horses analyzed by the TAMU cytogenetic lab over the last 20 years [3]. However, large-scale cytogenetic surveys analyzing whole populations are scarce. Here, we establish the incidence of the most important chromosomal abnormalities in a population of 25,237 Pura Raza Español horses.

Several reasons have been suggested for the lack of large cytogenetic surveys in horses, such as the inability of field practitioners to establish an association between infertility and chromosomal failures, but also the scarce availability of commercial laboratories providing karyotyping services in the species [4]. In addition, the classical cytogenetics techniques are slow, expensive, and require the collection and shipping of biological samples for cell culture. However, 20 years ago, Kakoi, et al. [7] proposed the viability of the use of STR parentage genotyping as a screening tool to detect chromosomal abnormalities in horses. Based on this idea, we established a relationship with the Pura Raza Español breeding association, which allowed us to detect several individuals carrying chromosomal abnormalities in the PRE breed, including chimerisms [11], DSD sex reversal horses [18] and complex karyotypes [19], such as normal and abnormal cell lines originating in two different individuals in the same horse. However, it was not until more recently, with the validation of an SNP-based methodology [12], that chromosomal screening was integrated into the breeding program. Since then, more than 25,000 horses have been screened, thus producing one of the most comprehensive datasets analyzed to date.

So far, two large-scale studies have been performed to analyze horse populations from a cytogenetic point of view. In 2005, Kakoi, et al. [7] analyzed 17,471 light-breed Japanese horses using a molecular approach based on STR parentage genotyping, similar to the one used in our study. The authors were able to detect 18 individuals with presumptive chromosomal abnormalities in the sex pair, establishing a prevalence of close to 1/1000.

In the second step, they employed an extended ECAX/ECAY STR panel, thus confirming the existence of 13 Turner mares (63,X, 0.07%), 4 Klinefelter horses (65,XXY, 0.02%), and a 65,XXX mare (0.006%). However, only four of these cases could be confirmed by karyotyping due to the lack of availability of biological samples. Two years later, Bugno, et al. [6] karyotyped 500 young Polish horses from several breeds, reporting 10 individuals (2%) with chromosomal abnormalities. Among them, the authors reported a male/female chimera (64,XX/64,XY, 0.2%), one Turner syndrome (63,X, 0.2%), 7 ECAX mosaicisms (63,X/64,XX, 1.4%), and an individual with an autosomal trisomy in the ECA31. Our study agrees with the results provided by Kakoi, et al. [7], probably due to the fact that they were performed using a similar diagnostic methodology in which a large population was first screened, and then only the horses suspected of carrying chromosomal abnormalities were re-analyzed for confirmation. In contrast, the later study karyotyped all 500 horses, with the most common abnormality reported being 63,X/64,XX mosaicism, which was barely detected in our study (just one case). Since this later syndrome is difficult to screen using molecular methods [20], it may explain the differences observed between these different approaches. However, it is also worth mentioning that an overall incidence equal to that reported by [6] would be equivalent to detecting nearly 500 horses carrying chromosomal abnormalities in our PRE dataset, which is far higher than the figure we found. In this context, Shah, et al. [21] reported a 33% discordance rate between karyotyping and molecular methods in the analysis of human miscarriages, with the appearance of chromosomal artifacts due to prolonged culture being one of the potential causes. This lack of consensus was even more noticeable when the growth of cultured cells was stimulated with phytohaemagglutinin (PHA), which resulted in an increased number of abnormalities in comparison with the same samples analyzed from unstimulated cultures [22]. However, we previously compared the results of karyotyping and molecular methods in 30 PRE horses (10 carrying chromosomal abnormalities) with 100% of accuracy [23]. For this reason, we hypothesize that this could be an additional cause to explain the differences observed among studies, such as the breed effect, or maybe another, which was not analyzed in the present report.

One of the key questions about the use of molecular methods for detecting chromosomal abnormalities is their accuracy in determining the presence of mosaicisms/chimerisms in which the percentage of abnormal cells is low [24]. This is particularly important since Power [2] reported that 15% of the 401 individuals carrying chromosomal abnormalities were 63,X/64,XX mares. These limitations were recently demonstrated in horses by Szczerbal, et al. [20] using a highly accurate technique (ddPCR) and by our group using an array-based SNP methodology [12]. However, this is also a well-known problem in humans, where the ability to detect low levels of mosaicism with less than 10% of abnormal cells is questionable [25]. Pienkowska-Schelling, et al. [26] recently reported a substantial increase in the rate of ECAX mosaicism (63,X/64,XX) in fertile mares by performing classical karyotyping. This finding agrees (although the incidence was different) with the reports demonstrating that some of these individuals can develop normally and produce offspring [6,27]. Nevertheless, most of the cases of 63,X/64,XX described are associated with subfertility [17]. Despite the fact that the use of molecular methods for this particular syndrome is still in its infancy, its importance may be limited in the whole population in comparison with the other chromosomal abnormalities reported to date [2,3].

One important limitation of this study worth mentioning is the inability to screen Turner's syndrome (63,X; ECAX monosomy) using parentage STR markers since the standardized test performed worldwide includes only one STR located in the ECAX. In this context, Bugno, et al. [6] reported a prevalence value of 0.2% (1 in 500) for this abnormality in the whole population (1/500). Similarly, several reports [2,17,28,29] have established that ECAX monosomy can account for approximately 30/40% of all the chromosomal abnormalities detected in horses. In contrast, Kakoi, et al. [7] reported a prevalence of this abnormality close to 0.075% (13/17,471), although they were only able to determine those individuals in which the foal's ECAX was of paternal origin. Since we are not able to estimate the incidence of such aberrations in PRE, we extrapolated the results obtained in other breeds (30% of incidence among chromosomal abnormalities) to our dataset to obtain a hypothetical incidence for comparison purposes. This estimated value (0.016%) is ten times lower than that reported by Bugno, et al. [6] but more in line with that reported by Kakoi, et al. [7] as well as with the incidence of this syndrome in humans (0.04%, according to Bondy [30]), although a large-scale study would be needed to determine the real prevalence of this syndrome in horse populations. Here, SNP-based methodologies can accurately detect mares carrying 63,X complements [12]. Given the introduction of SNP arrays and the genotyping of horses in breeding programs [31–33]), we should expect to find an increase in the detection of this chromosomal abnormality in large populations of horses within the next few years.

Disorders in sex development constitute another common syndrome in horses [5]. Among these, the most common is the sex-reversal mare (64,XY DSD), in which the individual shows a mare phenotype instead of carrying male chromosomal complements [34], even in the PRE breed [18]. In our study, we used a combination of an ECAX STR marker (LEX003) together with determining two fragments of the sex-related *amelogenin* gene (AME), which reduces the possibility of misdiagnosis close to null. However, Martinez, et al. [35] reported a very low percentage of inconsistencies in the AMEY testing (8/100,000) in a large population of horses, suggesting that they might be the result of a translocation (in the paternal line) from the Y to another chromosome. In our case, all the individuals showing a positive AMEY amplification were confirmed as 64,XY DSD by SNP genotyping (6/6). In terms of prevalence, Power [2] stated that 28% of the individuals carrying chromosomal abnormalities showed a 64,XY karyotype, whereas Bowling, et al. [28] detected 22 individuals showing this chromosomal arrangement among 98 mares. In our case, the incidence of 64,XY DSD among individuals showing abnormal karyotypes was ~40%. However, 63,X was not included in this study, and therefore, results may, to some extent, be overestimated. Interestingly, we determined a prevalence of 0.02% 64,XY DSD in the whole population analyzed. This result is in full agreement with the largest screening programs performed for this syndrome in the species, which reported the same value [35]. Interestingly, neither [6] nor [7] reported any individual showing a DSD in their populational studies. However, the fact is that analysis of almost 235,000 individuals, together with the long history of cases reported during the last 40 years, suggest that 64,XY DSD is one of the major chromosomal abnormalities in the domestic horse.

Finally, we were able to detect four cases of blood chimerism in the whole population (4/25,237; 0.016%). These results agree with Anaya, et al. [8] (0.024%), who analyzed ~21,000 PRE foalings but were ten times lower than those reported by Bugno, et al. [6]. Since the electropherogram pattern obtained in STR parentage can only be caused either by blood chimerism or by cross-contamination of the sample, the use of an additional DNA sample is mandatory. In our case, the results obtained using DNA obtained from hair bulbs were normal (not chimeric). However, SNP genotyping also allowed us to detect 64,XX/64,XX blood chimeras, which cannot be detected using classical or molecular karyotyping. Since it is perfectly possible to misdiagnose chimeric samples as normal, we believe that the occurrence of one case of blood chimerism every 5000 foals is a reasonable and reliable estimation for the domestic horse.

5. Conclusions

Chromosomal abnormalities associated with the sex pair are a noticeable problem in horse fertility. In this study, we demonstrated that the use of a combined technique, including STR and SNP genotyping, can detect most of these genetic abnormalities in horses at an early age. We detected the existence of chromosomal abnormalities in 0.05% of the 25,237 PRE individuals analyzed over the 24-month period within the official breeding program. However, we were also able to estimate a reliable prevalence for specific chromosomal abnormalities, such as 64,XY DSD and blood chimerism, by analyzing one of the largest datasets to date. However, the overall prevalence could be underestimated because of our inability to screen 63,X individuals due to methodological limitations. Finally, we suggest that the increasing use of SNP genotyping within breeding programs will allow us to detect most of the individuals carrying chromosomal abnormalities in the next few years in a reliable, systematic way.

Author Contributions: Conceptualization: S.D.-P., M.V. and A.M.; Methodology: N.L., G.A. and S.D.-P.; Formal analysis: G.A., N.L., B.K.-M. and A.K.; Data curation: G.A., N.L., B.K.-M. and A.K.; Investigation: G.A., N.L., B.K.-M. and A.K.; Resources: A.M., M.V. and S.D.-P. Writing—original draft preparation, B.K.-M., G.A. and N.L.; Writing—review and editing: B.K.-M., G.A., S.D.-P. and M.V.; Supervision: A.M. and S.D.-P.; Project management: S.D.-P. All authors have read and agreed to the published version of the manuscript.

Funding: This study was funded by UCO-FEDER 20 REF. 1380999-R (Junta de Andalucía, Spain), and AGL2017-84217-P (Mineco, Spain) grants (Antonio Molina, PI) and PICTA-2021-0063 grant (FONCyT, ANPCyT, Argentina), Sebastián Demyda-Peyrás (PI). Ayelén Karlau is a CONICET fellow.

Institutional Review Board Statement: Ethical review and approval were waived for this study since the data analyzed was produced by the ANCCE as a part of their official breeding program.

Informed Consent Statement: Not applicable.

Data Availability Statement: All the data analyzed in this study belong to the Pura Raza Española breeding program. The ANCCE allows its use for scientific purposes under a specific agreement of collaboration with the MERAGEM group. Therefore, the genetic and genomic data employed in this study is not publicly available. Access can be granted by request to the corresponding authors upon agreement with the ANCCE.

Acknowledgments: We would like to thank the ANCCE for providing STR data and samples.

Conflicts of Interest: The authors declare no conflict of interest.

References

1. Bouters, R.; Vandeplassche, M.; De Moor, A. An intersex (male pseudohermaphrodite) horse with 64 XX-65 XXY mosaicism. *Equine Vet. J.* **1972**, *4*, 150–153. [CrossRef] [PubMed]
2. Power, M.M. Chromosomes of the horse. *Adv. Vet. Sci. Comp. Med.* **1990**, *34*, 131–167.
3. Bugno-Poniewierska, M.; Raudsepp, T. Horse Clinical Cytogenetics: Recurrent Themes and Novel Findings. *Animals* **2021**, *11*, 831. [CrossRef] [PubMed]
4. Laseca, N.; Anaya, G.; Peña, Z.; Pirosanto, Y.; Molina, A.; Demyda Peyrás, S. Impaired Reproductive Function in Equines: From Genetics to Genomics. *Animals* **2021**, *11*, 393. [CrossRef] [PubMed]
5. Lear, T.L.; McGee, R.B. Disorders of sexual development in the domestic horse, Equus caballus. *Sex. Dev.* **2012**, *6*, 61–71. [CrossRef]
6. Bugno, M.; Słota, E.; Kościelny, M. Karyotype evaluation among young horse populations in Poland. *Schweiz. Arch. Tierheilkd.* **2007**, *149*, 227–232. [CrossRef]
7. Kakoi, H.; Hirota, K.; Gawahara, H.; Kurosawa, M.; Kuwajima, M. Genetic diagnosis of sex chromosome aberrations in horses based on parentage test by microsatellite DNA and analysis of X- and Y-linked markers. *Equine Vet. J.* **2005**, *37*, 143–147. [CrossRef]
8. Anaya, G.; Fernandez, M.E.; Valera, M.; Molina, A.; Azcona, F.; Azor, P.; Sole, M.; Moreno-Millan, M.; Demyda-Peyras, S. Prevalence of twin foaling and blood chimaerism in purebred Spanish horses. *Vet. J.* **2018**, *234*, 142–144. [CrossRef]
9. Bowling, A.T.; Stott, M.L.; Bickel, L. Silent blood chimaerism in a mare confirmed by DNA marker analysis of hair bulbs. *Anim. Genet.* **1993**, *24*, 323–324. [CrossRef]
10. Juras, R.; Raudsepp, T.; Das, P.J.; Conant, E.; Cothran, E.G. XX/XY Blood Lymphocyte Chimerism in Heterosexual Dizygotic Twins from an American Bashkir Curly Horse. Case Report. *J. Equine Vet. Sci.* **2010**, *30*, 575–580. [CrossRef]
11. Demyda-Peyras, S.; Bugno-Poniewierska, M.; Pawlina, K.; Anaya, G.; Moreno-Millán, M. The use of molecular and cytogenetic methods as a valuable tool in the detection of chromosomal abnormalities in horses: A Case of sex chromosome chimerism in a Spanish Purebred colt. *Cytogenet. Genome Res.* **2013**, *141*, 277–283. [CrossRef] [PubMed]
12. Pirosanto, Y.; Laseca, N.; Valera, M.; Molina, A.; Moreno-Millan, M.; Bugno-Poniewierska, M.; Ross, P.; Azor, P.; Demyda-Peyras, S. Screening and detection of chromosomal copy number alterations in the domestic horse using SNP-array genotyping data. *Anim. Genet.* **2021**, *52*, 431–439. [CrossRef] [PubMed]
13. Poyato-Bonilla, J.; Laseca, N.; Demyda-Peyras, S.; Molina, A.; Valera, M. 500 years of breeding in the Carthusian Strain of Pura Raza Espanol horse: An evolutional analysis using genealogical and genomic data. *J. Anim. Breed. Genet.* **2022**, *139*, 84–99. [CrossRef] [PubMed]
14. Perdomo-Gonzalez, D.I.; Laseca, N.; Demyda-Peyras, S.; Valera, M.; Cervantes, I.; Molina, A. Fine-tuning genomic and pedigree inbreeding rates in equine population with a deep and reliable stud book: The case of the Pura Raza Espanola horse. *J. Anim. Sci. Biotechnol.* **2022**, *13*, 127. [CrossRef]

15. Hasegawa, T.; Sato, F.; Ishida, N.; Fukushima, Y.; Mukoyama, H. Sex Determination by Simultaneous Amplification of Equine SRY and Amelogenin Genes. *J. Vet. Med. Sci.* **2000**, *62*, 1109–1110. [CrossRef]
16. Anaya-Calvo, G. Chromosomal alterations in the equine species: Development of a diagnostic screening tool in the purebred Spanish horse using molecular markers associated with the sex chromosomes. Ph.D. Thesis, University of Córdoba, Córdoba, Spain, 2014.
17. Raudsepp, T.; Chowdhary, B.P. Chromosome aberrations and fertility disorders in domestic animals. *Annu. Rev. Anim. Biosci.* **2016**, *4*, 15–43. [CrossRef]
18. Anaya, G.; Moreno-Millan, M.; Bugno-Poniewierska, M.; Pawlina, K.; Membrillo, A.; Molina, A.; Demyda-Peyras, S. Sex reversal syndrome in the horse: Four new cases of feminization in individuals carrying a 64,XY SRY negative chromosomal complement. *Anim. Reprod. Sci.* **2014**, *151*, 22–27. [CrossRef]
19. Demyda-Peyras, S.; Anaya, G.; Bugno-Poniewierska, M.; Pawlina, K.; Membrillo, A.; Valera, M.; Moreno-Millan, M. The use of a novel combination of diagnostic molecular and cytogenetic approaches in horses with sexual karyotype abnormalities: A rare case with an abnormal cellular chimerism. *Theriogenology* **2014**, *81*, 1116–1122. [CrossRef]
20. Szczerbal, I.; Nowacka-Woszuk, J.; Kopp-Kuhlman, C.; Mackowski, M.; Switonski, M. Application of droplet digital PCR in diagnosing of X monosomy in mares. *Equine Vet. J.* **2020**, *52*, 627–631. [CrossRef]
21. Shah, M.S.; Cinnioglu, C.; Maisenbacher, M.; Comstock, I.; Kort, J.; Lathi, R.B. Comparison of cytogenetics and molecular karyotyping for chromosome testing of miscarriage specimens. *Fertil. Steril.* **2017**, *107*, 1028–1033. [CrossRef]
22. Ballif, B.C.; Rorem, E.A.; Sundin, K.; Lincicum, M.; Gaskin, S.; Coppinger, J.; Kashork, C.D.; Shaffer, L.G.; Bejjani, B.A. Detection of low-level mosaicism by array CGH in routine diagnostic specimens. *Am. J. Med. Genet. Part A* **2006**, *140*, 2757–2767. [CrossRef] [PubMed]
23. Anaya, G.; Molina, A.; Valera, M.; Moreno-Millan, M.; Azor, P.; Peral-Garcia, P.; Demyda-Peyras, S. Sex chromosomal abnormalities associated with equine infertility: Validation of a simple molecular screening tool in the Purebred Spanish Horse. *Anim. Genet.* **2017**, *48*, 412–419. [CrossRef]
24. Albarella, S.; De Lorenzi, L.; Catone, G.; Magi, G.E.; Petrucci, L.; Vullo, C.; D'Anza, E.; Parma, P.; Raudsepp, T.; Ciotola, F.; et al. Diagnosis of XX/XY Blood Cell Chimerism at a Low Percentage in Horses. *J. Equine Vet. Sci.* **2018**, *69*, 129–135. [CrossRef]
25. Hall, G.K.; Mackie, F.L.; Hamilton, S.; Evans, A.; McMullan, D.J.; Williams, D.; Allen, S.; Kilby, M.D. Chromosomal microarray analysis allows prenatal detection of low level mosaic autosomal aneuploidy. *Prenat. Diagn.* **2014**, *34*, 505–507. [CrossRef]
26. Pienkowska-Schelling, A.; Kaul, A.; Schelling, C. X chromosome aneuploidy and micronuclei in fertile mares. *Theriogenology* **2020**, *147*, 34–38. [CrossRef] [PubMed]
27. Wieczorek, M.; Switoński, M.; Yang, F. A low-level X chromosome mosaicism in mares, detected by chromosome painting. *J. Appl. Genet.* **2001**, *42*, 205–209.
28. Bowling, A.T.; Millon, L.; Hughes, J.P. An update of chromosomal abnormalities in mares. *J. Reprod. Fertil. Suppl.* **1987**, *35*, 149–155.
29. Villagómez, D.A.F.; Parma, P.; Radi, O.; Di Meo, G.; Pinton, A.; Iannuzzi, L.; King, W.A. Classical and molecular cytogenetics of disorders of sex development in domestic animals. *Cytogenet. Genome Res.* **2009**, *126*, 110–131. [CrossRef]
30. Bondy, C.A. Care of girls and women with Turner syndrome: A guideline of the Turner Syndrome Study Group. *J. Clin. Endocrinol. Metab.* **2007**, *92*, 10–25. [CrossRef]
31. Schaefer, R.J.; McCue, M.E. Equine Genotyping Arrays. *Vet. Clin. N. Am. Equine Pract.* **2020**, *36*, 183–193. [CrossRef]
32. MacLeod, J.N.; Kalbfleisch, T.S. Genetics, Genomics, and Emergent Precision Medicine 12 Years After the Equine Reference Genome Was Published. *Vet. Clin. N. Am. Equine Pract.* **2020**, *36*, 173–181. [CrossRef] [PubMed]
33. McGivney, B.A.; Han, H.; Corduff, L.R.; Katz, L.M.; Tozaki, T.; MacHugh, D.E.; Hill, E.W. Genomic inbreeding trends, influential sire lines and selection in the global Thoroughbred horse population. *Sci. Rep.* **2020**, *10*, 466. [CrossRef] [PubMed]
34. Villagómez, D.A.F.; Lear, T.L.; Chenier, T.; Lee, S.; McGee, R.B.; Cahill, J.; Foster, R.A.; Reyes, E.; St John, E.; King, W.A. Equine disorders of sexual development in 17 mares including XX, SRY-negative, XY, SRY-negative and XY, SRY-positive genotypes. *Sex. Dev.* **2011**, *5*, 16–25. [CrossRef] [PubMed]
35. Martinez, M.M.; Costa, M.; Ratti, C. Molecular screening of XY SRY-negative sex reversal cases in horses revealed anomalies in amelogenin testing. *J. Vet. Diagn. Investig.* **2020**, *32*, 938–941. [CrossRef] [PubMed]

Disclaimer/Publisher's Note: The statements, opinions and data contained in all publications are solely those of the individual author(s) and contributor(s) and not of MDPI and/or the editor(s). MDPI and/or the editor(s) disclaim responsibility for any injury to people or property resulting from any ideas, methods, instructions or products referred to in the content.

Review

Molecular Cytogenetics in Domestic Bovids: A Review

Alessandra Iannuzzi [1], Leopoldo Iannuzzi [1] and Pietro Parma [2,*]

[1] Institute for Animal Production System in Mediterranean Environment, National Research Council, 80055 Portici, Italy
[2] Department of Agricultural and Environmental Sciences, University of Milan, 20133 Milan, Italy
* Correspondence: pietro.parma@unimi.it; Tel.: +39-0250316454

Simple Summary: Molecular cytogenetics, and particularly the use of fluorescence in situ hybridization (FISH), has allowed deeper investigation of the chromosomes of domestic animals in order to: (a) create physical maps of specific DNA sequences on chromosome regions; (b) use specific chromosome markers to confirm the identification of chromosomes or chromosome regions involved in chromosome abnormalities, especially when poor banding patterns are produced; (c) better anchor radiation hybrid and genetic maps to specific chromosome regions; (d) better compare related and unrelated species by comparative FISH mapping and/or Zoo-FISH techniques; (e) study meiotic segregation, especially by sperm-FISH, in some chromosome abnormalities; (f) better show conserved or lost DNA sequences in chromosome abnormalities; (g) use informatic and genomic reconstructions, in addition to CGH arrays in related species, to predict conserved or lost chromosome regions; and (h) study some chromosome abnormalities and genomic stability using PCR applications. This review summarizes the most important applications of molecular cytogenetics in domestic bovids, with an emphasis on FISH mapping applications.

Abstract: The discovery of the Robertsonian translocation (rob) involving cattle chromosomes 1 and 29 and the demonstration of its deleterious effects on fertility focused the interest of many scientific groups on using chromosome banding techniques to reveal chromosome abnormalities and verify their effects on fertility in domestic animals. At the same time, comparative banding studies among various species of domestic or wild animals were found useful for delineating chromosome evolution among species. The advent of molecular cytogenetics, particularly the use of fluorescence in situ hybridization (FISH), has allowed a deeper investigation of the chromosomes of domestic animals through: (a) the physical mapping of specific DNA sequences on chromosome regions; (b) the use of specific chromosome markers for the identification of the chromosomes or chromosome regions involved in chromosome abnormalities, especially when poor banding patterns are produced; (c) better anchoring of radiation hybrid and genetic maps to specific chromosome regions; (d) better comparisons of related and unrelated species by comparative FISH mapping and/or Zoo-FISH techniques; (e) the study of meiotic segregation, especially by sperm-FISH, in some chromosome abnormalities; (f) better demonstration of conserved or lost DNA sequences in chromosome abnormalities; (g) the use of informatic and genomic reconstructions, in addition to CGH arrays, to predict conserved or lost chromosome regions in related species; and (h) the study of some chromosome abnormalities and genomic stability using PCR applications. This review summarizes the most important applications of molecular cytogenetics in domestic bovids, with an emphasis on FISH mapping applications.

Keywords: animal cytogenetics; cattle; river buffalo; sheep; goat; FISH mapping; PCR

Citation: Iannuzzi, A.; Iannuzzi, L.; Parma, P. Molecular Cytogenetics in Domestic Bovids: A Review. *Animals* 2023, 13, 944. https://doi.org/10.3390/ani13050944

Academic Editor: Emilia Bagnicka

Received: 31 January 2023
Revised: 28 February 2023
Accepted: 2 March 2023
Published: 6 March 2023

Copyright: © 2023 by the authors. Licensee MDPI, Basel, Switzerland. This article is an open access article distributed under the terms and conditions of the Creative Commons Attribution (CC BY) license (https://creativecommons.org/licenses/by/4.0/).

1. Introduction

The application of cytogenetics to domestic animals emerged about 60 years ago with the study of normal stained chromosome preparations from some cases of domestic animals with reproductive defects [1–3]. However, the discovery of the Robertsonian translocation (rob) involving cattle chromosomes 1 and 29 [4,5] and the demonstration of its deleterious effects on fertility [6,7] was what piqued the interest of many scientific groups and focused their attention on studying the chromosomes of domestic animals. This approach was particularly useful for selecting bulls to be used for artificial insemination, as it could avoid the transmission of chromosome abnormalities (i.e., rob1;29) from bull carriers to their progeny. Evolutionary studies also benefitted from advancements beyond normal chromosome staining. Among the various studies, the most important was the study of the Bovidae family by Wurster and Benirske [8], who looked at the diploid number and shape of chromosomes. They concluded that while the diploid number varies from 38 to 60 among all bovid species, the number of chromosome arms (Fundamental Number = NF) varies only between 58 and 62, with three exceptions; therefore, they hypothesized a high degree of autosome arm conservation among all bovid species. This hypothesis was later confirmed with the application of chromosome banding techniques [9], which ushered in a new era of chromosome studies in various domestic animal species, allowing (a) the establishment of standard karyotypes of the most important domestic species as a point of reference for various applications; (b) better characterization and identification of the chromosomes involved in chromosome abnormalities of domestic animals [10], particularly domestic bovids [11], pigs [12], horses [13], and dogs [14]; (c) the study of the chromosome homologies between related and unrelated species [15–17]; and (d) the study of chromosome fragility in animals exposed in vivo or in vitro to particular mutagens [18,19]. The molecular cytogenetics, particularly the introduction of fluorescence in situ hybridization (FISH), offered a deeper investigation of the chromosomes of domestic animals through: (a) the physical mapping of specific DNA sequences on chromosome regions; (b) the use of specific chromosome markers for the identification of chromosomes or chromosome regions involved in chromosome abnormalities, especially when poor banding patterns are produced; (c) better anchoring of radiation hybrid (RH) and genetic maps to specific chromosome regions; (d) better comparisons of related and unrelated species by comparative FISH mapping and/or Zoo-FISH techniques; (e) the study of meiotic segregation, especially by sperm-FISH, in some chromosome abnormalities or aneuploidies in both oocytes and embryos; (f) better demonstration of conserved or lost DNA sequences in chromosome abnormalities by CGH (comparative genomic hybridization) or SNP (single-nucleotide polymorphism) arrays; (g) the use of informatic and genomic reconstructions, in addition to CGH arrays, for the prediction of conserved or lost chromosome regions in related species; and (h) the study of chromosome abnormalities and genomic stability using PCR (polymerase chain reaction).

This review summarizes the most important applications of molecular cytogenetics in domestic bovids, with particular emphasis on FISH mapping applications.

2. The Fluorescence In Situ Hybridization (FISH) Technique

The FISH mapping technique is based on two main principles: the target and the probe. The target can be a whole chromosome (or chromosome arms) or a specific chromosome region. The probe is prepared according to the size of the target and is typically: (a) cDNA (generally applied when the target gene is a multi-copy); (b) cosmids with DNA insert sizes of 20–40 kb; (c) bacterial artificial chromosomes (BACs) with DNA insert sizes of 100–300 kb; (d) yeast artificial chromosome (YAC) clones (these are actually not used because they have a low cloning efficiency and show a high level of chimerism); (e) chromosome painting probes (obtained by cell sorter or chromosome microdissection techniques) that can visualize parts of or entire chromosomes; and (f) CGH arrays to check for genomic gains or losses. The probes are labeled directly with fluorochromes or indirectly with molecules that bind to the probe via fluorochrome-conjugated antibodies. The probe is specific for the target, based on complementary DNA base pairing, which allows the fluorescence-labeled probes to hybridize and form specific fluorescent signals on specific chromosome regions.

The advent of the fluorescence in situ hybridization (FISH) technique, initially applied to human chromosomes [20,21], noticeably expanded cytogenetics research and investigations applied to domestic animals due to the possibility of revealing specific chromosome regions, entire chromosomes, or chromosome arms according to the choice of probe. One of the great advantages of the FISH technique is that it can be applied to interphase cell nuclei, meiotic preparations (sperm and oocytes), embryos, and elongated chromatin fibers, in addition to metaphase chromosomes, thereby allowing more complete cytogenetic investigations of animal cells. The following sections describe the main uses of FISH in domestic bovids.

2.1. FISH and Chromosome Abnormalities

The first study to apply FISH for the precise identification of the chromosomes involved in a chromosome abnormality was published by Gallagher et al. [22], who discovered an X-autosome translocation (X;23) using both Q-banding and a BoLA Class I cDNA probe. The probe shows hybridization signals to the normal chromosome 23 and to the translocated autosomal material present on the X chromosome, allowing a more precise localization of MHC (major histocompatibility complex) in cattle than was achieved earlier by genetic mapping. Several subsequent studies also applied FISH to obtain better confirmation of the chromosome(s) involved in abnormalities (especially when banding was poor) and identification of the break points, especially in reciprocal translocations. Table 1 shows the main studies that applied FISH mapping, either alone or in combination with other classical cytogenetic techniques (e.g., C-banding, G-banding, R-banding, and Ag-NORs), to study the chromosome abnormalities of domestic bovids in somatic cells at the metaphase (Figure 1) or interphase nuclei of germinal cells, such as sperm and oocytes, or embryos at different cell stages.

Table 1. FISH mapping approaches applied for the detection of chromosome abnormalities in domestic bovids. The type of chromosome abnormality, the techniques used (including FISH), the main results, and authors are reported.

Species	Chromosome Abnormality	Techniques Used	Main Results	References
Cattle	t(X-BTA23) in two normal cows	QBH, FISH	Better position of MHC-locus	[22]
	Minute fragment	Bovine SAT-DNA	Visualization of fragment	[23]
	rob(4;10)	Bovine bivariate flow painting probes on R-banded karyotype	Discovery of a new rob	[24]
	iso(Yp)	GTG, FISH with repeat sequences	Visualization of iso(Yp)	[25]
	Trisomy 20	QBH, FISH	Malformed calf with cranial defects	[26]
	rob(2;28)	Q-, R-banding, telomeric probe	Monocentric translocation	[27]
	rob(1;29), rob(6;8), rob(26;29)	GBG, RBG, CBA, FISH, HAS painting probe	correct identification of two of the three robs earlier published	[28]
	Mixoploidy	Dual-color FISH with BTA6/BTA7 painting probes	72% of IVP blastocysts were mixoploid, versus 25% in vivo	[29]
	Mixoploidy/polyploidy	Dual-color FISH with BTA6 and 7 painting probes on in vitro embryo cells	Numerical chromosome aberrations were detected as early as day 2 post insemination (pi)	[30]
	rcp(1;5)(q21;qter)(q11;q33)	CBA, GBG, RBG, FISH with HSA3 and HSA12 painting probes	Bull and dam carriers, the latter with poor fertility	[31]
	invY(Yq11-q12.2)	CBG, RBA, FISH	12 young males of which one (carrier) had female traits	[32]

Table 1. Cont.

Species	Chromosome Abnormality	Techniques Used	Main Results	References
	Trisomy 28	CBG, RBA, FISH	New chrom. identification of a previous studied case of abnormal calf	[33]
	t(Xp+;23q-)	FISH with painting probe, SCA	Oligospermic bull	[34]
	rcp(Y;9)(q12.3;q21.1).	CBA, RBG, FISH	Azoospermic bull	[35]
	Polyploidy	Painting probe BTA6 and BTA7 by microdissection on in vitro embryos	Polyploidy was significantly higher in trophectoderm (TE) cells than in embryonic disc (ED) cells	[36]
	rob(1;29)	FISH with SAT-I, III, IV	Different pattern of satellite DNA families in several chromosomes, model of rob(1;29) origin	[37]
	Mosaicism 2 n = 60/2 n = 60 t (2q−;5p+)	FISH with painting probes BTA2 and BTA5	Translocation mosaicism in a bull	[38]
	XXY-Trisomy	X-Y painting probes	Testicular hypoplasia	[39]
	fragm/hypoploidy/hypoploidy-mixoploidy; hyperploidy/hyperploidy-mixoploidy	Karyotyping, FISH with X-Y painting probes in nuclear transfer embryos	Anomalies occurred in NT embryos varied according to the donor cell culture and paralleled the frequency of anomalies in donor cells	[40]
	rob(1;29)	CBG, GTG, FISH with a rob(1;29) painting probe	Presence of rob(1;29) in Gaur (Bos gaurus)	[41]
	rob(1;29)	CBA, RBA, FISH	Origin of rob(1;29) by complex chromosome rearrangements	[42]
	rob(1;29)	Sperm-FISH	Low percentage of abnormal sperm in two carriers	[43]
	rcp(9;11)(q27;q11)	RBG and FISH	De novo origin of the rcp	[44]
	Mosaicism XX/XY cells	FISH with a male-specific BC1.2 DNA sequence in interphase cell nuclei	Diagnosis of freemartin	[45]
	rcp(11;21)(q28-q12)	CBA, RBA, Ag-NORs, FISH	Normal bull but with absence of libido; reduced fertility (very low presence of spermatozoa in germinal elements)	[46]
	rob(1;29)	microdissection, DOP-PCR, cloning and sequencing, sperm-FISH	Detection of sperm-carrying rob(1;29)	[47]
	rcp(2;4)(q45;q34)	G-banding, SCA, and chromosome painting	Detection of a new rcp in bull	[48]
	Aneuploidy	Dual-color FISH with Xcen/Y painting probes in sperm	Study the aneuploidy in different breeds	[49]
	rcp(4;7)	RBG, FISH (painting probe), aCGH	Normal male and no genomic loss in the rcp	[50]
	Aneuploidy	Dual-color FISH with Xcen and BTA5 painting probes	Study of aneuploidy in oocytes of two breeds	[51]
	Aneuploidy	FISH with BTAX, BTAY, and BTA6 painting probes on sperm of several young bulls	Aneuploidy frequencies in young fertile bull spermatozoa were relatively low	[52]
	rcp(Y;21)(p11;q11)	G-banding, FISH	Normal young bull but lower testosterone level at 12 months	[53]

Table 1. Cont.

Species	Chromosome Abnormality	Techniques Used	Main Results	References
	rcp(11;25)(q11, q14~21)	CBA, RBA, FISH, NOR	der11 with two C-bands for a break at the centromere of BTA25; cow with reduced fertility	[54]
	Aberrant oocytes	Dual-color FISH of X-cent/BTA5 painting probes	Similar rate of aneuploidy in different cattle breeds	[55]
	rob(1;29)	FISH, aCGH	New results of the origin of this rob by transposition, inversion; no gene-coding regions were disrupted during the rearrangements	[56]
	Xp-del (inactive X)	CBA, RBA, FISH	del found in both dam and calf (normal cow)	[57]
	X-Y aneuploidy	Dual-color FISH with Xcen-BTAY painting probes	Testing X-Y ratio and aneuploidy	[58]
	Aneuploidy	Dual-color FISH with Xcen and five autosome painting probes	Similar rates of chromosomal aberrant secondary oocytes in two indigenous cattle breeds	[59]
	Mixoploidy	FISH with BTAX and BTA6 painting probes	First zygotic cleavage (FZC) is a marker of embryo quality by demonstrating a significantly lower incidence of aberrations in early embryos	[60]
	Aneuploidy/polyploidy	CA, SCE, MN, MI, FISH	Effect of the tebuconazole-based fungicide: monosomies and trisomies on BTA5 and 7	[61]
	rcp(5;6)(q13;q34)	RBG, FISH, aCGH	Normal young bull with balanced rcp	[62]
	rcp(13;26)(q24;q11)	CBG, GTG, painting probes BTA13 and 26, telomeric probe	De novo rcp in both dam and calf	[63]
	der(11)t(11;25)(q11;q14–21)	CBA, RBA, FISH	Abnormal female calf	[64]
	Chromosome damages	SCE, MN, FISH with BTA1, 5, 7 painting probes	No significant chromosome fragility with use of thiacloprid	[65]
	Abnormal BTA17 in a young bull	CBA, R-banding, FISH, PNA-telomeric probe, aCGH, SNP array	Centromere repositioning	[66]
	X-monosomy	Karyotyping, FISH, SNP genotype data	Sterile for abnormal internal sex adducts	[67]
	rob(3;16)	Sperm-FISH	Low rate of unbalanced gametes produced by adjacent segregation (5.87%) and interchromosomal effect (ICE) on BTA17 and BTA20	[68]
	Trisomy 20	Q-banding, FISH	Malformed fetus, cranial defects	[69]
	Trisomy 29	FISH/genomic analysis	Malformed female calf showing dwarfism with severe facial anomalies	[70]
	rob(1;29); rcp(12;23)	FISH, use of BAC clones mapping prox- and dist- regions of all cattle autosomes and X	Identification of chromosome abnormalities in all autosomes and BTAX	[71]
	tan(18;27)	CBA, RBA, FISH	Male calf with congenital hypospadias and a ventricular septal defect	[72]

Table 1. *Cont.*

Species	Chromosome Abnormality	Techniques Used	Main Results	References
River buffalo	X-monosomy	CBA, RBA, FISH	Normal body conformation and external genitalia, ovaries not detectable, sterile	[73]
	rob(1p;23)	CBA, RBA, Ag-NORS, FISH	Complex chromosome abnormality with fission on BBU1 and centric fusion of BBU1p with BBU23 in both dam and female calf; reduced fertility in the dam	[74]
	rob(1p;18)	CBA, RBA, FISH	Famous bull eliminated from reproduction for the presence of the same chrom. abnormality in part of progeny	[75]
	Chromosome abnormalities	Zoo-FISH	Sequential approach with 13 chromosome river buffalo painting probes to detect river buffalo chromosome abnormalities	[76]
	rob(1p;18)	Sperm-FISH in motile and total fraction sperm	Limited effects on the aneuploidy in gametes on the motile fraction sperm	[77]
River/Swamp buffalo	Aneuploidy	M-FISH	Study of aneuploidy in river and swamp buffalo oocytes	[78]
Sheep	Chromosome abnormality	Production of all sheep chromosome painting probes from cell sorter technique	Easy identification of chromosome abnormalities	[79]
	rob(8;11)	G-bands, painting probes 8 and 11, SAT-I and SAT-II	SAT-I proximal on both arms with SAT-II covering the centromere	[80]
	Diploid-polyploid mosaicism	Zoo-FISH with bovine painting probes X/Y and 1;29 on nuclei of in vivo and in vitro embryos	In vitro embryos showed significant higher number of abnormal embryos than in vivo ones	[81]
	del(10q22)	Use of ovine BAC clone in addition to genetic analyses	Micro-chromosomal deletion responsible for EDNRB gene lack	[82]
	rcp(4q;12q)(q13;q25)	CBA, RBA, FISH with both specific markers and PNA-telomeric probe	Characterization of a new rcp in a young sheep	[83]
	rcp(18;23)(q14;q26).	CBA, RBA, FISH with bovine painting probe	Reduced fertility	[84]
	Chromosome abnormalities in bovids	Partial river buffalo chromosome painting probes from microdissection	Detection of chromosome abnormalities in bovids	[85]

A more complete classification of all chromosome abnormalities studied by classical cytogenetic techniques alone or (in some cases) with other molecular cytogenetic techniques is provided by Iannuzzi et al. [11].

Two examples of the importance of the use of FISH for the correct identification of the chromosomes involved in chromosome abnormalities of cattle were a case of autosome trisomy and two types of Robertsonian translocations. A case of autosome trisomy 28 in an abnormal calf, revealed by both R-banding and FISH mapping with a specific molecular marker [33], was identified, and the same abnormality was reported earlier as trisomy 22 using only the banding technique [86]. Two robs earlier reported as rob (4;8) [87] and rob (25;27) [88] in cattle were later corrected as rob (6;8) and rob (26;29), respectively, using C-, G-, and R-banding and FISH mapping with specific molecular markers and the use of HSA painting probes [28].

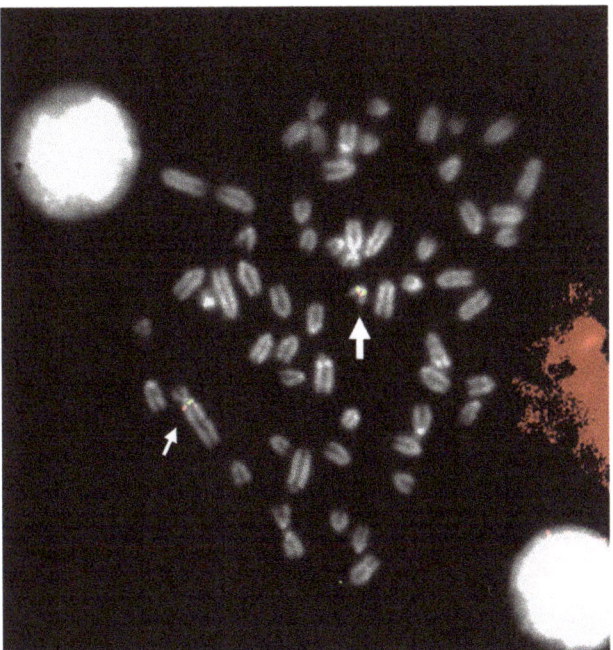

Figure 1. FISH mapping with a BAC clone mapping proximal to BTA29 (large arrow) and proximal to q-arms (BTA1) of rob (1;29) (small arrows). Indeed, a small chromosome region of 5,4 Mb translocated from proximal BTA29 to the proximal region of BTA1 (with an inversion), originating rob (1;29) [56]. Different colors indicate different BACs.

Table 1 shows that FISH mapping applications were used for the diagnosis of chromosome abnormalities in both metaphase (the majority) and interphase cells, the latter applied to lymphocyte nuclei (Figure 2), sperm (Figure 3), oocytes, and embryos.

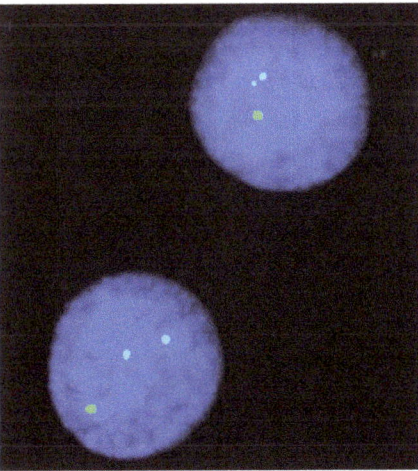

Figure 2. FISH mapping in an interphase nucleus of a female river buffalo affected by X-trisomy. Note the three hybridization signals due to the X chromosome PGK marker.

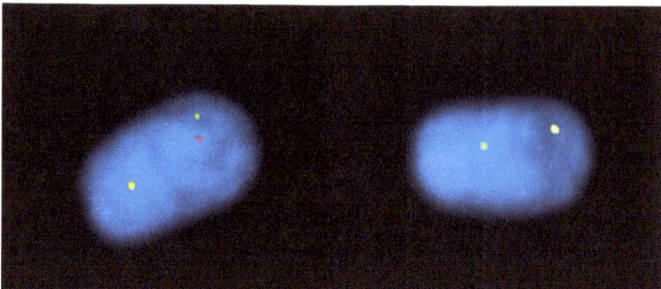

Figure 3. Sperm-FISH in a river buffalo bull carrying a rob (1p;18) using BAC probes for BBU 1p (red), BBU 1q (green), and BBU 18q (yellow) chromosomes. Normal sperm nucleus with 1/1/1 fluorescent phenotype and separate signals on left. Unbalanced sperm nucleus with 1/0/1 fluorescent phenotype on right.

Concerning the studies on meiotic preparations, those performed on the synaptonemal complexes (SCs), especially in spermatocytes, were particularly important for establishing the regularity of the pairing processes during the pachytene substage of meiotic prophase in animals carrying chromosome abnormalities (reviewed in [89]). Recent analyses of meiotic preparations have been performed using immune fluorescence approaches and have provided more detailed information on SCs [90–92]. Other studies have addressed the fragile sites in the chromosomes of domestic animals (reviewed by [93]), and limited studies have used CGH and SNP arrays to establish possible genomic losses occurring during chromosome rearrangements (Table 1).

FISH mapping was also very important for the definitive establishment of the agreement between various chromosome nomenclatures due to some discrepancies found during the Reading conference [94] and the subsequent ISCNDA1989 [95] (the inverted position between BTA4 and BTA6, as well as the correct position of BTA25, BTA27, and BTA29). This aspect was vital for the clinical cytogenetics of domestic bovids, as it allowed a correct identification of the chromosomes involved in chromosome abnormalities. During the Texas conference [96], specific molecular markers (only type I loci) were selected for each bovine syntenic group and each cattle chromosome based on previous standard chromosome nomenclatures.

The next advance was the application of FISH mapping by two labs that used 31 selected BAC clones (from the Texas Conference) on RBG- and QBH-banded cattle preparations [97]. The chromosome-banding homologies among bovids (cattle, sheep, goats, and river buffalo) were then used to establish a definitive standard chromosome nomenclature for the main domestic bovid species [98]. Subsequent studies using FISH mapping and the same Texas markers on river buffalo, sheep, and goat R-banded chromosomes [99,100] definitively confirmed the chromosome homologies among domestic bovids, as established at the ISCNDB2000 [98].

2.2. FISH in Physical Mapping

The identification of the DNA structure [101] paved the way for the development of in situ hybridization technology. In the early days of its development, this technology allowed the localization of genes using radioactive probes [102]. It was also used in studies of domestic animals [103,104], but the greatest diffusion of the physical mapping of genes awaited the development of fluorescent probes [105]. At that moment, we entered the golden years of gene mapping, and domestic animals were not excluded. One of the first examples was the localization of bovine alpha and beta interferon genes [106], and this localization was rapidly replicated in buffalos, goats, and sheep [107,108]. Subsequently, many other localizations were obtained using this technology (Figure 4).

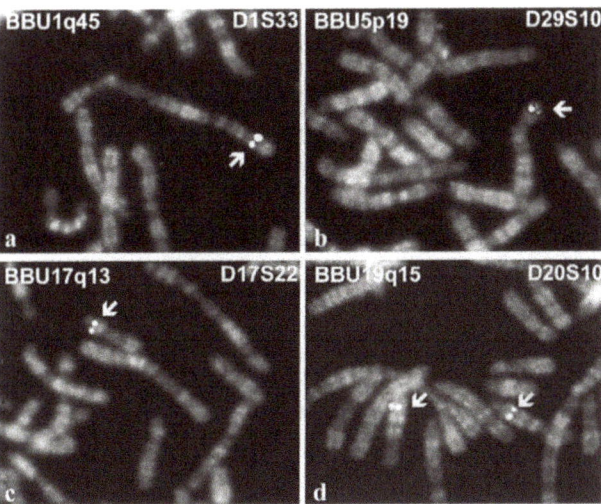

Figure 4. FISH mapping of type II loci in river buffalo R-banded chromosomes. FITC signals (arrows) of the markers and RBH banding were separately acquired by two different microscope filter combinations. Then signals were precisely superimposed to R-banded chromosomes (*Drawn from Iannuzzi et al., Cytogenet Cell Genet. 102, 65–75, 2003, DOI: 1 0.1159/000075727, S. Karger AG, Basel* [109]).

Considering the practical impossibility of compiling a complete list of all gene localizations obtained using this technology, some significant examples are listed in Table 2.

Table 2. Gene mapping obtained with FISH in domestic bovids. Type I and type II markers are expressed with polymorphic (SSRs, microsatellite, STSs) sequences, respectively.

Gene/Genes/Marker	Species	Reference
Lysozyme gene cluster	BBU	[110]
Uridine monophosphate synthase	BTA	[111]
Uridine monophosphate synthase	BBU	[112]
BTA1 to 7	BTA	[113]
Microsatellites	BTA	[114]
Microsatellites	BTA	[115]
Beta-defensin genes	BTA; OAR	[116]
Alpha-S2 casein	BTA; BBU	[117]
Fas/APO-1	BTA	[118]
Interferon gamma	OAR	[119]
Interleukin-2 receptor gamma	BTA	[120]
Beta-lactoglobulin pseudogene	BTA, OAR, CHI	[121]
Bone morphogenetic protein 1	BTA	[122]
TSPY	BTA, OAR, CHI	[123]
VIL	OAR, CHI, BBU	[124]
Type I markers	BTA	[125]

Table 2. *Cont.*

Gene/Genes/Marker	Species	Reference
Prion protein gene	BTA, OAR, CHI, BBU	[126]
IL2RA, VIM, THBD, PLC-II, CSNK2A1, TOP1	BTA	[127]
NF1, CRYB1, CHRNB1, TP53, P4HB, GH1	OAR, BBU	[128]
PAX8	BTA, OAR, CHI	[129]
Type I markers	BTA	[97]
PREF1	BTA	[130]
PRKCI	BTA	[131]
MHC	BTA	[132]
Type I markers	OAR, CHI	[100]
CACNA2D1	BTA	[133]
SLC26a2	BTA	[134]
SMN	BTA, OAR, CHI, BBU	[135]
Type I markers	BBU	[109]
Type I and II markers	OAR	[136]
PRPH	BTA	[137]
CYP11b/CYHR1	BTA	[138]
SRY, ANT3, CSF2RA	BTA	[139]
Autosomal loci (11)	BTA, OAR, CHI, BBU	[140]
Autosomal loci (88)	OAR	[141]
Autosomal loci (68)	BBU	[142]
BMPR1B, BMP15, GDF9	BTA, OAR, CHI, BBU	[143]

Localization sometimes involved a single gene [124,129] or a family of genes [132]. Other reports, however, mapped many genomic markers [100,141]. A point to remember is that FISH technology has significantly benefited from the availability of BAC genomic libraries—elements that represent the ideal source for the construction of the probes. Among these, the INRA library [144] and the CHORI-240 have played relevant roles. The publication of genomes [145–148] has since inevitably diminished interest in using this technology for mapping genetic factors, although genetic factor mapping continued for species whose genomes were sequenced later, such as the water buffalo [149]. However, this technology has proved useful in several aspects, including: a) the identification of errors in genomic assembly [150]; b) the refinement of genome assembly [151]; and c) the mapping of sequences not included in genomic assemblages [152]. Clearly, the interest today is very limited in locating a genetic factor in a species whose genomic sequence is available, but this does not mean that FISH technology is no longer indispensable for solving other problems related to the organization of genomes.

The mapping of genomic elements by FISH has also been used successfully for the physical mapping of data obtained by other technologies. The first examples concerned the physical anchoring of a genetic map to a chromosome [153–155] and the mapping of a synteny group to a specific chromosome [114]. Subsequent examples of the combined use of FISH and genetic maps followed [127,156].

2.3. Comparative FISH Mapping

Two main methods have been applied thus far to obtain a FISH mapping comparison between related and unrelated species: Zoo-FISH, which uses chromosome painting probes, and FISH mapping, which uses specific molecular markers of both type I and type II. Zoo-

FISH is a molecular technique that provides an easier comparison between related and unrelated species from a macro point of view. The term was first reported by [157], based on earlier studies that used genomic chromosome painting probes, obtained by cell sorter chromosomes, to compare related species [158–160].

Zoo-FISH was first applied in domestic animals when human chromosome painting probes became commercially available. This approach demonstrated the conservation of several human chromosome segments in both domestic bovids (Table 3) and other domestic species (reviewed in [161]).

Table 3. Comparative FISH mapping in domestic bovids with related and unrelated species.

Author/s	Results
[107]	Mapping omega and trophoblast interferon genes in cattle and river buffalo
[162]	Mapping of lactoperoxidase, retinoblastoma, and alpha-lactalbumin genes in cattle, sheep, and goats
[108]	Mapping omega and trophoblast interferon genes in sheep and goats
[163]	Mapping LGB and IGHML in cattle, sheep, and goats
[164]	Mapping CASAS2 gene to the cattle, sheep, and goat chromosome 4
[165]	Mapping MHC-complex in cattle and river buffalo
[166]	Mapping inhibin-alpha (INHA) to OAR2 and BTA2
[167]	Mapping inhibin subunit beta b to OAR2 and BTA2
[121]	Mapping beta-lactoglobulin pseudogene in sheep, goats, and cattle
[168]	Mapping ZNF164, ZNF146, GGTA1, SOX2, PRLR, and EEF2 in bovids
[117]	Mapping of the alpha-S2 casein gene on river buffalo and cattle
[116]	Mapping of beta-defensin genes to river buffalo and sheep chromosomes suggest a chromosome discrepancy in cattle standard karyotypes
[169]	Mapping STAT5A gene maps to BTA19, CHI19, and ORA11
[170]	Mapping in Y chromosomes of cattle and zebu by microdissected painting probes
[124]	Mapping of villin (VIL) gene in river buffalo, sheep, and goats
[126]	Mapping prion protein gene (PRNP) on cattle, river buffalo, sheep, and goats
[171]	Mapping BCAT2 gene to cattle, sheep, and goats
[172]	Comparative mapping in X chromosomes of bovids
[173]	Comparative mapping between BTA-X and CHI-X
[174]	Survey of chromosome rearrangements between ruminants and humans
[175]	Comparative mapping between cattle and pig chromosomes using pig painting probes
[176]	Extensive conservation of human chromosome regions in euchromatic regions of river buffalo chromosomes
[128]	Mapping of six expressed gene loci (NF1, CRYB1, CHRNB1, TP53, P4HB, and GH1) to river buffalo and sheep chromosomes
[177]	Comparison of human and sheep chromosomes using human chromosome painting probes
[178]	Mapping four HSA2 type I loci in river buffalo chromosomes 2q and 12
[179]	Mapping BCAT1 in cattle, sheep, and goats
[180]	Comparative mapping in bovid X chromosomes reveals homologies and divergences between the subfamilies *Bovinae* and *Caprinae*
[181]	Mapping 16 type I loci in river buffalo and sheep
[182]	Mapping 13 type I loci from HSA4q, HSA6p, HSA7q, and HSA12q on in river buffalo
[183]	Mapping forty autosomal type I loci in river buffalo and sheep chromosomes and assignment from sixteen human chromosomes
[184]	Mapping eight genes from HSA11 to bovine chromosomes 15 and 29

Table 3. *Cont.*

Author/s	Results
[98]	International chromosome nomenclature in domestic bovids based on Q-, G-, and R-banding and FISH with 31 specific Texas marker chromosomes
[185]	Mapping 28 loci in river buffalo and sheep chromosomes
[186]	Sheep/human comparative map in a chromosome region involved in scrapie incubation time shows multiple breakpoints between human chromosomes 14 and 15 and sheep chromosomes 7 and 18
[135]	Physical map of the survival of motor neuron gene (SMN) in domestic bovids
[100]	Assignment of the 31 type I Texas bovine markers in sheep and goat chromosomes by comparative FISH mapping and R-banding
[187]	Mapping 195 genes in cattle and updated comparative map with humans, mice, rats, and pigs
[188]	Mapping of F9, HPRT, and XIST in BTAX and HSAX clarifies breakpoints between the two species
[189]	15 gene loci were mapped in the telomeric region of BTA18q and HSA19q
[190]	Comparative G- and Q-banding of saola and cattle chromosomes as well as FISH mapping of 32 type I Texas markers
[191]	Mapping of fragile histidine triad (FHIT) gene in bovids
[192]	Chromosome evolution and improved cytogenetic maps of the Y chromosome in cattle, zebu, river buffalo, sheep, and goats
[193]	Physical map of mucin 1, transmembrane (MUC1) among cattle, river buffalo, sheep, and goat chromosomes and comparison with HSA1
[194]	Mapping of LEP and SLC26A2 in *bovidae* chrom. 4 (BTA4/OAR4/CHI4) and HSA7
[140]	Mapping 11 genes to BTA2, BBU2q, OAR2q, and CHI2, and comparison with HSA2q
[195]	Mapping among humans, cattle, and mice suggests a role for repeat sequences in mammalian genome evolution
[196]	Mapping sheep and goat BAC clones identifies the transcriptional orientation of T cell receptor gamma genes on chromosome 4 in bovids
[197]	Mapping of twelve loci in river buffalo and sheep chromosomes: comparison with HSA8p and HSA4q
[198]	Mapping 25 new loci in BTA27 and comparison with both human and mouse chromosomes
[141]	An advanced sheep cytogenetic map and assignment of 88 new autosomal loci
[199]	Cross-species FISH with cattle whole-chromosome paints and satellite DNA I probes was used to identify the chromosomes involved in the translocations of some tribe *Bovinae* species
[142]	Extended river buffalo cytogenetic map, assignment of 68 autosomal loci and comparison with human chromosomes
[200]	FISH with 28S and telomeric probes in 17 bovid species. NORs are an important and frequently overlooked source of additional phylogenetic information within the *Bovidae*
[201]	Mapping DMRT1 genes to BTA8 and HSA9
[202]	Comparative DM domain genes between cattle and pigs
[203]	Assignments of new loci to BBU7 and OAR6 and comparison with HSA4
[204]	Mapping 22 ovine BAC clones in sheep, cattle, and human X chromosome
[205]	Mapping and genomic annotation of bovine oncosuppressor gene in domestic bovids
[206]	Cytogenetic map in sheep as anchor of genomic maps also using different genomic resources from other species
[207]	Molecular cytogenetics in goats and comparative mapping with human maps
[208]	Mapping of 6 loci containing genes involved in the dioxin metabolism of domestic bovids
[209]	Extended cytogenetic maps of sheep chromosome 1 and their cattle and river buffalo homologues: comparison with the OAR1 RH-map and HSA2, 3, 21, and 1q
[210]	Mapping between BTA5 and some *Antilopinae* species using Sat-I and SAT-II sequence and BTA-painting probes
[211]	Comparison of centromeric repeats between cattle and other *Bovidae* species

Table 3. *Cont.*

Author/s	Results
[212]	Advanced comparative map in X chromosome of *Bovidae*
[143]	Physical map of BMPR1B, BMP15, and GDF9 fecundity genes on cattle, river buffalo, sheep, and goat chromosomes
[152]	Physical mapping of 20 unmapped fragments in Btau 4.0 Genome Assembly in cattle, sheep, and river buffalo
[213]	Physical map of LCA5L gene in cattle, sheep, and goats
[214]	New cryptic difference between cattle and goat karyotypes
[215]	Small evolutionary rearrangement between BTA21 and homologous OAR18
[216]	Assignment of 23 endogenous retrovirus to both sheep and homologous chromosomes regions of river buffalo

The use of human-chromosome painting probes allowed the identification of a substantial number of human chromosome segments (around 50) in bovid chromosomes [175,176,217–219]. Zoo-FISH has also been applied to correctly identify some chromosomes involved in the chromosome abnormalities shown in Table 1. The availability of specific painting probes obtained by both cell sorting and/or by the microdissection of specific chromosomes (or chromosome arms) from domestic animals extended these studies to investigations between related species (Table 3). For example, in cattle, Zoo-FISH was applied to study X-Y aneuploidy in sperm [55] and in oocytes [58] (Table 1). An interesting approach was demonstrated in two studies characterizing two cases of goat/sheep [220] and donkey/zebra [221] hybrids using multicolor FISH (M-FISH), starting from painting probes obtained from microdissected river buffalo chromosomes (or chromosome arms) and from flow-sorted donkey chromosomes, respectively.

Chromosome painting probes allow the delineation of large, conserved chromosome regions between related and unrelated species, as reported above. The use of comparative FISH mapping using several chromosome markers to map a single type I or type II locus along the chromosomes allows a more accurate establishment of the gene order within chromosome regions, thereby confirming that chromosome rearrangements occurred to differentiate related or unrelated species in key evolutionary studies (Table 3). These detailed comparisons have confirmed a high degree of autosome (or chromosome arm) conservation among all bovid species. The main autosome difference found thus far in bovids was a chromosome translocation of a proximal chromosome region from *Bovinae* chromosome 9 to *Caprinae* chromosome 14, as demonstrated by both chromosome banding and, in particular, by a molecular marker (COL9A1) mapping proximal to *Bovinae* chromosome 9 and proximal to *Caprinae* chromosome 14 (reviewed in [9]). This translocation involved a genome region of about 13 MB and was followed by an inversion in *Caprinae* chromosome 14, as demonstrated earlier [213]. This chromosome event was common to all remaining *Bovidae* subfamilies, leading to the conclusion that the *Bovinae* subfamily is an ancestor to the remaining *Bovidae* subfamilies (reviewed in [9]).

In contrast to autosomes, sex chromosomes are differentiated by more complex chromosome rearrangements. Indeed, the *Caprinae* X chromosome (as for all remaining X chromosomes of the other *Bovidae* subfamilies) is differentiated from the ancestor *Bovinae* X (very probably a large acrocentric chromosome, such as that of the water buffalo) by at least three chromosome transpositions and one inversion (reviewed in [9]). Detailed FISH mapping data are also useful for better anchoring of both genetic and RH maps [203,222–224]. The availability of detailed cytogenetic maps in bovid species allowed a better comparison of the bovid and human chromosomes, especially using type I loci. These comparisons facilitated the translation of genomic information from the human genome to the genomes of domestic animals, especially in those with no genome sequencing available. These comparisons also revealed a very high number of chromosome rearrangements that differentiate bovid species from humans. Indeed, the conservation of entire chromosomes or large regions of them between bovid and human chromosomes, as revealed by Zoo-FISH, was the result of complex chromosome rearrangements that differentiated human and bovid species according to their gene order. An example is presented in Figure 5 which illustrates the comparison of FISH mapping between HSA2q and BTA2. As seen, when utilizing the Zoo-FISH technique with the HSA2q painting probe, almost all BTA2 is painted [217], indicating a high degree of chromosome conservation between the chromosomes of the two species. By conducting the same comparison using comparative FISH mapping and examining the gene order along the chromosomes of the two species, we observe a distinct gene order between the two species, thus revealing complex chromosome rearrangements that differentiated the chromosomes of the two species during their evolution.

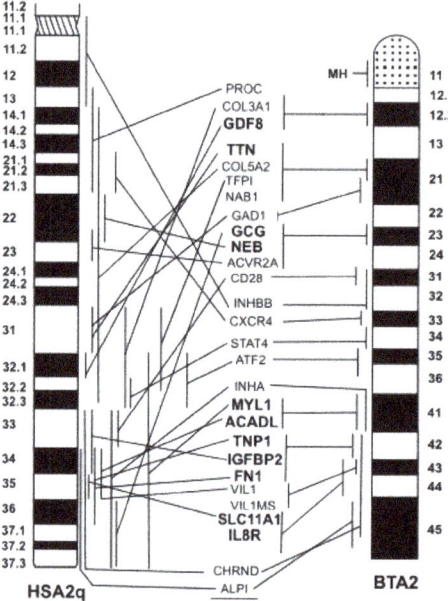

Figure 5. Comparative FISH mapping between HSA2q and BTA2. Note the different gene order between the two chromosomes due to complex chromosome rearrangements occurred during the chromosome evolution of the two species (*Drawn from Di Meo et al., Animal Genetics 37, 299–300, 2006, Wiley Online Library* [140]).

2.4. Fiber-FISH

The various FISH mapping techniques developed for human cytogenetics (reviewed by [225]) include SKY-FISH (spectral karyotyping FISH), Q-FISH (quantitative FISH), M-FISH (multicolor FISH), heterochromatin-M-FISH, COBRA-FISH (combined binary ratio labeling FISH), cenM-FISH (centromere-specific M-FISH), and fiber-FISH. Among these techniques, only fiber-FISH and M-FISH have been applied to domestic bovids. The use of fiber-FISH yields high-resolution maps of chromosomal regions and related genes on a single DNA fiber. This approach establishes the physical location of DNA probes with a resolution of 1000 bp. It is particularly useful for detecting gene duplications, gaps, and variations in the nuclear genome. The DNA fibers are obtained from nucleated cells by releasing the DNA fibers from the nucleus, stretching them mechanically, and then fixing them on slides [226] (Figure 6). Table 4 summarizes the studies that have used this technique in domestic bovids.

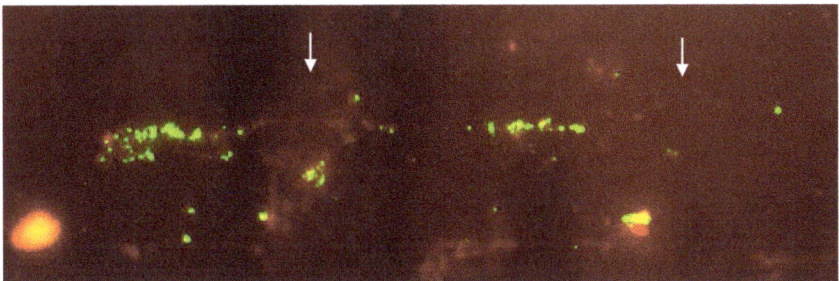

Figure 6. Details of the fiber-FISH performed on a lymphocyte nucleus of cattle affected by arthrogryposis using a BAC clone containing the survival of motor neuron gene (SMN). The presence of two groups of linear hybridization signals (arrows) supports the hypothesis that SMN was at least duplicated [135].

Table 4. Studies using the fiber-FISH on domestic bovids.

Species	Author/s	Results
Cattle	[227]	Genomic organization of the bovine aromatase
	[228]	Molecular characterization of STAT5A- and STAT5B-encoding genes
	[135]	Demonstration of survival of motor neuron gene (SMN) duplication in a calf affected by arthrogryposis
	[229]	Demonstration of multiple TSPY copies on the Y chromosome
Sheep	[230]	DNA fiber barcodes indicated a chromosomal deletion

2.5. CGH Arrays

The CGH array technology, an evolution of in situ comparative genomic hybridization (CGH), is a method of cytogenetic investigation that emerged in the 1990s to overcome the limitations of common banding cytogenetic analyses, especially those involving the presence of genomic imbalances, such as duplications or deletions [231,232]. In situ CGH technology has many similarities to FISH: the support used is the same, i.e., denatured metaphases fixed on slides and the approaches to label the probes are identical. However, in this case, the probes are produced using complete genomic DNA deriving from two subjects: typically, one healthy and one relating to the subject being investigated. The two DNAs are labeled with two different fluorochromes and then hybridized simultaneously on the slide. In the hybridization phase, a competition is therefore created between the probes, and in the presence of a normal chromosomal segment, an intermediate color is obtained, while in the presence of chromosomal alterations, a fluorescence closer to one of the two colors used is obtained. Although this technology has been widely used and has provided important results, its major limitation lies in the resolution. CGH array technology follows the same principle, but the support is no longer represented by slides but by synthetic DNA fixed on slides. Initially, the chips for CGH array analyses contained DNA extracted from BAC to provide as uniform a representation of the genome as possible [233]. Current CGH array analyses are performed using devices containing oligonucleotides chosen that uniformly cover the whole genome and achieve resolutions of 5–10 kb [234,235]. More information about this technology and its use is provided by [236]. In species of zootechnical interest, CGH array analyses (Figure 7) became common following the appearance of the first commercial arrays, and these analyses are conducted essentially for two purposes: the identification of copy number variation (CNV) polymorphisms and the characterization of chromosome anomalies. CNVs are polymorphic variations present very frequently in the genomes of higher organisms [237–239]. In humans, approximately 4.8–9.7% of the genome contains CNVs [240]. The introduction of commercial arrays has allowed the use of this technology to obtain a great amount of information about the distribution of CNVs in species differences and how these variations are related to phenotypic traits. The transfer of this technology to the animal field and the availability of commercial arrays has led to the publication of several reports (Table 5).

Table 5. Identification of CNV.

Specie	Reference	Note
Cattle	[241]	3 Holstein bulls
Cattle	[242]	90 animals: 11 Bos taurus breeds, 3 Bos indicus breeds, and 3 composite breeds for beef, dairy, or dual purpose
Cattle	[243]	20 animals: 14 Holsteins, 3 Simmental 2 Red Danish and 1 Hereford
Cattle	[244]	47 Holstein bulls
Cattle	[245]	24 animals from Chianese breeds
Cattle	[246]	3 Angus, 6 Brahman, and 1 composite animal
Sheep	[247]	36 animals
Sheep	[248]	12 animals
Goat	[249]	10 animals

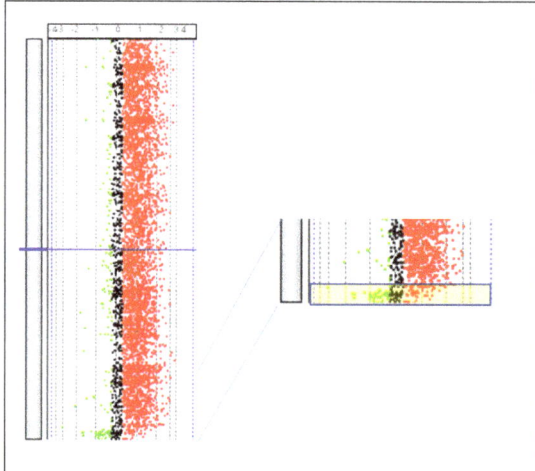

Figure 7. Identification of the PAR region present on BTAX and BTAY. The PAR region (yellow box) is identified by comparing DNA obtained from a male subject and that obtained from a female subject using a SurePrint G3 Bovine CGH Microarray 180 k (Agilent Technologies, Santa Clara, CA, USA). Parma P. Personal communication.

3. Combined Informatic and Genomic Information

The publication of animal genomes [145–149,250] has made available a very large series of data that required the development of sophisticated analysis techniques and often required the use of computers with large processing capacities. The first bio-informatic analyses were used to assemble thousands of short genomic sequences, produced by modern high-throughput sequencing technologies, into genomes. Today, most of these programs are available free of charge through web pages that function as interfaces between the user and calculation tools [251]. Currently, dozens of bio-informatics programs are available to analyze the data contained in genomic assemblies, and many of these are accessible through various web platforms. Making a complete list is very complicated, in part because this is a rapidly evolving discipline that introduces, almost daily, new analytical tools.

3.1. Visualization of Genomes

The genomic sequences produced by the various assemblies can be visualized using one of the available websites available, including Genome Data viewer [252], UCSC Genome Browser [253], and Ensembl [254]. Currently, these websites provide the ability to view and process data relating to several genome assemblies (Table 6).

Table 6. Independent genomic assemblies that can be analyzed through the main genomic visualization sites.

Specie [1]	Genome Assembly [2]	Origin	GDW [3]	UCSC [4]	ENS [5]
BTA	ARS-UCD1.3	USDA ARS	yes	no	no
	ARS-UCD1.2	USDA ARS	no	yes	no
	Btau_5.0.1	Cattle Gen. Seq. Int. Consortium	yes	no	no
	Btau_4.6.1	Cattle Gen. Seq. Int. Consortium	no	yes	no
	Btau_4.0	Cattle Gen. Seq. Int. Consortium	no	no	yes
	UMD_3.1.1	University of Maryland	yes	yes	no
	UMD_3.1	University of Maryland	no	no	yes
	Baylor 4.0	Baylor College of Medicine	no	yes	no
OAR	ARS-UI_Ramb_v2.0	University of Idaho	yes	no	no
	Oar_rambouillet_v1.0	Baylor College of Medicine	yes	no	yes
	Oar_v4.0	Int. Sheep Gen. Consortium	yes	yes	no
	CAU_O.aries_1.0	China Agricultural University	yes	no	no
CHI	ARS1.2	USDA ARS	yes	no	no
	ARS1	USDA ARS	no	no	yes
	CHIR_1.0	Int. Goat Gen. Consortium	yes	no	no
BBU	NDDB_SH_1	Nat. Dairy Dev. Board, India	yes	no	no
	UOA_WB_1	University of Adelaide	yes	no	no
BIN	Bos_indicus_1.0	Genoa Biotecnologia SA	yes	no	no

[1] BTA = cattle; OAR = sheep; CHI = goat; BBU = water buffalo and BIN = Zebu. [2] Only genomic assemblages at the chromosomal level were considered and not those limited to scaffolds. [3] Genome data viewer. [4] USCS genome browser. [5] Ensembl genome browser.

These genome viewers are constantly evolving and contain several tools within them that allow the user to obtain highly relevant genetic data and information. This includes, but is not limited to, the possibility of: (a) identifying the structure of genetic factors (in terms of exon–intron boundaries); (b) identifying SNP polymorphisms in a particular region of the genome; (c) identifying the position of BACs by mapping the BES (Bac Ends Sequences, particularly useful when the user wants to choose the BACs to use in FISH analysis); (d) observing the genomic regions expressed in particular types of tissues; (e) analyzing the relationships between different assemblies of the same species; (f) visualizing the relationships between similar regions in different species (comparative genomics); and (g) viewing the repeating regions. In this review, we do not specify a best genome viewer, as this will often depend on personal needs and experience. However, as each genome viewer has its own specific analysis tools, sometimes the best solution is to use all three to obtain more complete information.

3.2. Use of Genomic Assemblies

The availability of genomic assemblages has, on the one hand, limited the interest in the physical mapping of genomic elements, but has, on the other hand, allowed the evolution of a very large number of genetic and genomic analyses. Probably one of the most common uses (even if not directly related to cytogenetics) is to design primers for use in PCR amplifications. This operation can be performed using different software, both available for free and for a fee. Among those available free of charge, the most frequently used is Primer3 [255]. The availability of genomic assemblages also makes rapid evolutionary investigation possible (i.e., visualizing, in a simple and rapid way, the similarities that exist between the various genomic regions of different species). The publication of genomes has certainly had a great impact on cytogenetics (both negatively and positively). If the golden era of gene mapping has ended, the possibility of rapidly identifying BACs for use as probes in FISH experiments has certainly provided great benefits to cytogenetics, as it avoids long and tedious testing of BAC libraries. This aspect has allowed the rapid characterization of some chromosomal anomalies, such as a centromere repositioning event in cattle [66], a reciprocal translocation, also in cattle [62], and cryptic evolutionary rearrangements between cattle and sheep [213]. Finally, the rapid localization of BACs on genomes has allowed the development of complex approaches for the identification of chromosomal abnormalities, which are also difficult to identify [71]. Obviously, these are not all the possible uses of genomic assemblies, but they represent the best examples in relation to cytogenetics. Each genomic assembly contains substantial information that can be used for very specific purposes and avoids the need for probes that would be complex to synthesize. The continuous evolution of these data analysis tools creates difficulty in any attempt to compile their possible uses.

3.3. Tools for Genomic Data Analyses

Simultaneously with the publication of the genomes, bio-informatics tools were developed for the analysis of the vast amount of data generated—data that are characterized by both their great variety and their large quantity. One of the main repositories of tools for analyzing genomic data is Galaxy [251]. This repository provides access to bio-informatic analysis tools, which are constantly updated. SNP variations represent the major source of variation in genomes, and the genomes of the species covered in this review are no exception. Currently, identifying these sources of variation is quite simple (through modern high-throughput sequencing techniques at ever-lower cost), but this does not characterize the effect that these variations can cause. For this scenario, the variant effect predictor (VEP available on the Ensembl website) software is helpful [256].

Without a doubt, BACs represent one of the most useful tools for molecular cytogenetics, and, as previously mentioned, their identification in genomes is currently greatly facilitated. However, the current situation would not be possible without the existence of two important institutions that have dedicated part of their activities to the construction, maintenance, and distribution of BAC libraries: the BACPAC Resources Center (BPRC, https://bacpacresources.org/ (accessed on 2 March 2023)) and INRA (http://abridge.inra.fr/index.php?option=com_flexicontent&view=item&cid=17&id=61&Itemid=202&lang=fr (accessed on 2 March 2023)). Through these two institutes, BACs belonging to different libraries can be obtained.

3.4. Whole-Genome Sequencing

In recent years, the decreasing costs of sequencing have made it possible to analyze many subjects. The purposes of these sequencings are different; in many cases, the aim is the identification of signatures of selection [257–259], but other purposes are represented, such as: (a) the identification of genetic variants in specific genes [260]; (b) the verification of data obtained regarding the identification of SNPs with chip arrays [261]; (c) the identification of the run of homozygosity in breeds intended for different productions [262]; (d) prediction and QTL mapping [263]; and (e) the identification of copy number variants [264] and transcriptome characterization [265]. Similar analyses were performed on sheep [266,267] and goats [268,269]. Additionally, in this case, the water buffalo seems to be slightly behind, as there are very few papers available on it [265].

4. PCR-Based Methods and Molecular Cytogenetics

The polymerase chain reaction (PCR) [270] is a method largely used to make millions of copies of a specific DNA sample in a fast and economical way for the detection, quantification, and typing of infectious diseases and genetic changes. Current PCR-based methods are distinguished as: (a) first-generation PCR, (b) second-generation quantitative PCR (qPCR), and (c) third-generation droplet-based digital PCR (dPCR). PCR detects endpoint, qualitative, or semi-quantitative assays by gel electrophoresis, separating DNA fragments according to size. The qPCR measures DNA/RNA in real time using PCR methods, fluorescent dyes, and fluorometry for relative quantification and quantitative assays with standard curves. The dPCR splits a PCR sample labeled with fluorescent dye into millions of microsamples to digitize the pool of DNA molecules with a single or no copy in each droplet. It quantifies the DNA/RNA copy number faster than qPCR based on standard curves [271].

In recent years, PCR-based methods have replaced the classic cytogenetic techniques for detecting chromosome abnormalities and aneuploidy due to greater precision, lower cost, and faster data than are possible with cytogenetic methods, because of the small quantities of DNA (30 ng) required from any stored or fresh biological samples. PCR-based approaches are most commonly used in bovid studies to examine sex chromosomes in early-sex-determination assays to detect aberrations (Table 7).

Table 7. PCR-based approaches on bovids for the detection of chromosomal aberrations.

Species	Objective	Sample	PCR-Based Method	Reference
Cattle	Sex-determination	Embryos	PCR	[272]
Cattle	Freemartinism diagnosis	Blood	PCR	[273]
Cattle	Sex-determination	Embryos	PCR	[274]
Cattle	Sex-determination	Spermatozoa	PCR	[275]
Cattle	Chimerism diagnosis	Blood	qPCR	[276]
Cattle	XX/XY chimerism diagnosis	Blood	PCR	[277]
Cattle	SRY-positive hermaphrodite diagnosis	Blood	PCR	[278]
Cattle	XY (SRY-positive) diagnosis	Blood	PCR	[279]
Cattle	Freemartinism diagnosis	Blood	qPCR	[280]
Cattle	Freemartinism diagnosis	Blood	dPCR	[281]
Cattle	Sex-determination	Spermatozoa	dPCR	[282]
Cattle	Mosaic karyotype (60,XX/60,XX,+mar) diagnosis	Skin tissue	PCR	[283]
Cattle	Mosaicism (60,XX/90,XXY) diagnosis	Blood, skin, buccal epithelial cells, and hair follicles	dPCR	[284]
Cattle	XX/XY chimerism diagnosis	Blood and hair follicles	dPCR	[285]

Telomere assessment is another critical goal of cytogenetics research due to the central roles of telomeres in chromosome stability, aging, cancer development, apoptosis, and senescence. The telomeres consist of thousands of noncoding repetitive sequences of DNA composed of six nucleotide motifs (TTAGGG)n localized at the ends of chromosomes and are responsible for maintaining DNA integrity during each cell division. They are associated with several proteins, with the most abundant being the shelterin complex, which is made up of six different polypeptides. Telomeres also contain other genomic structures, such as T-loops, D-loops, G-quadruplexes (G4), R-loops, and long noncoding RNA (TERRA) [286].

In farm animals, telomere length (TL) did not receive much interest initially due to the difficulty in determining the natural limits of their lifespans. However, a recent study related TL to health, genome stability, and aging in cattle aged between 2 and 13 years and transformed TL into a sensitive biomarker for longevity and wellness (critical traits of selective breeding), responding to the "One Health" approach (improving animal welfare) [287]. TL is not often used as a unique marker of aging in humans because of its poor predictive accuracy due to increased telomere shortening in elderly humans as a consequence of age-related diseases (e.g., cancer, atherosclerosis, autoimmune disorders, obesity, chronic obstructive pulmonary disease, diabetes, hematological disorders, and neurodegenerative diseases) [288]. By contrast, TL proved to be a relevant biomarker of the general state of farm animals due to their lack of age-related pathologies [289,290].

Approaches for measuring TL include: (a) telomere restriction fragment (TRF) length [291]; (b) length analysis by Southern blotting; (c) fluorescent in situ hybridization (FISH) by flow cytometry (flow-FISH) or in metaphase cells (Q-FISH) [292,293]; and (d) PCR-based methods. Most of these methods have several limitations. For example, TRF and flow-FISH are labor-intensive and expensive; Southern blot analysis requires large amounts of genomic DNA, and Q-FISH works only on chromosomes (metaphase stage). Of the available methods, the PCR-based ones are the fastest, most recent, and least costly and require only small quantities of DNA (30 ng) from stored or fresh biological samples [294]. The qPCR method amplifies telomere repeats relative to a single-copy gene (reference gene) according to a method described by Cawthon et al. [295] and follows the MIQE guidelines [296]. One limitation of qPCR is the inconsistent repeatability and reproducibility of different TL measurement methods, producing a high variation in results [297]. Several studies on humans and animals indicated that the DNA extraction method might affect TL measurements using q-PCR, as DNA yields were higher using the non-silica membrane kit (salting-out method), and DNA integrity on electrophoresis gels varied [298,299]. A recent study showed comparable results for DNA quality and purity (tested using a NanoDrop instrument and electrophoresis gels) in cattle blood and milk samples using two different extraction kits (a salting-out kit for blood and a silica membrane kit for milk samples) due to the difficulty of extracting DNA from milk matrices. The DNA quality results were similar in both matrices, demonstrating a synchronous trend between them for the first time [287].

5. Current Developments and Knowledge Gaps

Molecular cytogenetics is approaching its first 30 years of history and during this period, it performed important functions that evolved over time. It therefore seems normal that in the coming years, we will witness further developments; however, some approaches will always be current and irreplaceable. The FISH technology represents, and will represent, the main methodology for the verification of chromosomal anomalies eventually identified with other approaches, just as the CGH array technology that will be increasingly used for the identification of genomic variants linked to a particular phenotype. Molecular cytogenetics could be very useful for the study of those species which have not yet benefited from the genomic revolution, or which are still in its early stages: in this sense, the water buffalo (Bubalus bubalis) is the main example. Despite possessing a great economic importance, its genome has been decrypted and made available only recently, and the application of other technologies is very late. A further gap that can be filled is the development of a technological approach that can allow the identification of all chromosomal types identifiable by cytogenetic analyses. A similar approach has already been published [71], but only the transfer of SKY-FISH technologies [300] from humans to bovids will bridge this gap. Finally, the certain decrease in costs will mean that even the species considered in this review will be able to benefit from long-read genomic sequencing, such as PacBio [301] and Oxford Nanopore [302].

6. Conclusions

The study of the chromosomes of domestic bovids is about to enter its seventh decade, and, as expected, it has undergone a notable evolution along the way. This evolutionary process for this discipline is mainly a result of the appearance of technologies that have significantly increased the potential of applied cytogenetics. Banding techniques, FISH, CGH arrays, and PCR have radically changed animal cytogenetics, making them irreplaceable tools for understanding the genetics of bred animals. Therefore, considering the history of cytogenetics, a quite easy prediction is that even the next evolutions will be dictated by technological advances. Predicting the next technological leap is difficult, but if we were to make a prediction, it would be that long-read genomic sequencing technologies will have important impacts on cytogenetics. Cytogenetics will likely retain its functionality, particularly in the confirmation of genomic results and the characterization of cytogenetic anomalies, as well as in evolutionary studies. This is because the most significant genetic mutations have accumulated at the chromosome level during the evolution of species. Finally, the implication and progresses from animal cytogenetics can be summarized as follows:

- In the pre-genomic era, FISH technology represented the almost exclusive technology available for the localization of genes in genomes.
- Prior to the availability of low-cost genomic sequencing, molecular cytogenetics was the only approach for identifying similarities between karyotypes of different species.
- The technologies of molecular cytogenetics represent the best approach for the characterization of chromosomal abnormalities.
- Despite scientific progress in similar disciplines, molecular cytogenetics will always find its place and represent an inescapable investigation methodology.

Author Contributions: Conceptualization, P.P, L.I. and A.I.; writing—original draft preparation, P.P, L.I. and A.I.; writing—review and editing, P.P. and L.I. All authors have read and agreed to the published version of the manuscript.

Funding: The study was supported by the project PON1_486 GENOBU.

Institutional Review Board Statement: An Institutional Review Board statement was not required.

Data Availability Statement: Data sharing is not applicable to this article as no new data were created or analyzed in this study.

Conflicts of Interest: The authors declare no conflict of interest.

Abbreviations

aCGH	array Comparative Genomic Hybridization
BAC	Bacterial Artificial Chromosome
BBU	*Bubalus bubalis* Chromosome, 2 n = 50
BES	Bac Ends Sequences
BIN	*Bos indicus* Chromosomes, 2 n = 60
BTA	*Bos taurus* Chromosome, 2 n = 60
CA	Chromosome Abnormalities (chromosome breaks)
CBA	C-banding by Acrine Orange Staining
CHI	*Capra hircus* Chromosomes, 2 n = 60
Fiber-FISH	Extended Chromatin Fiber-FISH
FISH	Fluorescence In Situ Hybridization
GBG	G-banding by Early BrdU-Incorporation and Giemsa Staining

HSA	*Human sapiens* Chromosome, 2 n = 46	
IVP	In Vitro Production	
MHC	Major Histocompatibility Complex	
MI	Mitotic Index	
MN	Micronuclei	
OAR	*Ovis aries* Chromosomes, 2 n = 54	
PCR	Polymerase Chain Reaction	
PNA	Peptide Nucleic Acids	
QBH	Q-banding by Early BrdU-Incorporation and Hoescht Staining	
RBA	R-banding by Late BrdU-Incorporation and Acridine Orange Staining	
RBG	R-banding by Late BrdU-Incorporation and Giemsa Staining	
RH	Radiation Hybrids	
SCA	Synaptonemal Complex Analysis	
SCE	Sister Chromatid Exchange	
SKY-FISH	Spectral Karyotyping	
SNP	Single Nucleotide Polymorphism	
YAC	Yeast Artificial Chromosome	

References

1. Kanagawa, H.; Kawata, K.; Ishikawa, T.; Odajima, T.; Inoue, T. Chromosome studies on heterosexual twins in cattle. 3. Sex chromosome chimerism (XX/XY) in bone marrow specimens. *Jpn. J. Vet. Res.* **1966**, *14*, 123–126.
2. Mcfeely, R.; Hare, W.; Biggers, J.; Diggers, J. Chromosome Studies in 14 Cases of Intersex in Domestic Mammals. *Cytogenet. Genome Res.* **1967**, *6*, 242–253. [CrossRef]
3. Mcfeely, R.A. Chromosome abnormalities in early embryos of the pig. *J. Reprod. Fertil.* **1967**, *13*, 579–581. [CrossRef]
4. Gustavsson, I.; Rockborn, G. Chromosome abnormality in three cases of lymphatic leukemia in cattle. *Nature* **1964**, *203*, 990. [CrossRef]
5. Gustavsson, I. Distribution of the 1/29 translocation in the A.I. Bull population of Swedish Red and White cattle. *Hereditas* **1971**, *69*, 101–106. [CrossRef] [PubMed]
6. Gustavsson, I. Cytogenetics, distribution and phenotypic effects of a translocation in swedish cattle. *Hereditas* **1979**, *63*, 68–169. [CrossRef]
7. Dyrendahl, I.; Gustavsson, I. Sexual functions, semen characteristics and fertility of bulls carrying the 1/29 chromosome translocation. *Hereditas* **1979**, *90*, 281–289. [CrossRef]
8. Wurster, D.H.; Benirschke, K. Chromosome studies in the superfamily Bovoidea. *Chromosoma* **1968**, *25*, 152–171. [CrossRef] [PubMed]
9. Iannuzzi, L.; King, W.A.; Di Berardino, D. Chromosome evolution in domestic bovids as revealed by chromosome banding and FISH-mapping techniques. *Cytogenet. Genome Res.* **2009**, *126*, 49–62. [CrossRef]
10. Ducos, A.; Revay, T.; Kovacs, A.; Hidas, A.; Pinton, A.; Bonnet-Garnier, A.; Molteni, L.; Slota, E.; Switonski, M.; Arruga, M.V.; et al. Cytogenetic screening of livestock populations in Europe: An overview. *Cytogenet. Genome Res.* **2008**, *120*, 26–41. [CrossRef] [PubMed]
11. Iannuzzi, A.; Parma, P.; Iannuzzi, L. Chromosome Abnormalities and Fertility in Domestic Bovids: A Review. *Animals* **2021**, *11*, 802. [CrossRef] [PubMed]
12. Donaldson, B.; Villagomez, D.; King, W.A. Classical, Molecular, and Genomic Cytogenetics of the Pig, a Clinical Perspective. *Animals* **2021**, *11*, 1257. [CrossRef] [PubMed]
13. Bugno-Poniewierska, M.; Raudsepp, T. Horse Clinical Cytogenetics: Recurrent Themes and Novel Findings. *Animals* **2021**, *11*, 831. [CrossRef]
14. Szczerbal, I.; Switonski, M. Clinical Cytogenetics of the Dog: A Review. *Animals* **2021**, *11*, 947. [CrossRef]
15. Hayes, H.; Petit, E.; Dutrillaux, B. Comparison of RBG-banded karyotypes of cattle, sheep, and goats. *Cytogenet. Cell Genet.* **1991**, *57*, 51–55. [CrossRef] [PubMed]
16. Gallagher, D.S.; Womack, J.E. Chromosome conservation in the Bovidae. *J. Hered.* **1992**, *83*, 287–298. [CrossRef]
17. Iannuzzi, L.; Di Meo, G.P. Chromosomal evolution in bovids: A comparison of cattle, sheep and goat G- and R-banded chromosomes and cytogenetic divergences among cattle, goat and river buffalo sex chromosomes. *Chromosome Res.* **1995**, *3*, 291–299. [CrossRef]
18. Perucatti, A.; Genualdo, V.; Pauciullo, A.; Iorio, C.; Incarnato, D.; Rossetti, C.; Vizzarri, F.; Palazzo, M.; Casamassima, D.; Iannuzzi, L.; et al. Cytogenetic tests reveal no toxicity in lymphocytes of rabbit (*Oryctolagus cuniculus*, 2n=44) feed in presence of verbascoside and/or lycopene. *Food. Chem. Toxicol.* **2018**, *114*, 311–315. [CrossRef]
19. Iannuzzi, A.; Perucatti, A.; Genualdo, V.; Pauciullo, A.; Melis, R.; Porqueddu, C.; Marchetti, M.; Usai, M.; Iannuzzi, L. Sister chromatid exchange test in river buffalo lymphocytes treated in vitro with furocoumarin extracts. *Mutagenesis* **2016**, *31*, 547–551. [CrossRef]

20. Pinkel, D.; Gray, J.W.; Trask, B.; Van Den Engh, G.; Fuscoe, J.; Van Dekken, H. Cytogenetic analysis by in situ hybridization with fluorescently labeled nucleic acid probes. *Cold Spring Harb. Symp. Quant. Biol.* **1986**, *51*, 151–157. [CrossRef]
21. Trask, B.; Pinkel, D. Fluorescence in situ hybridization with DNA probes. *Methods Cell Biol.* **1990**, *33*, 383–400. [CrossRef] [PubMed]
22. Gallagher, D.S.; Basrur, P.K.; Womack, J.E. Identification of an autosome to X chromosome translocation in the domestic cow. *J. Hered.* **1992**, *83*, 451–453. [CrossRef] [PubMed]
23. Miyake, Y.I.; Kawakura, K.; Murakami, R.K.; Kaneda, Y. Minute fragment observed in a bovine pedigree with Robertsonian translocation. *J. Hered.* **1994**, *85*, 488–490. [CrossRef] [PubMed]
24. Schmitz, A.; Oustry, A.; Chaput, B.; Bahri-Darwich, I.; Yerle, M.; Millan, D.; Frelat, G.; Cribiu, E.P. The bovine bivariate flow karyotype and peak identification by chromosome painting with PCR-generated probes. *Mamm. Genome* **1995**, *6*, 415–420. [CrossRef] [PubMed]
25. Kawakura, K.; Miyake, Y.; Murakami, R.K.; Kondoh, S.; Hirata, T.I.; Kaneda, Y. Abnormal structure of the Y chromosome detected in bovine gonadal hypoplasia (XY female) by FISH. *Cytogenet. Cell Genet.* **1997**, *76*, 36–38. [CrossRef]
26. Gallagher, D.S.; Lewis, B.C.; De Donato, M.; Davis, S.K.; Taylor, J.F.; Edwards, J.F. Autosomal trisomy 20 (61,XX,+20) in a malformed bovine foetus. *Vet. Pathol.* **1999**, *36*, 448–451. [CrossRef]
27. Tanaka, K.; Yamamoto, Y.; Amano, T.; Yamagata, T.; Dang, V.B.; Matsuda, Y.; Namikawa, T. A Robertsonian translocation, rob(2;28), found in Vietnamese cattle. *Hereditas* **2000**, *133*, 19–23. [CrossRef]
28. Di Meo, G.P.; Molteni, L.; Perucatti, A.; De Giovanni, A.; Incarnato, D.; Succi, G.; Schibler, L.; Cribiu, E.P.; Iannuzzi, L. Chromosomal characterization of three centric fusion translocations in cattle using G-, R- and C-banding and FISH technique. *Caryologia* **2000**, *53*, 213–218. [CrossRef]
29. Hyttel, P.; Viuff, D.; Laurincik, J.; Schmidt, M.; Thomsen, P.D.; Avery, B.; Callesen, H.; Rath, D.; Niemann, H.; Rosenkranz, C.; et al. Risks of in-vitro production of cattle and swine embryos: Aberrations in chromosome numbers, ribosomal RNA gene activation and perinatal physiology. *Hum. Reprod.* **2000**, *5*, 87–97. [CrossRef] [PubMed]
30. Viuff, D.; Greve, T.; Avery, B.; Hyttel, P.; Brockhoff, P.B.; Thomsen, P.D. Chromosome aberrations in in vitro-produced bovine embryos at days 2–5 post-insemination. *Biol. Reprod.* **2000**, *63*, 1143–1148. [CrossRef] [PubMed]
31. Iannuzzi, L.; Molteni, L.; Di Meo, G.P.; Perucatti, A.; Lorenzi, L.; Incarnato, D.; De Giovanni, A.; Succi, G.; Gustavsson, I. A new balanced autosomal reciprocal translocation in cattle revealed by banding techniques and human-painting probes. *Cytogenet. Cell Genet.* **2000**, *94*, 225–228. [CrossRef] [PubMed]
32. Iannuzzi, L.; Di Meo, G.P.; Perucatti, A.; Eggen, A.; Incarnato, D.; Sarubbi, F.; Cribiu, E. A pericentric inversion in the cattle Y chromosome. *Cytogenet. Cell Genet.* **2001**, *94*, 202–205. [CrossRef] [PubMed]
33. Iannuzzi, L.; Di Meo, G.P.; Leifsson, P.S.; Eggen, A.; Christensen, K. A case of trisomy 28 in cattle revealed by both banding and FISH-mapping techniques. *Hereditas* **2001**, *134*, 147–151. [CrossRef] [PubMed]
34. Basrur, P.K.; Reyes, E.R.; Farazmand, A.; King, W.A.; Popescu, P.C. X-autosome translocation and low fertility in a family of crossbred cattle. *Anim. Reprod. Sci.* **2001**, *67*, 1–16. [CrossRef]
35. Iannuzzi, L.; Molteni, L.; Di Meo, G.P.; De Giovanni, A.; Perucatti, A.; Succi, G.; Incarnato, D.; Eggen, A.; Cribiu, E.P. A case of azoospermia in a bull carrying a Y-autosome reciprocal translocation. *Cytogenet. Cell Genet.* **2001**, *95*, 225–227. [CrossRef]
36. Viuff, D.; Palsgaard, A.; Rickords, L.; Lawson, L.G.; Greve, T.; Schmidt, M.; Avery, B.; Hyttel, P.; Thomsen, P.D. Bovine embryos contain a higher proportion of polyploid cells in the trophectoderm than in the embryonic disc. *Mol. Reprod. Dev.* **2002**, *62*, 483–488. [CrossRef]
37. Chaves, R.; Adega, F.; Heslop-Harrison, J.S.; Guedes-Pinto, H.; Wienberg, J. Complex satellite DNA reshuffling in the polymorphic t(1;29) Robertsonian translocation and evolutionarily derived chromosomes in cattle. *Chromosome Res.* **2003**, *11*, 641–648. [CrossRef]
38. Pinton, A.; Ducos, A.; Yerle, M. Chromosomal rearrangements in cattle and pigs revealed by chromosome microdissection and chromosome painting. *Genet. Sel. Evol.* **2003**, *35*, 685–696. [CrossRef]
39. Słota, E.; Kozubska-Sobocińska, A.; Kościelny, M.; Danielak-Czech, B.; Rejduch, B. Detection of the XXY trisomy in a bull by using sex chromosome painting probes. *J. Appl. Genet.* **2003**, *44*, 379–382.
40. Bureau, W.S.; Bordignon, V.; Léveillée, C.; Smith, L.C.; King, W.A. Assessment of chromosomal abnormalities in bovine nuclear transfer embryos and in their donor cells. *Cloning Stem Cells* **2003**, *5*, 123–132. [CrossRef]
41. Mastromonaco, G.F.; Coppola, G.; Crawshaw, G.; Di Berardino, D.; King, W.A. Identification of the homologue of the bovine Rob(1;29) in a captive gaur (*Bos gaurus*). *Chromosome Res.* **2004**, *12*, 725–731. [CrossRef]
42. Di Meo, G.P.; Perucatti, A.; Chaves, R.; Adega, F.; De Lorenzi, L.; Molteni, L.; De Giovanni, A.; Incarnato, D.; Guedes-Pinto, H.; Eggen, A.; et al. Cattle rob(1;29) originating from complex chromosome rearrangements as revealed by both banding and FISH-mapping techniques. *Chromosome Res.* **2006**, *14*, 649–655. [CrossRef]
43. Bonnet-Garnier, A.; Pinton, A.; Berland, H.M.; Khireddine, B.; Eggen, A.; Yerle, M.; Darré, R.; Ducos, A. Sperm nuclei analysis of 1/29 Robertsonian translocation carrier bulls using fluorescence in situ hybridization. *Cytogenet. Genome Res.* **2006**, *12*, 241–247. [CrossRef]
44. De Lorenzi, L.; De Giovanni, A.; Molteni, L.; Denis, C.; Eggen, A.; Parma, P. Characterization of a balanced reciprocal translocation, rcp(9;11)(q27;q11) in cattle. *Cytogenet. Genome Res.* **2007**, *19*, 231–234. [CrossRef]

45. Sohn, S.H.; Cho, E.J.; Son, W.J.; Lee, C.Y. Diagnosis of bovine freemartinism by fluorescence in situ hybridization on interphase nuclei using a bovine Y chromosome-specific DNA probe. *Theriogenology* **2007**, *68*, 1003–1011. [CrossRef]
46. Molteni, L.; Perucatti, A.; Iannuzzi, A.; Di Meo, G.P.; De Lorenzi, L.; De Giovanni, A.; Incarnato, D.; Succi, G.; Cribiu, E.; Eggen, A.; et al. A new case of reciprocal translocation in a young bull: Rcp(11;21)(q28;q12). *Cytogenet. Genome Res.* **2007**, *116*, 80–84. [CrossRef]
47. Vozdova, M.; Kubickova, S.; Cernohorska, H.; Rubes, J. Detection of translocation rob(1;29) in bull sperm using a specific DNA probe. *Cytogenet. Genome Res.* **2008**, *120*, 102–105. [CrossRef]
48. Switonski, M.; Andersson, M.; Nowacka-Woszuk, J.; Szczerbal, I.; Sosnowski, J.; Kopp, C.; Cernohorska, H.; Rubes, J. Identification of a new reciprocal translocation in an AI bull by synaptonemal complex analysis, followed by chromosome painting. *Cytogenet. Genome Res.* **2008**, *121*, 245–248. [CrossRef] [PubMed]
49. Nicodemo, D.; Pauciullo, A.; Castello, A.; Roldan, E.; Gomendio, M.; Cosenza, G.; Peretti, V.; Perucatti, A.; Di Meo, G.P.; Ramunno, L.; et al. X-Y sperm aneuploidy in 2 cattle (*Bos taurus*) breeds as determined by dual color fluorescent in situ hybridization (FISH). *Cytogenet. Genome Res.* **2009**, *126*, 217–225. [CrossRef] [PubMed]
50. De Lorenzi, L.; Kopecna, O.; Gimelli, S.; Cernohorska, H.; Zannotti, M.; Béna, F.; Molteni, L.; Rubes, J.; Parma, P. Reciprocal translocation t(4;7)(q14;q28) in cattle: Molecular characterization. *Cytogenet. Genome Res.* **2010**, *129*, 298–304. [CrossRef] [PubMed]
51. Nicodemo, D.; Pauciullo, A.; Cosenza, G.; Peretti, V.; Perucatti, A.; Di Meo, G.P.; Ramunno, L.; Iannuzzi, L.; Rubes, J.; Di Berardino, D. Frequency of aneuploidy in in vitro-matured MII oocytes and corresponding first polar bodies in two dairy cattle (*Bos taurus*) breeds as determined by dual-color fluorescent in situ hybridization. *Theriogenology* **2010**, *73*, 523–529. [CrossRef] [PubMed]
52. Rybar, R.; Kopecka, V.; Prinosilova, P.; Kubickova, S.; Veznik, Z.; Rubes, J. Fertile bull sperm aneuploidy and chromatin integrity in relationship to fertility. *Int. J. Androl.* **2010**, *33*, 613–622. [CrossRef] [PubMed]
53. Switonski, M.; Szczerbal, I.; Krumrych, W.; Nowacka-Woszuk, J. A case of Y-autosome reciprocal translocation in a Holstein-Friesian bull. *Cytogenet. Genome Res.* **2011**, *132*, 22–25. [CrossRef]
54. Perucatti, A.; Genualdo, V.; Iannuzzi, A.; De Lorenzi, L.; Matassino, D.; Parma, P.; Di Berardino, D.; Iannuzzi, L.; Di Meo, G.P. A new and unusual reciprocal translocation in cattle: Rcp(11;25)(q11;q14-21). *Cytogenet. Genome Res.* **2011**, *134*, 96–100. [CrossRef] [PubMed]
55. Pauciullo, A.; Cosenza, G.; Peretti, V.; Iannuzzi, A.; Di Meo, G.P.; Ramunno, L.; Iannuzzi, L.; Rubes, J.; Di Berardino, D. Incidence of X-Y aneuploidy in sperm of two indigenous cattle breeds by using dual color fluorescent in situ hybridization (FISH). *Theriogenology* **2011**, *76*, 328–333. [CrossRef]
56. De Lorenzi, L.; Genualdo, V.; Gimelli, S.; Rossi, E.; Perucatti, A.; Iannuzzi, A.; Zannotti, M.; Malagutti, L.; Molteni, L.; Iannuzzi, L.; et al. Genomic analysis of cattle rob(1;29). *Chromosome Res.* **2012**, *20*, 815–823. [CrossRef]
57. De Lorenzi, L.; Rossi, E.; Genualdo, V.; Gimelli, S.; Lasagna, E.; Perucatti, A.; Iannuzzi, A.; Parma, P. Molecular characterization of Xp chromosome deletion in a fertile cow. *Sex Dev.* **2012**, *6*, 298–302. [CrossRef]
58. Pauciullo, A.; Nicodemo, D.; Peretti, V.; Marino, G.; Iannuzzi, A.; Cosenza, G.; Di Meo, G.P.; Ramunno, L.; Iannuzzi, L.; Rubes, J.; et al. X-Y aneuploidy rate in sperm of two "minor" breeds of cattle (*Bos taurus*) by using dual color fluorescent in situ hybridization (FISH). *Theriogenology* **2012**, *78*, 688–695. [CrossRef] [PubMed]
59. Pauciullo, A.; Nicodemo, D.; Cosenza, G.; Peretti, V.; Iannuzzi, A.; Di Meo, G.P.; Ramunno, L.; Iannuzzi, L.; Rubes, J.; Di Berardino, D. Similar rates of chromosomal aberrant secondary oocytes in two indigenous cattle (*Bos taurus*) breeds as determined by dual-color FISH. *Theriogenology* **2012**, *77*, 675–683. [CrossRef] [PubMed]
60. Pers-Kamczyc, E.; Pawlak, P.; Rubes, J.; Lechniak, D. Early cleaved bovine embryos show reduced incidence of chromosomal aberrations and higher developmental potential on day 4.5 post-insemination. *Reprod. Domest. Anim.* **2012**, *47*, 899–906. [CrossRef]
61. Šiviková, K.; Dianovský, J.; Holečková, B.; Galdíková, M.; Kolesárová, V. Assessment of cytogenetic damage in bovine peripheral lymphocytes exposed to in vitro tebuconazole-based fungicide. *Chemosphere* **2013**, *92*, 555–562. [CrossRef]
62. De Lorenzi, L.; Rossi, E.; Gimelli, S.; Parma, P. De novo reciprocal translocation t(5;6)(q13;q34) in cattle: Cytogenetic and molecular characterization. *Cytogenet. Genome Res.* **2014**, *142*, 95–100. [CrossRef]
63. Biltueva, L.; Kulemzina, A.; Vorobieva, N.; Perelman, P.; Kochneva, M.; Zhidenova, A.; Graphodatsky, A. A new case of an inherited reciprocal translocation in cattle: Rcp(13;26) (q24;q11). *Cytogenet. Genome Res.* **2014**, *144*, 208–211. [CrossRef] [PubMed]
64. Iannuzzi, A.; Genualdo, V.; Perucatti, A.; Pauciullo, A.; Varricchio, G.; Incarnato, D.; Matassino, D.; Iannuzzi, L. Fatal outcome in a newborn calf associated with partial trisomy 25q and partial monosomy 11q, 60,XX,der(11)t(11;25)(q11;q14~21). *Cytogenet. Genome Res.* **2015**, *146*, 222–229. [CrossRef] [PubMed]
65. Galdíková, M.; Šiviková, K.; Holečková, B.; Dianovský, J.; Drážovská, M.; Schwarzbacherová, V. The effect of thiacloprid formulation on DNA/chromosome damage and changes in GST activity in bovine peripheral lymphocytes. *J. Environ. Sci. Health B* **2015**, *50*, 698–707. [CrossRef]
66. De Lorenzi, L.; Iannuzzi, A.; Rossi, E.; Bonacina, S.; Parma, P. Centromere repositioning in cattle (*Bos taurus*) chromosome 17. *Cytogenet. Genome Res.* **2017**, *151*, 191–197. [CrossRef]
67. Berry, D.P.; Wolfe, A.; O'Donovan, J.; Byrne, N.; Sayers, R.G.; Dodds, K.G.; McEwan, J.C.; O'Connor, R.E.; McClure, M.; Purfield, D.C. Characterization of an X-chromosomal non-mosaic monosomy (59, X0) dairy heifer detected using routinely available single nucleotide polymorphism genotype data. *J. Anim. Sci.* **2017**, *95*, 1042–1049. [CrossRef]

68. Barasc, H.; Mouney-Bonnet, N.; Peigney, C.; Calgaro, A.; Revel, C.; Mary, N.; Ducos, A.; Pinton, A. Analysis of meiotic segregation pattern and interchromosomal effects in a bull heterozygous for a 3/16 robertsonian translocation. *Cytogenet. Genome Res.* **2018**, *156*, 197–203. [CrossRef]
69. Häfliger, I.M.; Agerholm, J.S.; Drögemüller, C. Constitutional trisomy 20 in an aborted Holstein fetus with pulmonary hypoplasia and anasarca syndrome. *Anim. Genet.* **2020**, *51*, 988–989. [CrossRef]
70. Häfliger, I.M.; Seefried, F.; Drögemüller, C. Trisomy 29 in a stillborn swiss original braunvieh calf. *Anim. Genet.* **2020**, *51*, 483–484. [CrossRef] [PubMed]
71. Jennings, R.L.; Griffin, D.K.; O'Connor, R.E. A new approach for accurate detection of chromosome rearrangements that affect fertility in cattle. *Animals* **2020**, *10*, 114. [CrossRef]
72. Iannuzzi, A.; Braun, M.; Genualdo, V.; Perucatti, A.; Reinartz, S.; Proios, I.; Heppelmann, M.; Rehage, J.; Hülsköter, K.; Beineke, A.; et al. Clinical, cytogenetic and molecular genetic characterization of a tandem fusion translocation in a male Holstein cattle with congenital hypospadias and a ventricular septal defect. *PLoS ONE* **2020**, *15*, e0227117. [CrossRef]
73. Iannuzzi, L.; Di Meo, G.P.; Perucatti, A.; Zicarelli, L. Sex chromosome monosomy (2n=49,X) in a river buffalo (*Bubalus bubalis*). *Vet. Rec.* **2000**, *147*, 690–691.
74. Di Meo, G.P.; Perucatti, A.; Genualdo, V.; Iannuzzi, A.; Sarubbi, F.; Caputi-Jambrenghi, A.; Incarnato, D.; Peretti, V.; Vonghia, G.; Iannuzzi, L. A rare case of centric fission and fusion in a river buffalo (*Bubalus bubalis*, 2n = 50) cow with reduced fertility. *Cytogenet. Genome Res.* **2011**, *132*, 26–30. [CrossRef]
75. Albarella, S.; Ciotola, F.; Coletta, A.; Genualdo, V.; Iannuzzi, L.; Peretti, V. A new translocation t(1p;18) in an Italian Mediterranean river buffalo (*Bubalus bubalis*, 2n = 50) bull: Cytogenetic, fertility and inheritance studies. *Cytogenet. Genome Res.* **2013**, *139*, 17–21. [CrossRef]
76. Pauciullo, A.; Perucatti, A.; Iannuzzi, A.; Incarnato, D.; Genualdo, V.; Di Berardino, D.; Iannuzzi, L. Development of a sequential multicolor-FISH approach with 13 chromosome-specific painting probes for the rapid identification of river buffalo (*Bubalus bubalis*, 2n = 50) chromosomes. *J. Appl. Genet.* **2014**, *55*, 397–401. [CrossRef]
77. Di Dio, C.; Longobardi, V.; Zullo, G.; Parma, P.; Pauciullo, A.; Perucatti, A.; Higgins, J.; Iannuzzi, A. Analysis of meiotic segregation by triple-color fish on both total and motile sperm fractions in a t(1p;18) river buffalo bull. *PLoS ONE* **2020**, *15*, e0232592. [CrossRef] [PubMed]
78. Pauciullo, A.; Versace, C.; Perucatti, A.; Gaspa, G.; Li, L.Y.; Yang, C.Y.; Zheng, H.Y.; Liu, Q.; Shang, J.H. Oocyte aneuploidy rates in river and swamp buffalo types (*Bubalus bubalis*) determined by Multi-color Fluorescence In Situ Hybridization (M-FISH). *Sci. Rep.* **2022**, *12*, 8440. [CrossRef] [PubMed]
79. Burkin, D.J.; O'Brien, P.C.; Broad, T.E.; Hill, D.F.; Jones, C.A.; Wienberg, J.; Ferguson-Smith, M.A. Isolation of chromosome-specific paints from high-resolution flow karyotypes of the sheep (*Ovis aries*). *Chromosome Res.* **1997**, *5*, 102–108. [CrossRef] [PubMed]
80. Chaves, R.; Adega, F.; Wienberg, J.; Guedes-Pinto, H.; Heslop-Harrison, J.S. Molecular cytogenetic analysis and centromeric satellite organization of a novel 8;11 translocation in sheep: A possible intermediate in biarmed chromosome evolution. *Mamm. Genome* **2003**, *14*, 706–710. [CrossRef] [PubMed]
81. Coppola, G.; Alexander, B.; Di Berardino, D.; St John, E.; Basrur, P.K.; King, W.A. Use of cross-species in-situ hybridization (ZOO-FISH) to assess chromosome abnormalities in day-6 in-vivo- or in-vitro-produced sheep embryos. *Chromosome Res.* **2007**, *15*, 399–408. [CrossRef]
82. Lühken, G.; Fleck, K.; Pauciullo, A.; Huisinga, M.; Erhardt, G. Familiar hypopigmentation syndrome in sheep associated with homozygous deletion of the entire endothelin type-B receptor gene. *PLoS ONE* **2012**, *7*, e53020. [CrossRef]
83. Iannuzzi, A.; Perucatti, A.; Genualdo, V.; De Lorenzi, L.; Di Berardino, D.; Parma, P.; Iannuzzi, L. Cytogenetic elaboration of a novel reciprocal translocation in sheep. *Cytogenet. Genome Res.* **2013**, *139*, 97–101. [CrossRef] [PubMed]
84. Iannuzzi, A.; Perucatti, A.; Genualdo, V.; Pauciullo, A.; Incarnato, D.; Musilova, P.; Rubes, J.; Iannuzzi, C. The utility of chromosome microdissection in clinical cytogenetics: A new reciprocal translocation in sheep. *Cytogenet. Genome Res.* **2014**, *142*, 174–178. [CrossRef] [PubMed]
85. Pauciullo, A.; Perucatti, A.; Cosenza, G.; Iannuzzi, A.; Incarnato, D.; Genualdo, V.; Di Berardino, D.; Iannuzzi, L. Sequential cross-species chromosome painting among river buffalo, cattle, sheep and goat: A useful tool for chromosome abnormalities diagnosis within the family Bovidae. *PLoS ONE* **2014**, *9*, e110297. [CrossRef]
86. Christensen, K.; Juul, L. A case of trisomy 22 in a live hereford calf. *Acta Vet. Scand.* **1999**, *40*, 85–88. [CrossRef] [PubMed]
87. De Giovanni, A.; Molteni, L.; Succi, G.; Galliani, C.; Bocher, J.; Popescu, C.P. A new type of Robertsonian translocation in cattle. In Proceedings of the 8th European Colloquium on Cytogenetics of Domestic Animals, Bristol, UK, 19–22 July 1988; pp. 53–59.
88. De Giovanni, A.; Succi, G.; Molteni, L.; Castiglioni, M. A new autosomal translocation in "Alpine grey cattle". *Ann. Genet. Sel. Anim.* **1979**, *11*, 115–120. [CrossRef]
89. Switoński, M.; Stranzinger, G. Studies of synaptonemal complexes in farm mammals- a review. *J. Hered.* **1998**, *89*, 473–480. [CrossRef] [PubMed]
90. Hart, E.J.; Pinton, A.; Powell, A.; Wall, R.; King, W.A. Meiotic recombination in normal and clone bulls and their offspring. *Cytogenet. Genome Res.* **2008**, *120*, 97–101. [CrossRef]
91. Sebestova, H.; Vozdova, M.; Kubickova, S.; Cernohorska, H.; Kotrba, R.; Rubes, J. Effect of species-specific differences in chromosome morphology on chromatin compaction and the frequency and distribution of RAD51 and MLH1 foci in two bovid species: Cattle (*Bos taurus*) and the common eland (*Tautragus oryx*). *Chromosoma* **2016**, *125*, 137–149. [CrossRef]

92. Villagómez, D.A.; Pinton, A. Chromosomal abnormalities, meiotic behaviour and fertility in domestic animals. *Cytogenet. Genome Res.* **2008**, *120*, 69–80. [CrossRef]
93. Riggs, P.K.; Rønne, M. Fragile sites in domestic animal chromosomes: Molecular insights and challenges. *Cytogenet. Genome Res.* **2009**, *126*, 97–109. [CrossRef] [PubMed]
94. Ford, C.E.; Pollock, D.L.; Gustavsson, I. Proceedings of the First International Conference for the Standardisation of Banded Karyotypes of Domestic Animals. University of Reading, Reading, England. 2nd–6th August 1976. *Hereditas* **1980**, *92*, 145–162. [CrossRef] [PubMed]
95. Di Berardino, D.; Hayes, H.; Fries, R.; Long, S. ISCNDA1989: International System for Cytogenetic Nomenclature of Domestic Animals. *Cytogenet. Cell Genet.* **1990**, *53*, 65–79. [CrossRef]
96. Popescu, C.P.; Long, S.; Riggs, P.; Womack, J.; Schmutz, S.; Fries, R.; Gallagher, D.S. Standardization of cattle karyotype nomenclature: Report of the committee for the standardization of the cattle karyotype. *Cytogenet. Cell Genet.* **1996**, *74*, 259–261. [CrossRef] [PubMed]
97. Hayes, H.; Di Meo, G.P.; Gautier, M.; Laurent, P.; Eggen, A.; Iannuzzi, L. Localization by FISH of the 31 Texas nomenclature type I markers to both Q- and R-banded bovine chromosomes. *Cytogenet. Cell Genet.* **2000**, *90*, 315–320. [CrossRef]
98. Cribiu, E.P.; Di Berardino, D.; Di Meo, G.P.; Gallagher, D.S.; Hayes, H.; Iannuzzi, L.; Popescu, C.P.; Rubes, J.; Schmutz, S.; Stranzinger, G.; et al. International System for Chromosome Nomenclature of Domestic Bovids (ISCNDB 2000). *Cytogenet. Cell Genet.* **2001**, *92*, 283–299. [CrossRef]
99. Iannuzzi, L.; Di Meo, G.P.; Hayes, H.; Perucatti, A.; Incarnato, D.; Gautier, M.; Eggen, A. FISH-mapping of 31 type I loci (Texas markers) to river buffalo chromosomes. *Chromosome Res.* **2001**, *9*, 339–342. [CrossRef]
100. Di Meo, G.P.; Perucatti, A.; Gautier, M.; Hayes, H.; Incarnato, D.; Eggen, A.; Iannuzzi, L. Chromosome localization of the 31 type I Texas bovine markers in sheep and goat chromosomes by comparative FISH-mapping and R-banding. *Anim. Genet.* **2003**, *34*, 294–296. [CrossRef]
101. Watson, J.D.; Crick, F.H. Molecular structure of nucleic acids; a structure for deoxyribose nucleic acid. *Nature* **1953**, *171*, 737–738. [CrossRef]
102. Pardue, M.L.; Gall, J.G. Molecular hybridization of radioactive DNA to the DNA of cytological preparations. *Proc. Natl. Acad. Sci. USA* **1969**, *64*, 600–604. [CrossRef] [PubMed]
103. Fries, R.; Hediger, R.; Stranzinger, G. Tentative chromosomal localization of the bovine major histocompatibility complex by in situ hybridization. *Anim. Genet.* **1986**, *17*, 287–294. [CrossRef]
104. Yerle, M.; Gellin, J.; Echard, G.; Lefevre, F.; Gillois, M. Chromosomal localization of leukocyte interferon gene in the pig (*Sus scrofa domestica* L.) by in situ hybridization. *Cytogenet. Cell Genet.* **1986**, *42*, 129–132. [CrossRef] [PubMed]
105. Rudkin, G.T.; Stollar, B.D. High resolution detection of DNA-RNA hybrids in situ by indirect immunofluorescence. *Nature* **1977**, *265*, 472–473. [CrossRef]
106. Ryan, A.M.; Gallagher, D.S.; Womack, J.E. Syntenic mapping and chromosomal localization of bovine alpha and beta interferon genes. *Mamm. Genome* **1992**, *3*, 575–578. [CrossRef]
107. Iannuzzi, L.; Gallagher, D.S.; Ryan, A.M.; Di Meo, G.P.; Womack, J.E. Chromosomal localization of omega and trophoblast interferon genes in cattle and river buffalo by sequential R-banding and fluorescent in situ hybridization. *Cytogenet. Cell Genet.* **1993**, *62*, 224–227. [CrossRef]
108. Iannuzzi, L.; Di Meo, G.P.; Gallagher, D.S.; Ryan, A.M.; Ferrara, L.; Womack, J.E. Chromosomal localization of omega and trophoblast interferon genes in goat and sheep by fluorescent in situ hybridization. *J. Hered.* **1993**, *84*, 301–304. [CrossRef] [PubMed]
109. Iannuzzi, L.; Di Meo, G.P.; Perucatti, A.; Schibler, L.; Incarnato, D.; Gallagher, D.; Eggen, A.; Ferretti, L.; Cribiu, E.P.; Womack, J. The river buffalo (*Bubalus bubalis*, 2n = 50) cytogenetic map: Assignment of 64 loci by fluorescence in situ hybridization and R-banding. *Cytogenet. Genome Res.* **2003**, *102*, 65–75. [CrossRef]
110. Iannuzzi, L.; Gallagher, D.S.; Di Meo, G.P.; Ryan, A.M.; Perucatti, A.; Ferrara, L.; Irwin, D.M.; Womack, J.E. Chromosomal localization of the lysozyme gene cluster in river buffalo (*Bubalus bubalis* L.). *Chromosome Res.* **1993**, *1*, 253–255. [CrossRef]
111. Friedl, R.; Rottmann, O.J. Assignment of the bovine uridine monophosphate synthase gene to the bovine chromosome region 1q34-36 by FISH. *Mamm. Genome* **1994**, *5*, 38–40. [CrossRef]
112. Iannuzzi, L.; Di Meo, G.P.; Ryan, A.M.; Gallagher, D.S.; Ferrara, L.; Womack, J.E. Localization of uridine monophosphate synthase (UMPS) gene to river buffalo chromosomes by FISH. *Chromosome Res.* **1994**, *2*, 255–256. [CrossRef]
113. Solinas-Toldo, S.; Mezzelani, A.; Hawkins, G.A.; Bishop, M.D.; Olsaker, I.; Mackinlay, A.; Ferretti, L.; Fries, R. Combined Q-banding and fluorescence in situ hybridization for the identification of bovine chromosomes 1 to 7. *Cytogenet. Cell Genet.* **1995**, *69*, 1–6. [CrossRef]
114. Hawkins, G.A.; Toldo, S.S.; Bishop, M.D.; Kappes, S.M.; Fries, R.; Beattie, C.W. Physical and linkage mapping of the bovine genome with cosmids. *Mamm Genome* **1995**, *6*, 249–254. [CrossRef]
115. Mezzelani, A.; Zhang, Y.; Redaelli, L.; Castiglioni, B.; Leone, P.; Williams, J.L.; Toldo, S.S.; Wigger, G.; Fries, R.; Ferretti, L. Chromosomal localization and molecular characterization of 53 cosmid-derived bovine microsatellites. *Mamm. Genome* **1995**, *6*, 629–635. [CrossRef] [PubMed]

116. Iannuzzi, L.; Gallagher, D.S.; Di Meo, G.P.; Diamond, G.; Bevins, C.L.; Womack, J.E. High-resolution FISH mapping of beta-defensin genes to river buffalo and sheep chromosomes suggests a chromosome discrepancy in ffttle standard karyotypes. *Cytogenet. Cell Genet.* **1996**, *75*, 10–13. [CrossRef] [PubMed]
117. Iannuzzi, L.; Gallagher, D.S.; Womack, J.E.; Meo, G.P.; Shelling, C.P.; Groenen, M.A. FISH mapping of the alpha-S2 casein gene on river buffalo and cattle chromosomes identifies a nomenclature discrepancy in the bovine karyotype. *Chromosome Res.* **1996**, *4*, 159–162. [CrossRef] [PubMed]
118. Yoo, J.; Stone, R.T.; Kappes, S.M.; Toldo, S.S.; Fries, R.; Beattie, C.W. Genomic organization and chromosomal mapping of the bovine Fas/APO-1 gene. *DNA Cell Biol.* **1996**, *15*, 377–385. [CrossRef]
119. Goldammer, T.; Brunner, R.M.; Schmidt, P.; Schwerin, M. Mapping of the interferon gamma gene (IFNG) to chromosomes 3 in sheep and 5 in goat by FISH. *Mamm. Genome* **1996**, *7*, 470–471. [CrossRef]
120. Yoo, J.; Stone, R.T.; Solinas-Toldo, S.; Fries, R.; Beattie, C.W. Cloning and chromosomal mapping of bovine interleukin-2 receptor gamma gene. *DNA Cell Biol.* **1996**, *15*, 453–459. [CrossRef] [PubMed]
121. Folch, J.M.; Coll, A.; Hayes, H.C.; Sànchez, A. Characterization of a caprine beta-lactoglobulin pseudogene, identification and chromosomal localization by in situ hybridization in goat, sheep and cow. *Gene* **1996**, *177*, 87–91. [CrossRef]
122. Martín-Burriel, I.; Goldammer, T.; Elduque, C.; Lundin, M.; Barendse, W.; Zaragoza, P.; Olsaker, I. Physical and linkage mapping of the bovine bone morphogenetic protein 1 on the evolutionary break region of BTA 8. *Cytogenet. Cell Genet.* **1997**, *79*, 179–183. [CrossRef]
123. Vogel, T.; Borgmann, S.; Dechend, F.; Hecht, W.; Schmidtke, J. Conserved Y-chromosomal location of TSPY in Bovidae. *Chromosome Res.* **1997**, *5*, 182–185. [CrossRef]
124. Iannuzzi, L.; Skow, L.; Di Meo, G.P.; Gallagher, D.S.; Womack, J.E. Comparative FISH-mapping of villin (VIL) gene in river buffalo, sheep and goat chromosomes. *Chromosome Res.* **1997**, *5*, 199–202. [CrossRef] [PubMed]
125. Gallagher, D.S.; Yang, Y.P.; Burzlaff, J.D.; Womack, J.E.; Stelly, D.M.; Davis, S.K.; Taylor, J.F. Physical assignment of six type I anchor loci to bovine chromosome 19 by fluorescence in situ hybridization. *Anim. Genet.* **1998**, *29*, 130–134. [CrossRef]
126. Iannuzzi, L.; Palomba, R.; Di Meo, G.P.; Perucatti, A.; Ferrara, L. Comparative FISH-mapping of the prion protein gene (PRNP) on cattle, river buffalo, sheep and goat chromosomes. *Cytogenet. Cell Genet.* **1998**, *81*, 202–204. [CrossRef] [PubMed]
127. Gallagher, D.S.; Schläpfer, J.; Burzlaff, J.D.; Womack, J.E.; Stelly, D.M.; Davis, S.K.; Taylor, J.F. Cytogenetic alignment of the bovine chromosome 13 genome map by fluorescence in-situ hybridization of human chromosome 10 and 20 comparative markers. *Chromosome Res.* **1999**, *7*, 115–119. [CrossRef] [PubMed]
128. Iannuzzi, L.; Gallagher, D.S.; Di Meo, G.P.; Yang, Y.; Womack, J.E.; Davis, S.K.; Taylor, J.F. Comparative FISH-mapping of six expressed gene loci to river buffalo and sheep. *Cytogenet. Cell Genet.* **1999**, *84*, 161–163. [CrossRef] [PubMed]
129. Lòpez-Corrales, N.L.; Sonstegard, T.S.; Smith, T.P. Physical mapping of the bovine, caprine and ovine homologues of the paired box gene PAX8. *Cytogenet. Cell Genet.* **1999**, *84*, 179–181. [CrossRef]
130. Minoshima, Y.; Taniguchi, Y.; Tanaka, K.; Yamada, T.; Sasaki, Y. Molecular cloning, expression analysis, promoter characterization, and chromosomal localization of the bovine PREF1 gene. *Anim. Genet.* **2001**, *32*, 333–339. [CrossRef]
131. De Donato, M.; Gallagher, D.S.; Davis, S.K.; Stelly, D.M.; Taylor, J.F. The assignment of PRKCI to bovine chromosome 1q34->q36 by FISH suggests a new assignment to human chromosome 3. *Cytogenet. Cell Genet.* **2001**, *95*, 79–81. [CrossRef]
132. McShane, R.D.; Gallagher, D.S., Jr.; Newkirk, H.; Taylor, J.F.; Burzlaff, J.D.; Davis, S.K.; Skow, L.C. Physical localization and order of genes in the class I region of the bovine MHC. *Anim. Genet.* **2001**, *32*, 235–239. [CrossRef] [PubMed]
133. Buitkamp, J.; Ewald, D.; Masabanda, J.; Bishop, M.D.; Fries, R. FISH and RH mapping of the bovine alpha (2)/delta calcium channel subunit gene (CACNA2D1). *Anim. Genet.* **2003**, *34*, 309–310. [CrossRef]
134. Brenig, B.; Baumgartner, B.G.; Kriegesmann, B.; Habermann, F.; Fries, R.; Swalve, H.H. Molecular cloning, mapping, and functional analysis of the bovine sulfate transporter SLC26a2 gene. *Gene* **2003**, *319*, 161–166. [CrossRef] [PubMed]
135. Iannuzzi, L.; Di Meo, G.P.; Perucatti, A.; Rullo, R.; Incarnato, D.; Longeri, M.; Bongioni, G.; Molteni, L.; Galli, A.; Zanotti, M.; et al. Comparative FISH-mapping of the survival of motor neuron gene (SMN) in domestic bovids. *Cytogenet. Genome Res.* **2003**, *102*, 39–41. [CrossRef]
136. Iannuzzi, L.; Perucatti, A.; Di Meo, G.P.; Schibler, L.; Incarnato, D.; Cribiu, E.P. Chromosomal localization of sixty autosomal loci in sheep (*Ovis aries*, 2n = 54) by fluorescence in situ hybridization and R-banding. *Cytogenet. Genome Res.* **2003**, *103*, 135–138. [CrossRef]
137. Mömke, S.; Kuiper, H.; Spötter, A.; Drögemüller, C.; Williams, J.L.; Distl, O. Assignment of the PRPH gene to bovine chromosome 5q1.4 by FISH and confirmation by RH mapping. *Anim. Genet.* **2004**, *35*, 477–478. [CrossRef] [PubMed]
138. Kaupe, B.; Kollers, S.; Fries, R.; Erhardt, G. Mapping of CYP11B and a putative CYHR1 paralogous gene to bovine chromosome 14 by FISH. *Anim. Genet.* **2004**, *35*, 478–479. [CrossRef] [PubMed]
139. Liu, W.S.; de León, F.A. Assignment of SRY, ANT3, and CSF2RA to the bovine Y chromosome by FISH and RH mapping. *Anim. Biotechnol.* **2004**, *15*, 103–109. [CrossRef]
140. Di Meo, G.P.; Gallagher, D.; Perucatti, A.; Wu, X.; Incarnato, D.; Mohammadi, G.; Taylor, J.F.; Iannuzzi, L. Mapping of 11 genes by FISH to BTA2, BBU2q, OAR2q and CHI2, and comparison with HSA2q. *Anim. Genet.* **2006**, *37*, 299–300. [CrossRef]
141. Di Meo, G.P.; Perucatti, A.; Floriot, S.; Hayes, H.; Schibler, L.; Rullo, R.; Incarnato, D.; Ferretti, L.; Cockett, N.; Cribiu, E.; et al. An advanced sheep (*Ovis aries*, 2n = 54) cytogenetic map and assignment of 88 new autosomal loci by fluorescence in situ hybridization and R-banding. *Anim. Genet.* **2007**, *38*, 233–240. [CrossRef]

142. Di Meo, G.P.; Perucatti, A.; Floriot, S.; Hayes, H.; Schibler, L.; Incarnato, D.; Di Berardino, D.; Williams, J.; Cribiu, E.; Eggen, A.; et al. An extended river buffalo (*Bubalus bubalis*, 2n = 50) cytogenetic map: Assignment of 68 autosomal loci by FISH-mapping and R-banding and comparison with human chromosomes. *Chromosome Res.* **2008**, *16*, 827–837. [CrossRef] [PubMed]
143. Farhadi, A.; Genualdo, V.; Perucatti, A.; Hafezian, S.H.; Rahimi-Mianji, G.; De Lorenzi, L.; Parma, P.; Iannuzzi, L.; Iannuzzi, A. Comparative FISH mapping of BMPR1B, BMP15 and GDF9 fecundity genes on cattle, river buffalo, sheep and goat chromosomes. *J. Genet.* **2013**, *92*, 595–597. [CrossRef] [PubMed]
144. Eggen, A.; Gautier, M.; Billaut, A.; Petit, E.; Hayes, H.; Laurent, P.; Urban, C.; Pfister-Genskow, M.; Eilertsen, K.; Bishop, M.D. Construction and characterization of a bovine BAC library with four genome-equivalent coverage. Genetics, selection, evolution. *Genet. Sel. Evol.* **2001**, *33*, 543–548. [CrossRef]
145. Bovine Genome Sequencing and Analysis Consortium; Elsik, C.G.; Tellam, R.L.; Worley, K.C.; Gibbs, R.A.; Muzny, D.M.; Weinstock, G.M.; Adelson, D.L.; Eichler, E.E.; Elnitski, L.; et al. The genome sequence of taurine cattle: A window to ruminant biology and evolution. *Science* **2009**, *324*, 522–528. [CrossRef]
146. Zimin, A.V.; Delcher, A.L.; Florea, L.; Kelley, D.R.; Schatz, M.C.; Puiu, D.; Hanrahan, F.; Pertea, G.; Van Tassell, C.P.; Sonstegard, T.S.; et al. A whole-genome assembly of the domestic cow, *Bos taurus*. *Genome Biol.* **2009**, *10*, R42. [CrossRef]
147. Dong, Y.; Xie, M.; Jiang, Y.; Xiao, N.; Du, X.; Zhang, W.; Tosser-Klopp, G.; Wang, J.; Yang, S.; Liang, J.; et al. Sequencing and automated whole-genome optical mapping of the genome of a domestic goat (*Capra hircus*). *Nat. Biotechnol.* **2013**, *31*, 135–141. [CrossRef]
148. International Sheep Genomics Consortium; Archibald, A.L.; Cockett, N.E.; Dalrymple, B.P.; Faraut, T.; Kijas, J.W.; Maddox, J.F.; McEwan, J.C.; Hutton Oddy, V.; Raadsma, H.W.; et al. The sheep genome reference sequence: A work in progress. *Anim. Genet.* **2010**, *41*, 449–453. [CrossRef] [PubMed]
149. Mintoo, A.A.; Zhang, H.; Chen, C.; Moniruzzaman, M.; Deng, T.; Anam, M.; Emdadul Huque, Q.M.; Guang, X.; Wang, P.; Zhong, Z.; et al. Draft genome of the river water buffalo. *Ecol. Evol.* **2019**, *9*, 3378–3388. [CrossRef]
150. De Lorenzi, L.; Molteni, L.; Parma, P. FISH mapping in cattle (*Bos taurus* L.) is not yet out of fashion. *J. Appl. Genet.* **2010**, *51*, 497–499. [CrossRef]
151. Partipilo, G.; D'Addabbo, P.; Lacalandra, G.M.; Liu, G.E.; Rocchi, M. Refinement of *Bos taurus* sequence assembly based on BAC-FISH experiments. *BMC Genom.* **2011**, *12*, 639. [CrossRef] [PubMed]
152. De Lorenzi, L.; Genualdo, V.; Perucatti, A.; Iannuzzi, A.; Iannuzzi, L.; Parma, P. Physical mapping of 20 unmapped fragments of the btau_4.0 genome assembly in cattle, sheep and river buffalo. *Cytogenet. Genome Res.* **2013**, *140*, 29–35. [CrossRef] [PubMed]
153. Toldo, S.S.; Fries, R.; Steffen, P.; Neibergs, H.L.; Barendse, W.; Womack, J.E.; Hetzel, D.J.; Stranzinger, G. Physically mapped, cosmid-derived microsatellite markers as anchor loci on bovine chromosomes. *Mamm. Genome* **1993**, *4*, 720–727. [CrossRef] [PubMed]
154. Vaiman, D.; Mercier, D.; Eggen, A.; Bahri-Darwich, I.; Grohs, C.; Cribiu, E.P.; Dolf, G.; Oustry, A.; Guérin, G.; Levéziel, H. A genetic and physical map of bovine chromosome 11. *Mamm. Genome* **1994**, *5*, 553–556. [CrossRef] [PubMed]
155. Drögemüller, C.; Bader, A.; Wöhlke, A.; Kuiper, H.; Leeb, T.; Distl, O. A high-resolution comparative RH map of the proximal part of bovine chromosome 1. *Anim. Genet.* **2002**, *33*, 271–279. [CrossRef]
156. Smith, T.P.; Lopez-Corrales, N.; Grosz, M.D.; Beattie, C.W.; Kappes, S.M. Anchoring of bovine chromosomes 4, 6, 7, 10, and 14 linkage group telomeric ends via FISH analysis of lambda clones. *Mamm. Genome* **1997**, *8*, 333–336. [CrossRef]
157. Scherthan, H.; Cremer, T.; Arnason, U.; Weier, H.U.; Lima-de-Faria, A.; Frönicke, L. Comparative chromosome painting discloses homologous segments in distantly related mammals. *Nat. Genet.* **1994**, *6*, 342–347. [CrossRef] [PubMed]
158. Jauch, A.; Wienberg, J.; Stanyon, R.; Arnold, N.; Tofanelli, S.; Ishida, T.; Cremer, T. Reconstruction of genomic rearrangements in great apes and gibbons by chromosome painting. *Proc. Natl. Acad. Sci. USA* **1992**, *89*, 8611–8615. [CrossRef]
159. Lengauer, C.; Wienberg, J.; Cremer, T.; Lüdecke, H.J.; Horstehmke, B. Comparative chromosome band mapping in primates by in situ suppression hybridization of band specific DNA microlibraries. *Hum. Evol.* **1991**, *6*, 67–71. [CrossRef]
160. Wienberg, J.; Stanyon, R.; Jauch, A.; Cremer, T. Homologies in human and Macaca fuscata chromosomes revealed by in situ suppression hybridization with human chromosome specific DNA libraries. *Chromosoma* **1992**, *101*, 265–270. [CrossRef]
161. Chowdhary, B.P.; Raudsepp, T.; Frönicke, L.; Scherthan, H. Emerging patterns of comparative genome organization in some mammalian species as revealed by Zoo-FISH. *Genome Res.* **1998**, *8*, 577–589. [CrossRef]
162. Hayes, H.C.; Popescu, P.; Dutrillaux, B. Comparative gene mapping of lactoperoxidase, retinoblastoma, and alpha-lactalbumin genes in cattle, sheep, and goats. *Mamm. Genome* **1993**, *4*, 593–597. [CrossRef] [PubMed]
163. Hayes, H.C.; Petit, E.J. Mapping of the beta-lactoglobulin gene and of an immunoglobulin M heavy chain-like sequence to homoeologous cattle, sheep, and goat chromosomes. *Mamm. Genome* **1993**, *4*, 207–210. [CrossRef]
164. Hayes, H.; Petit, E.; Bouniol, C.; Popescu, P. Localization of the alpha-S2-casein gene (CASAS2) to the homoeologous cattle, sheep, and goat chromosomes 4 by in situ hybridization. *Cytogenet. Cell Genet.* **1993**, *64*, 281–285. [CrossRef]
165. Iannuzzi, L.; Gallagher, D.S.; Womack, J.E.; Di Meo, G.P.; Skow, L.C.; Ferrara, L. Chromosomal localization of the major histocompatibility complex in cattle and river buffalo by fluorescent in situ hybridization. *Hereditas* **1993**, *118*, 187–190. [CrossRef]
166. Brunner, R.M.; Goldammer, T.; Hiendleder, S.; Jäger, C.; Schwerin, M. Comparative mapping of the gene coding for inhibin-alpha (INHA) to chromosome 2 in sheep and cattle. *Mamm. Genome* **1995**, *6*, 309. [CrossRef]
167. Goldammer, T.; Brunner, R.M.; Hiendleder, S.; Schwerin, M. Comparative mapping of sheep inhibin subunit beta b to chromosome 2 in sheep and cattle by fluorescence in situ hybridization. *Anim. Genet.* **1995**, *26*, 199–200. [CrossRef] [PubMed]

168. Hayes, H.; Le Chalony, C.; Goubin, G.; Mercier, D.; Payen, E.; Bignon, C.; Kohno, K. Localization of ZNF164, ZNF146, GGTA1, SOX2, PRLR and EEF2 on homoeologous cattle, sheep and goat chromosomes by fluorescent in situ hybridization and comparison with the human gene map. *Cytogenet. Cell Genet.* **1996**, *72*, 342–346. [CrossRef]
169. Goldammer, T.; Meyer, L.; Seyfert, H.M.; Brunner, R.M.; Schwerin, M. STAT5A encoding gene maps to chromosome 19 in cattle and goat and to chromosome 11 in sheep. *Mamm. Genome* **1997**, *8*, 705–706. [CrossRef] [PubMed]
170. Goldammer, T.; Brunner, R.M.; Schwerin, M. Comparative analysis of Y chromosome structure in *Bos taurus* and *B. indicus* by FISH using region-specific, microdissected, and locus-specific DNA probes. *Cytogenet. Cell Genet.* **1997**, *77*, 238–241. [CrossRef]
171. Faure, M.; Hayes, H.; Bledsoe, R.K.; Hutson, S.M.; Papet, I. Assignment of the gene of mitochondrial branched chain aminotransferase (BCAT2) to sheep chromosome band 14q24 and to cattle and goat chromosome bands 18q24 by in situ hybridization. *Cytogenet. Cell Genet.* **1998**, *83*, 96–97. [CrossRef]
172. Robinson, T.J.; Harrison, W.R.; Ponce de León, F.A.; Davis, S.K.; Elder, F.F. A molecular cytogenetic analysis of X-chromosome repatterning in the Bovidae: Transpositions, inversions, and phylogenetic inference. *Cytogenet. Cell Genet.* **1998**, *80*, 179–184. [CrossRef] [PubMed]
173. Piumi, F.; Schibler, L.; Vaiman, D.; Oustry, A.; Cribiu, E.P. Comparative cytogenetic mapping reveals chromosome rearrangements between the X chromosomes of two closely related mammalian species (cattle and goats). *Cytogenet. Cell Genet.* **1998**, *81*, 36–41. [CrossRef]
174. Schibler, L.; Vaiman, D.; Oustry, A.; Giraud-Delville, C.; Cribiu, E.P. Comparative gene mapping: A fine-scale survey of chromosome rearrangements between ruminants and humans. *Genome Res.* **1998**, *8*, 901–915. [CrossRef]
175. Schmitz, A.; Oustry, A.; Vaiman, D.; Chaput, B.; Frelat, G.; Cribiu, E.P. Comparative karyotype of pig and cattle using whole chromosome painting probes. *Hereditas* **1998**, *128*, 257–263. [CrossRef] [PubMed]
176. Iannuzzi, L.; Di Meo, G.P.; Perucatti, A.; Bardaro, T. ZOO-FISH and R-banding reveal extensive conservation of human chromosome regions in euchromatic regions of river buffalo chromosomes. *Cytogenet. Cell Genet.* **1998**, *82*, 210–214. [CrossRef] [PubMed]
177. Iannuzzi, L.; Di Meo, G.P.; Perucatti, A.; Incarnato, D. Comparison of the human with the sheep genomes by use of human chromosome-specific painting probes. *Mammal. Genome* **1999**, *10*, 719–723. [CrossRef]
178. Iannuzzi, L.; Di Meo, G.P.; Perucatti, A.; Incarnato, D.; Lopez-Corrales, N.; Smith, J. Chromosomal localization of four HSA2 type I loci in river buffalo (*Bubalus bubalis*, 2n=50) chromosomes 2q and 12. *Mammal. Genome* **2000**, *11*, 241–242. [CrossRef]
179. Hayes, H.; Bonfils, J.; Faure, M.; Papet, I. Assignment of BCAT1, the gene encoding cytosolic branched chain aminotransferase, to sheep chromosome band 3q33 and to cattle and goat chromosome bands 5q33 by in situ hybridization. *Cytogenet. Cell Genet.* **2000**, *90*, 84–85. [CrossRef] [PubMed]
180. Iannuzzi, L.; Di Meo, G.P.; Perucatti, A.; Incarnato, D.; Schibler, L.; Cribiu, E.P. Comparative FISH-mapping of bovid X chromosomes reveals homologies and divergences between the subfamilies Bovinae and Caprinae. *Cytogenet. Cell Genet.* **2000**, *89*, 171–176. [CrossRef]
181. Iannuzzi, L.; Di Meo, G.P.; Perucatti, A.; Schibler, L.; Incarnato, D.; Ferrara, L.; Bardaro, T.; Cribiu, E.P. Sixteen type I loci from six chromosomes were comparatively fluorescent in-situ mapped to river buffalo (*Bubalus bubalis*) and sheep (*Ovis aries*) chromosomes. *Chromosome Res.* **2000**, *8*, 447–450. [CrossRef] [PubMed]
182. Di Meo, G.P.; Perucatti, A.; Schibler, L.; Incarnato, D.; Ferrara, L.; Cribiu, E.P.; Iannuzzi, L. Thirteen type I loci from HSA4q, HSA6p, HSA7q and HSA12q were comparatively FISH-mapped in four river buffalo and sheep chromosomes. *Cytogenet. Cell Genet.* **2000**, *90*, 102–105. [CrossRef]
183. Iannuzzi, L.; Di Meo, G.P.; Perucatti, A.; Schibler, L.; Incarnato, D.; Cribiu, E.P. Comparative FISH-mapping in river buffalo and sheep chromosomes: Assignment of forty autosomal type I loci from sixteen human chromosomes. *Cytogenet. Cell Genet.* **2001**, *94*, 43–48. [CrossRef]
184. Gautier, M.; Hayes, H.; Taourit, S.; Laurent, P.; Eggen, A. Assignment of eight additional genes from human chromosome 11 to bovine chromosomes 15 and 29: Refinement of the comparative map. *Cytogenet. Cell Genet.* **2001**, *93*, 60–64. [CrossRef]
185. Di Meo, G.P.; Perucatti, A.; Incarnato, D.; Ferretti, L.; Di Berardino, D.; Caputi Jambrenghi, A.; Vonghia, G.; Iannuzzi, L. Comparative mapping of twenty-eight bovine loci in sheep (*Ovis aries*, 2n = 54) and river buffalo (*Bubalus bubalis*, 2n = 50) by FISH. *Cytogenet. Genome Res.* **2002**, *98*, 262–264. [CrossRef]
186. Cosseddu, G.M.; Oustry-Vaiman, A.; Jego, B.; Moreno, C.; Taourit, S.; Cribiu, E.P.; Elsen, J.M.; Vaiman, D. Sheep/human comparative map in a chromosome region involved in scrapie incubation time shows multiple breakpoints between human chromosomes 14 and 15 and sheep chromosomes 7 and 18. *Chromosome Res.* **2002**, *10*, 369–378. [CrossRef] [PubMed]
187. Hayes, H.; Elduque, C.; Gautier, M.; Schibler, L.; Cribiu, E.; Eggen, A. Mapping of 195 genes in cattle and updated comparative map with man, mouse, rat and pig. *Cytogenet. Genome Res.* **2003**, *102*, 16–24. [CrossRef]
188. Goldammer, T.; Amaral, M.E.; Brunner, R.M.; Owens, E.; Kata, S.R.; Schwerin, M.; Womack, J.E. Clarifications on breakpoints in HSAX and BTAX by comparative mapping of F9, HPRT, and XIST in cattle. *Cytogenet. Genome Res.* **2003**, *101*, 39–42. [CrossRef] [PubMed]
189. Brunner, R.M.; Sanftleben, H.; Goldammer, T.; Kühn, C.; Weikard, R.; Kata, S.R.; Womack, J.E.; Schwerin, M. The telomeric region of BTA18 containing a potential QTL region for health in cattle exhibits high similarity to the HSA19q region in humans. *Genomics* **2003**, *81*, 270–278. [CrossRef]

190. Ahrens, E.; Graphodatskaya, D.; Nguyen, B.X.; Stranzinger, G. Cytogenetic comparison of saola (*Pseudoryx nghetinhensis*) and cattle (*Bos taurus*) using G- and Q-banding and FISH. *Cytogenet. Genome Res.* **2005**, *111*, 147–151. [CrossRef] [PubMed]
191. Di Meo, G.P.; Perucatti, A.; Uboldi, C.; Roperto, S.; Incarnato, D.; Roperto, F.; Williams, J.; Eggen, A.; Ferretti, L.; Iannuzzi, L. Comparative mapping of the fragile histidine triad (FHIT) gene in cattle, river buffalo, sheep and goat by FISH and assignment to BTA22 by RH-mapping: A comparison with HSA3. *Anim. Genet* **2005**, *36*, 363–364. [CrossRef]
192. Di Meo, G.P.; Perucatti, A.; Floriot, S.; Incarnato, D.; Rullo, R.; Caputi Jambrenghi, A.; Ferretti, L.; Vonghia, G.; Cribiu, E.; Eggen, A.; et al. Chromosome evolution and improved cytogenetic maps of the Y chromosome in cattle, zebu, river buffalo, sheep and goat. *Chromosome Res.* **2005**, *13*, 349–355. [CrossRef] [PubMed]
193. Perucatti, A.; Floriot, S.; Di Meo, G.P.; Soglia, D.; Rullo, R.; Maione, S.; Incarnato, D.; Eggen, A.; Sacchi, P.; Rasero, R.; et al. Comparative FISH mapping of mucin 1, transmembrane (MUC1) among cattle, river buffalo, sheep and goat chromosomes: Comparison between bovine chromosome 3 and human chromosome 1. *Cytogenet. Genome Res.* **2006**, *112*, 103–105. [CrossRef]
194. Perucatti, A.; Di Meo, G.P.; Vallinoto, M.; Kierstein, G.; Schneider, M.P.; Incarnato, D.; Caputi Jambrenghi, A.; Mohammadi, G.; Vonghia, G.; Silva, A.; et al. FISH-mapping of LEP and SLC26A2 genes in sheep, goat and cattle R-banded chromosomes: Comparison between bovine, ovine and caprine chromosome 4 (BTA4/OAR4/CHI4) and human chromosome 7 (HSA7). *Cytogenet. Genome Res.* **2006**, *115*, 7–9. [CrossRef] [PubMed]
195. Schibler, L.; Roig, A.; Mahe, M.-F.; Laurent, P.; Hayes, H.; Rodolphe, F.; Cribiu, E.P. High-resolution comparative mapping among man, cattle and mouse suggests a role for repeat sequences in mammalian genome evolution. *BMC Genom.* **2006**, *7*, 194. [CrossRef] [PubMed]
196. Antonacci, R.; Vaccarelli, G.; Di Meo, G.P.; Piccinni, B.; Miccoli, M.C.; Cribiu, E.P.; Perucatti, A.; Iannuzzi, L.; Ciccarese, S. Molecular in situ hybridization analysis of sheep and goat BAC clones identifies the transcriptional orientation of T cell receptor gamma genes on chromosome 4 in bovids. *Vet. Res. Comm.* **2007**, *31*, 977–983. [CrossRef]
197. Perucatti, A.; Di Meo, G.P.; Goldammer, T.; Incarnato, D.; Brunner, R.; Iannuzzi, L. Comparative FISH-mapping of twelve loci in river buffalo and sheep chromosomes: Comparison with HSA8p and HSA4q. *Cytogenet. Genome. Res.* **2007**, *119*, 242–244. [CrossRef]
198. Goldammer, T.; Brunner, R.M.; Weikard, R.; Kuehn, C.; Wimmers, K. Generation of an improved cytogenetic and comparative map of *Bos taurus* chromosome BTA27. *Chromosome Res.* **2007**, *15*, 203–213. [CrossRef]
199. Ropiquet, A.; Gerbault-Seureau, M.; Deuve, J.L.; Gilbert, C.; Pagacova, E.; Chai, N.; Rubes, J.; Hassanin, A. Chromosome evolution in the subtribe Bovina (Mammalia, Bovidae): The karyotype of the Cambodian banteng (*Bos javanicus birmanicus*) suggests that Robertsonian translocations are related to interspecific hybridization. *Chromosome Res.* **2008**, *16*, 1107–1118. [CrossRef]
200. Nguyen, T.T.; Aniskin, V.M.; Gerbault-Seureau, M.; Planton, H.; Renard, J.P.; Nguyen, B.X.; Hassanin, A.; Volobouev, V.T. Phylogenetic position of the saola (*Pseudoryx nghetinhensis*) inferred from cytogenetic analysis of eleven species of Bovidae. *Cytogenet. Genome Res.* **2008**, *122*, 41–54. [CrossRef]
201. Bratuś, A.; Bugno, M.; Klukowska-Rötzler, J.; Sawińska, M.; Eggen, A.; Słota, E. Chromosomal homology between the human and the bovine DMRT1 genes. *Folia Biol.* **2009**, *57*, 29–32. [CrossRef]
202. Bratuś, A.; Słota, E. Comparative cytogenetic and molecular studies of DM domain genes in pig and cattle. *Cytogenet. Genome Res.* **2009**, *126*, 180–185. [CrossRef] [PubMed]
203. Perucatti, A.; Di Meo, G.P.; Goldammer, T.; Incarnato, D.; Nicolae, I.; Brunner, R.; Iannuzzi, L. FISH-mapping comparison between river buffalo chromosome 7 and sheep chromosome 6: Assignment of new loci and comparison with HSA4. *Cytogenet. Genome Res.* **2009**, *124*, 106–111. [CrossRef]
204. Goldammer, T.; Brunner, R.M.; Rebl, A.; Wu, C.H.; Nomura, K.; Hadfield, T.; Maddox, J.F.; Cockett, N.E. Cytogenetic anchoring of radiation hybrid and virtual maps of sheep chromosome X and comparison of X chromosomes in sheep, cattle, and human. *Chromosome Res.* **2009**, *17*, 497–506. [CrossRef]
205. Manera, S.; Bonfiglio, S.; Malusà, A.; Denis, C.; Boussaha, M.; Russo, V.; Roperto, F.; Perucatti, A.; Di Meo, G.P.; Eggen, A.; et al. Comparative mapping and genomic annotation of the bovine oncosuppressor gene WWOX. *Cytogenet. Genome Res.* **2009**, *126*, 186–193. [CrossRef] [PubMed]
206. Goldammer, T.; Di Meo, G.P.; Lühken, G.; Drögemüller, C.; Wu, C.H.; Kijas, J.; Dalrymple, B.P.; Nicholas, F.W.; Maddox, J.F.; Iannuzzi, L.; et al. Molecular cytogenetics and gene mapping in sheep (*Ovis aries*, 2n = 54). *Cytogenet. Genome Res.* **2009**, *126*, 63–76. [CrossRef]
207. Schibler, L.; Di Meo, G.P.; Cribiu, E.P.; Iannuzzi, L. Molecular cytogenetics and comparative mapping in goats (*Capra hircus*, 2n = 60). *Cytogenet. Genome Res.* **2009**, *126*, 77–85. [CrossRef]
208. Genualdo, V.; Spalenza, V.; Perucatti, A.; Iannuzzi, A.; Di Meo, G.P.; Caputi-Jambrenghi, A.; Vonghia, G.; Rasero, R.; Nebbia, C.; Sacchi, P.; et al. Fluorescence in situ hybridization mapping of six loci containing genes involved in the dioxin metabolism of domestic bovids. *J. Appl. Genet.* **2011**, *52*, 229–232. [CrossRef]
209. Di Meo, G.P.; Goldammer, T.; Perucatti, A.; Genualdo, V.; Iannuzzi, A.; Incarnato, D.; Rebl, A.; Di Berardino, D.; Iannuzzi, L. Extended cytogenetic maps of sheep chromosome 1 and their cattle and river buffalo homoeologues: Comparison with the OAR1 RH map and human chromosomes 2, 3, 21 and 1q. *Cytogenet. Genome Res.* **2011**, *133*, 16–24. [CrossRef]
210. Cernohorska, H.; Kubickova, S.; Vahala, J.; Rubes, J. Molecular insights into X;BTA5 chromosome rearrangements in the tribe Antilopini (*Bovidae*). *Cytogenet. Genome Res.* **2012**, *136*, 188–198. [CrossRef]

211. Kopecna, O.; Kubickova, S.; Cernohorska, H.; Cabelova, K.; Vahala, J.; Rubes, J. Isolation and comparison of tribe-specific centromeric repeats within Bovidae. *J. Appl. Genet.* **2012**, *53*, 193–202. [CrossRef] [PubMed]
212. Perucatti, A.; Genualdo, V.; Iannuzzi, A.; Rebl, A.; Di Berardino, D.; Goldammer, T.; Iannuzzi, L. Advanced comparative cytogenetic analysis of X chromosomes in river buffalo, cattle, sheep, and human. *Chromosome Res.* **2012**, *20*, 413–425. [CrossRef] [PubMed]
213. Kolesárová, V.; Šiviková, K.; Holečková, B.; Dianovský, J. A comparative FISH mapping of LCA5L gene in cattle, sheep, and goats. *Anim. Biotechnol.* **2015**, *26*, 37–39. [CrossRef] [PubMed]
214. De Lorenzi, L.; Planas, J.; Rossi, E.; Malagutti, L.; Parma, P. New cryptic karyotypic differences between cattle (*Bos taurus*) and goat (*Capra hircus*). *Chromosome Res.* **2015**, *23*, 225–235. [CrossRef] [PubMed]
215. De Lorenzi, L.; Pauciullo, A.; Iannuzzi, A.; Parma, P. Cytogenetic Characterization of a Small Evolutionary Rearrangement Involving Chromosomes BTA21 and OAR18. *Cytogenet. Genome Res.* **2020**, *160*, 193–198. [CrossRef]
216. Perucatti, A.; Iannuzzi, A.; Armezzani, A.; Palmarini, M.; Iannuzzi, L. Comparative fluorescence in Situ hybridization (FISH) mapping of twenty-three endogenous Jaagsiekte sheep retrovirus (enJSRVs) in sheep (*Ovis aries*) and river buffalo (*Bubalus bubalis*) chromosomes. *Animals* **2022**, *12*, 2834. [CrossRef]
217. Hayes, H. Chromosome painting with human chromosome-specific DNA libraries reveals the extent and distribution of conserved segments in bovine chromosomes. *Cytogenet. Cell Genet.* **1995**, *71*, 168–174. [CrossRef]
218. Chowdhary, B.P.; Frönicke, L.; Gustavsson, I.; Scherthan, H. Comparative analysis of the cattle and human genomes: Detection of ZOO-FISH and gene mapping-based chromosomal homologies. *Mamm. Genome* **1996**, *7*, 297–302. [CrossRef]
219. Solinas-Toldo, S.; Lengauer, C.; Fries, R. Comparative genome map of human and cattle. *Genomics* **1995**, *27*, 489–496. [CrossRef]
220. Pauciullo, A.; Knorr, C.; Perucatti, A.; Iannuzzi, A.; Iannuzzi, L.; Erhardt, G. Characterization of a very rare case of living ewe-buck hybrid using classical and molecular cytogenetics. *Sci. Rep.* **2016**, *6*, 34781. [CrossRef]
221. Iannuzzi, A.; Pereira, J.; Iannuzzi, C.; Fu, B.; Ferguson-Smith, M. Pooling strategy and chromosome painting characterize a living zebroid for the first time. *PLoS ONE* **2017**, *12*, e0180158. [CrossRef]
222. Amaral, M.E.; Grant, J.R.; Riggs, P.K.; Stafuzza, N.B.; Filho, E.A.; Goldammer, T.; Weikard, R.; Brunner, R.M.; Kochan, K.J.; Greco, A.J.; et al. A first generation whole genome RH map of the river buffalo with comparison to domestic cattle. *BMC Genom.* **2008**, *9*, 631. [CrossRef] [PubMed]
223. Faraut, T.; de Givry, S.; Hitte, C.; Lahbib-Mansais, Y.; Morisson, M.; Milan, D.; Schiex, T.; Servin, B.; Vignal, A.; Galibert, F.; et al. Contribution of radiation hybrids to genome mapping in domestic animals. *Cytogenet. Genome Res.* **2009**, *126*, 21–33. [CrossRef] [PubMed]
224. Stafuzza, N.B.; Abbassi, H.; Grant, J.R.; Rodrigues-Filho, E.A.; Ianella, P.; Kadri, S.M.; Amarante, M.V.; Stohard, P.; Womack, J.E.; de León, F.A.; et al. Comparative RH maps of the river buffalo and bovine Y chromosomes. *Cytogenet. Genome Res.* **2009**, *126*, 132–138. [CrossRef] [PubMed]
225. Durmaz, A.A.; Karaca, E.; Demkow, U.; Toruner, G.; Schoumans, J.; Cogulu, O. Evolution of Genetic Techniques: Past, Present, and Beyond. *BioMed. Res. Intern.* **2015**, *2015*, 461524. [CrossRef]
226. Fidlerova, H.; Senger, G.; Kost, M.; Sanseau, P.; Sheer, D. Two simple procedures for releasing chromatin from routinely fixed cells for fluorescence in situ hybridization. *Cytogenet. Cell Genet.* **1994**, *65*, 203–205. [CrossRef]
227. Brunner, R.M.; Goldammer, T.; Fürbass, R.; Vanselow, J.; Schwerin, M. Genomic organization of the bovine aromatase encoding gene and a homologous pseudogene as revealed by DNA fiber FISH. *Cytogenet. Cell Genet.* **1998**, *82*, 37–40. [CrossRef]
228. Seyfert, H.M.; Pitra, C.; Meyer, L.; Brunner, R.M.; Wheeler, T.T.; Molenaar, A.; McCracken, J.Y.; Herrmann, J.; Thiesen, H.J.; Schwerin, M. Molecular characterization of STAT5A- and STAT5B-encoding genes reveals extended intragenic sequence homogeneity in cattle and mouse and different degrees of divergent evolution of various domains. *J Mol. Evol.* **2000**, *50*, 550–561. [CrossRef]
229. Hamilton, C.K.; Favetta, L.A.; Di Meo, G.P.; Floriot, S.; Perucatti, A.; Peippo, J.; Kantanen, J.; Eggen, A.; Iannuzzi, L.; King, W.A. Copy number variation of testis-specific protein, Y-encoded (TSPY) in 14 different breeds of cattle (*Bos taurus*). *Sex Dev.* **2009**, *3*, 205–213. [CrossRef]
230. Pauciullo, A.; Fleck, K.; Lühken, G.; Di Berardino, D.; Erhardt, G. Dual-color high-resolution fiber-FISH analysis on lethal white syndrome carriers in sheep. *Cytogenet. Genome Res.* **2013**, *140*, 46–54. [CrossRef]
231. du Manoir, S.; Speicher, M.R.; Joos, S.; Schröck, E.; Popp, S.; Döhner, H.; Kovacs, G.; Robert-Nicoud, M.; Lichter, P.; Cremer, T. Detection of complete and partial chromosome gains and losses by comparative genomic in situ hybridization. *Hum. Gene.t* **1993**, *90*, 590–610. [CrossRef]
232. Kallioniemi, A.; Kallioniemi, O.P.; Sudar, D.; Rutovitz, D.; Gray, J.W.; Waldman, F.; Pinkel, D. Comparative genomic hybridization for molecular cytogenetic analysis of solid tumors. *N. Y. Sci. J.* **1992**, *258*, 818–821. [CrossRef]
233. Solinas-Toldo, S.; Lampel, S.; Stilgenbauer, S.; Nickolenko, J.; Benner, A.; Döhner, H.; Cremer, T.; Lichter, P. Matrix-based comparative genomic hybridization: Biochips to screen for genomic imbalances. *Genes Chromosom. Cancer.* **1997**, *20*, 399–407. [CrossRef]
234. Brennan, C.; Zhang, Y.; Leo, C.; Feng, B.; Cauwels, C.; Aguirre, A.J.; Kim, M.; Protopopov, A.; Chin, L. High-resolution global profiling of genomic alterations with long oligonucleotide microarray. *Cancer Res.* **2004**, *64*, 4744–4748. [CrossRef] [PubMed]
235. Carvalho, B.; Ouwerkerk, E.; Meijer, G.A.; Ylstra, B. High resolution microarray comparative genomic hybridisation analysis using spotted oligonucleotides. *J. Clin. Pathol.* **2004**, *57*, 644–646. [CrossRef]

236. Pinkel, D.; Albertson, D.G. Array comparative genomic hybridization and its applications in cancer. *Nat. Genet.* **2005**, *37*, S11–S17. [CrossRef]
237. Iafrate, A.J.; Feuk, L.; Rivera, M.N.; Listewnik, M.L.; Donahoe, P.K.; Qi, Y.; Scherer, S.W.; Lee, C. Detection of large-scale variation in the human genome. *Nat. Genet.* **2004**, *36*, 949–951. [CrossRef]
238. Redon, R.; Ishikawa, S.; Fitch, K.R.; Feuk, L.; Perry, G.H.; Andrews, T.D.; Fiegler, H.; Shapero, M.H.; Carson, A.R.; Hurles, M.E.; et al. Global variation in copy number in the human genome. *Nature* **2006**, *444*, 444–454. [CrossRef]
239. Sebat, J.; Lakshmi, B.; Troge, J.; Alexander, J.; Young, J.; Lundin, P.; Månér, S.; Massa, H.; Walker, M.; Chi, M.; et al. Large-scale copy number polymorphism in the human genome. *N. Y. Sci. J.* **2004**, *305*, 525–528. [CrossRef]
240. Zarrei, M.; MacDonald, J.R.; Merico, D.; Scherer, S.W. A copy number variation map of the human genome. *Nat. Rev. Genet.* **2015**, *16*, 172–183. [CrossRef] [PubMed]
241. Liu, G.E.; Van Tassell, C.P.; Sonstegard, T.S.; Li, R.W.; Alexander, L.J.; Keele, J.W.; Matukumalli, L.K.; Smith, T.P.; Gasbarre, L.C. Detection of germline and somatic copy number variations in cattle. *Dev. Biol.* **2008**, *132*, 231–237. [CrossRef]
242. Liu, G.E.; Hou, Y.L.; Zhu, B.; Cardone, M.F.; Jiang, L.; Cellamare, A.; Mitra, A.; Alexander, L.J.; Coutinho, L.L.; Dell'Aquila, M.E.; et al. Analysis of copy number variations among diverse cattle breeds. *Genome Res.* **2010**, *20*, 693–703. [CrossRef] [PubMed]
243. Fadista, J.; Thomsen, B.; Holm, L.E.; Bendixen, C. Copy number variation in the bovine genome. *BMC Genom.* **2010**, *11*, 284. [CrossRef]
244. Liu, M.; Fang, L.; Liu, S.; Pan, M.G.; Seroussi, E.; Cole, J.B.; Ma, L.; Chen, H.; Liu, G.E. Array CGH-based detection of CNV regions and their potential association with reproduction and other economic traits in Holsteins. *BMC Genom.* **2019**, *20*, 181. [CrossRef] [PubMed]
245. Zhang, L.; Jia, S.; Yang, M.; Xu, Y.; Li, C.; Sun, J.; Huang, Y.; Lan, X.; Lei, C.; Zhou, Y.; et al. Detection of copy number variations and their effects in Chinese bulls. *BMC Genom.* **2014**, *15*, 480. [CrossRef]
246. Kijas, J.W.; Barendse, W.; Barris, W.; Harrison, B.; McCulloch, R.; McWilliam, S.; Whan, V. Analysis of copy number variants in the cattle genome. *Gene* **2011**, *482*, 73–77. [CrossRef] [PubMed]
247. Jenkins, G.M.; Goddard, M.E.; Black, M.A.; Brauning, R.; Auvray, B.; Dodds, K.G.; Kijas, J.W.; Cockett, N.; McEwan, J.C. Copy number variants in the sheep genome detected using multiple approaches. *BMC Genom.* **2016**, *17*, 441. [CrossRef]
248. Fontanesi, L.; Beretti, F.; Martelli, P.L.; Colombo, M.; Dall'olio, S.; Occidente, M.; Portolano, B.; Casadio, R.; Matassino, D.; Russo, V. A first comparative map of copy number variations in the sheep genome. *Genomics* **2011**, *97*, 158–165. [CrossRef]
249. Fontanesi, L.; Martelli, P.L.; Beretti, F.; Riggio, V.; Dall'Olio, S.; Colombo, M.; Casadio, R.; Russo, V.; Portolano, B. An initial comparative map of copy number variations in the goat (*Capra hircus*) genome. *BMC Genom.* **2010**, *11*, 639. [CrossRef]
250. Canavez, F.C.; Luche, D.D.; Stothard, P.; Leite, K.R.; Sousa-Canavez, J.M.; Plastow, G.; Meidanis, J.; Souza, M.A.; Feijao, P.; Moore, S.S.; et al. Genome sequence and assembly of *Bos indicus*. *J. Hered.* **2012**, *103*, 342–348. [CrossRef]
251. Afgan, E.; Baker, D.; Batut, B.; van den Beek, M.; Bouvier, D.; Cech, M.; Chilton, J.; Clements, D.; Coraor, N.; Grüning, B.A.; et al. The Galaxy platform for accessible, reproducible and collaborative biomedical analyses: 2018 update. *Nucleic Acids Res.* **2018**, *46*, W537–W544. [CrossRef]
252. Rangwala, S.H.; Kuznetsov, A.; Ananiev, V.; Asztalos, A.; Borodin, E.; Evgeniev, V.; Joukov, V.; Lotov, V.; Pannu, R.; Rudnev, D.; et al. Accessing NCBI data using the NCBI Sequence Viewer and Genome Data Viewer (GDV). *Genome Res.* **2021**, *31*, 159–169. [CrossRef]
253. Kent, W.J.; Sugnet, C.W.; Furey, T.S.; Roskin, K.M.; Pringle, T.H.; Zahler, A.M.; Haussler, D. The human genome browser at UCSC. *Genome Res.* **2002**, *12*, 996–1006. [CrossRef] [PubMed]
254. Cunningham, F.; Allen, J.E.; Allen, J.; Alvarez-Jarreta, J.; Amode, M.R.; Armean, I.M.; Austine-Orimoloye, O.; Azov, A.G.; Barnes, I.; Bennett, R.; et al. Ensembl 2022. *Nucleic Acids Res.* **2022**, *50*, D988–D995. [CrossRef] [PubMed]
255. Untergasser, A.; Cutcutache, I.; Koressaar, T.; Ye, J.; Faircloth, B.C.; Remm, M.; Rozen, S.G. Primer3–new capabilities and interfaces. *Nucleic Acids Res.* **2012**, *40*, e115. [CrossRef]
256. McLaren, W.; Gil, L.; Hunt, S.E.; Riat, H.S.; Ritchie, G.R.; Thormann, A.; Flicek, P.; Cunningham, F. The Ensembl Variant Effect Predictor. *Genome Biol.* **2016**, *17*, 122. [CrossRef] [PubMed]
257. Hulsegge, I.; Oldenbroek, K.; Bouwman, A.; Veerkamp, R.; Windig, J. Selection and Drift: A Comparison between Historic and Recent Dutch Friesian Cattle and Recent Holstein Friesian Using WGS Data. *Animals* **2022**, *12*, 329. [CrossRef] [PubMed]
258. Xia, X.; Zhang, S.; Zhang, H.; Zhang, Z.; Chen, N.; Li, Z.; Sun, H.; Liu, X.; Lyu, S.; Wang, X.; et al. Assessing genomic diversity and signatures of selection in Jiaxian Red cattle using whole-genome sequencing data. *BMC Genom.* **2021**, *22*, 43. [CrossRef]
259. Sun, L.; Qu, K.; Liu, Y.; Ma, X.; Chen, N.; Zhang, J.; Huang, B.; Lei, C. Assessing genomic diversity and selective pressures in Bashan cattle by whole-genome sequencing data. *Anim Biotechnol.* **2021**, *11*, 1–12. [CrossRef]
260. Pollott, G.E.; Piercy, R.J.; Massey, C.; Salavati, M.; Cheng, Z.; Wathes, D.C. Locating a novel autosomal recessive genetic variant in the cattle glucokinase gene using only WGS data from three cases and six carriers. *Front Genet.* **2022**, *13*, 755693. [CrossRef]
261. Gershoni, M.; Shirak, A.; Raz, R.; Seroussi, E. Comparing BeadChip and WGS Genotyping: Non-Technical Failed Calling Is Attributable to Additional Variation within the Probe Target Sequence. *Genes* **2022**, *13*, 485. [CrossRef]
262. Mulim, H.A.; Brito, L.F.; Pinto, L.F.B.; Ferraz, J.B.S.; Grigoletto, L.; Silva, M.R.; Pedrosa, V.B. Characterization of runs of homozygosity, heterozygosity-enriched regions, and population structure in cattle populations selected for different breeding goals. *BMC Genom..* **2022**, *23*, 209. [CrossRef] [PubMed]

263. Meuwissen, T.; van den Berg, I.; Goddard, M. On the use of whole-genome sequence data for across-breed genomic prediction and fine-scale mapping of QTL. *Genet. Sel. Evol.* **2021**, *53*, 19. [CrossRef] [PubMed]
264. Butty, A.M.; Chud, T.C.S.; Miglior, F.; Schenkel, F.S.; Kommadath, A.; Krivushin, K.; Grant, J.R.; Häfliger, I.M.; Drögemüller, C.; Cánovas, A.; et al. High confidence copy number variants identified in Holstein dairy cattle from whole genome sequence and genotype array data. *Sci. Rep.* **2020**, *10*, 8044. [CrossRef] [PubMed]
265. Taylor, J.F.; Whitacre, L.K.; Hoff, J.L.; Tizioto, P.C.; Kim, J.; Decker, J.E.; Schnabel, R.D. Lessons for livestock genomics from genome and transcriptome sequencing in cattle and other mammals. *Genet. Sel. Evol.* **2016**, *48*, 59. [CrossRef]
266. Stegemiller, M.R.; Redden, R.R.; Notter, D.R.; Taylor, T.; Taylor, J.B.; Cockett, N.E.; Heaton, M.P.; Kalbfleisch, T.S.; Murdoch, B.M. Using whole genome sequence to compare variant callers and breed differences of US sheep. *Front. Genet.* **2023**, *13*, 1060882. [CrossRef]
267. Hirter, N.; Letko, A.; Häfliger, I.M.; Becker, D.; Greber, D.; Drögemüller, C. A genome-wide significant association on chromosome 15 for congenital entropion in Swiss White Alpine sheep. *Anim. Genet.* **2020**, *51*, 278–283. [CrossRef]
268. Saleh, A.A.; Xue, L.; Zhao, Y. Screening Indels from the whole genome to identify the candidates and their association with economic traits in several goat breeds. *Funct. Integr. Genom.* **2023**, *23*, 58. [CrossRef]
269. Signer-Hasler, H.; Henkel, J.; Bangerter, E.; Bulut, Z.; VarGoats Consortium; Drögemüller, C.; Leeb, T.; Flury, C. Runs of homozygosity in Swiss goats reveal genetic changes associated with domestication and modern selection. *Genet. Sel. Evol.* **2022**, *54*, 6. [CrossRef] [PubMed]
270. Mullis, K.B. Target amplification for DNA analysis by the polymerase chain reaction. *Ann. Biol. Clin.* **1990**, *48*, 579–582.
271. Zhu, H.; Zhang, H.; Xu, Y.; Laššáková, S.; Korabečná, M.; Neužil, P. PCR past, present and future. *Biotechniques* **2020**, *69*, 317–325. [CrossRef]
272. Ennis, S.; Gallagher, T.F.A. PCR-based sex-determination assay in cattle based on the bovine amelogenin locus. *Anim. Genet.* **1994**, *25*, 425–427. [CrossRef] [PubMed]
273. McNiel, E.A.; Madrill, N.J.; Treeful, A.E.; Buoen, L.C.; Weber, A.F. Comparison of cytogenetics and polymerase chain reaction based detection of the amelogenin gene polymorphism for the diagnosis of freemartinism in cattle. *J. Vet. Diagn. Investig.* **2006**, *18*, 469–472. [CrossRef] [PubMed]
274. Lu, W.; Rawlings, N.; Zhao, J.; Wang, H. Amplification and application of the HMG box of bovine SRY gene for sex determination. *Anim. Reprod. Sci.* **2007**, *100*, 186–191. [CrossRef]
275. Chandler, J.E.; Taylor, T.M.; Canal, A.L.; Cooper, R.K.; Moser, E.B.; McCormick, M.E.; Willard, S.T.; Rycroft, H.E.; Gilbert, G.R. Calving sex ratio as related to the predicted Y-chromosome-bearing spermatozoa ratio in bull ejaculates. *Theriogenology* **2007**, *67*, 563–571. [CrossRef]
276. Ron, M.; Porat, B.; Band, M.R.; Weller, J.I. Chimaerism detection in bovine twins, triplets and quadruplets using sex chromosome-linked markers. *Anim. Genet.* **2011**, *42*, 208–211. [CrossRef]
277. Szczerbal, I.; Kociucka, B.; Nowacka-Woszuk, J.; Lach, Z.; Jaskowski, J.M.; Switonski, M. A high incidence of leukocyte chimerism (60,XX/60,XY) in single born heifers culled due to underdevelopment of internal reproductive tracts. *Czech. J. Anim. Sci.* **2014**, *59*, 445–449. [CrossRef]
278. Bresciani, C.; Parma, P.; De Lorenzi, L.; Di Ianni, F.; Bertocchi, M.; Bertani, V.; Cantoni, A.M.; Parmigiani, E. A clinical case of an SRY-positive intersex/hermaphrodite holstein cattle. *Sex Dev.* **2015**, *9*, 229–238. [CrossRef]
279. De Lorenzi, L.; Arrighi, S.; Rossi, E.; Grignani, P.; Previderè, C.; Bonacina, S.; Cremonesi, F.; Parma, P. XY (SRY-positive) ovarian disorder of sex development in cattle. *Sex Dev.* **2018**, *12*, 196–203. [CrossRef] [PubMed]
280. Qiu, Q.; Shao, T.; He, Y.; Muhammad, A.U.; Cao, B.; Su, H. Applying real-time quantitative PCR to diagnosis of freemartin in Holstein cattle by quantifying SRY gene: A comparison experiment. *PeerJ* **2018**, *6*, e4616. [CrossRef]
281. Szczerbal, I.; Nowacka-Woszuk, J.; Albarella, S.; Switonski, M. Technical note: Droplet digital PCR as a new molecular method for a simple and reliable diagnosis of freemartinism in cattle. *J. Dairy. Sci.* **2019**, *102*, 10100–10104. [CrossRef]
282. Cray, N.; Wagner, M.; Hauer, J.; Roti Roti, E. Technical note: Droplet digital PCR precisely and accurately quantifies sex skew in bovine semen. *J. Dairy Sci.* **2020**, *103*, 6698–6705. [CrossRef]
283. Uzar, T.; Szczerbal, I.; Serwanska-Leja, K.; Nowacka-Woszuk, J.; Gogulski, M.; Bugaj, S.; Switonski, M.; Komosa, M. Congenital malformations in a Holstein-Fresian calf with a unique mosaic karyotype: A Case Report. *Animals* **2020**, *10*, 1615. [CrossRef]
284. Szczerbal, I.; Komosa, M.; Nowacka-Woszuk, J.; Uzar, T.; Houszka, M.; Semrau, J.; Musial, M.; Barczykowski, M.; Lukomska, A.; Switonski, M. A disorder of sex development in a Holstein-Friesian heifer with a rare Mosaicism (60,XX/90,XXY): A genetic, anatomical, and histological study. *Animals* **2021**, *11*, 285. [CrossRef] [PubMed]
285. Szczerbal, I.; Nowacka-Woszuk, J.; Stachowiak, M.; Lukomska, A.; Konieczny, K.; Tarnogrodzka, N.; Wozniak, J.; Switonski, M. XX/XY chimerism in internal genitalia of a virilized heifer. *Animals* **2022**, *12*, 2932. [CrossRef] [PubMed]
286. Lansdorp, P. Telomere Length Regulation. *Front. Oncol.* **2022**, *12*, 943622. [CrossRef]
287. Iannuzzi, A.; Albarella, S.; Parma, P.; Galdiero, G.; D'Anza, E.; Pistucci, R.; Peretti, V.; Ciotola, F. Characterization of telomere length in Agerolese cattle breed, correlating blood and milk samples. *Anim. Genet.* **2022**, *53*, 676–679. [CrossRef] [PubMed]
288. Kordinas, V.; Ioannidis, A.; Chatzipanagiotou, S. The telomere/telomerase system in chronic inflammatory diseases. Cause or effect? *Genes* **2016**, *7*, 60. [CrossRef] [PubMed]
289. Bateson, M. Cumulative stress in research animals: Telomere attrition as a biomarker in a welfare context? *Bioessays* **2016**, *38*, 201–212. [CrossRef]

290. Ilska-Warner, J.J.; Psifidi, A.; Seeker, L.A.; Wilbourn, R.V.; Underwood, S.L.; Fairlie, J.; Whitelaw, B.; Nussey, D.H.; Coffey, M.P.; Banos, G. The Genetic Architecture of Bovine Telomere Length in Early Life and Association With Animal Fitness. *Front. Genet.* **2019**, *10*, 1048. [CrossRef] [PubMed]
291. Tilesi, F.; Domenico, E.G.D.; Pariset, L.; Bosco, L.; Willems, D.; Valentini, A.; Ascenzioni, F. Telomere length diversity in cattle breeds. *Diversity* **2010**, *2*, 1118–1129. [CrossRef]
292. Lai, T.P.; Wright, W.E.; Shay, J.W. Comparison of telomere length measurement methods. *Philos. Trans. R. Soc. Lond. B Biol. Sci.* **2018**, *373*, 20160451. [CrossRef]
293. Ribas-Maynou, J.; Llavanera, M.; Mateo-Otero, Y.; Ruiz, N.; Muiño, R.; Bonet, S.; Yeste, M. Telomere length in bovine sperm is related to the production of reactive oxygen species, but not to reproductive performance. *Theriogenology* **2022**, *189*, 290–300. [CrossRef]
294. Iannuzzi, A.; Della Valle, G.; Russo, M.; Longobardi, V.; Albero, G.; De Canditiis, C.; Kosior, M.A.; Pistucci, R.; Gasparrini, B. Evaluation of bovine sperm telomere length and association with semen quality. *Theriogenology* **2020**, *158*, 227–232. [CrossRef] [PubMed]
295. Cawthon, R.M. Telomere length measurement by a novel monochrome multiplex quantitative PCR method. *Nucleic Acids Res.* **2009**, *37*, e21. [CrossRef]
296. Bustin, S.A.; Benes, V.; Garson, J.A.; Hellemans, J.; Huggett, J.; Kubista, M.; Mueller, R.; Nolan, T.; Pfaffl, M.W.; Shipley, G.L.; et al. The MIQE guidelines: Minimum information for publication of quantitative real-time PCR experiments. *Clin. Chem.* **2009**, *55*, 611–622. [CrossRef]
297. Lindrose, A.R.; McLester-Davis, L.W.Y.; Tristano, R.I.; Kataria, L.; Gadalla, S.M.; Eisenberg, D.T.A.; Verhulst, S.; Drury, S. Method comparison studies of telomere length measurement using qPCR approaches: A critical appraisal of the literature. *PLoS ONE* **2021**, *16*, e0245582. [CrossRef] [PubMed]
298. Seeker, L.A.; Holland, R.; Underwood, S.; Fairlie, J.; Psifidi, A.; Ilska, J.J.; Bagnall, A.; Whitelaw, B.; Coffey, M.; Banos, G.; et al. Method specific calibration corrects for DNA extraction method effects on relative telomere length measurements by quantitative PCR. *PLoS ONE* **2016**, *11*, e0164046. [CrossRef]
299. Tolios, A.; Teupser, D.; Holdt, L.M. Preanalytical Conditions and DNA Isolation Methods Affect Telomere Length Quantification in Whole Blood. *PLoS ONE* **2015**, *10*, e0143889. [CrossRef]
300. Macville, M.; Veldman, T.; Padilla-Nash, H.; Wangsa, D.; O'Brien, P.; Schröck, E.; Ried, T. Spectral karyotyping, a 24-colour FISH technique for the identification of chromosomal rearrangements. *Histochem. Cell Biol.* **1997**, *108*, 299–305. [CrossRef] [PubMed]
301. Rhoads, A.; Au, K.F. PacBio Sequencing and Its Applications. *Genom. Prot. Bioinf.* **2015**, *13*, 278–289. [CrossRef]
302. Branton, D.; Deamer, D.W.; Marziali, A.; Bayley, H.; Benner, S.A.; Butler, T.; Di Ventra, M.; Garaj, S.; Hibbs, A.; Huang, X.; et al. The potential and challenges of nanopore sequencing. *Nat. Biotechnol.* **2008**, *26*, 1146–1153. [CrossRef] [PubMed]

Disclaimer/Publisher's Note: The statements, opinions and data contained in all publications are solely those of the individual author(s) and contributor(s) and not of MDPI and/or the editor(s). MDPI and/or the editor(s) disclaim responsibility for any injury to people or property resulting from any ideas, methods, instructions or products referred to in the content.

MDPI
St. Alban-Anlage 66
4052 Basel
Switzerland
Tel. +41 61 683 77 34
Fax +41 61 302 89 18
www.mdpi.com

Animals Editorial Office
E-mail: animals@mdpi.com
www.mdpi.com/journal/animals

www.ingramcontent.com/pod-product-compliance
Lightning Source LLC
LaVergne TN
LVHW070417100526
838202LV00014B/1475